T0233981

Springer Monographs in Mathematics

More information about this series at
http://www.springer.com/series/3733

Vieri Benci • Donato Fortunato

Variational Methods in Nonlinear Field Equations

Solitary Waves, Hylomorphic Solitons and Vortices

 Springer

Vieri Benci
Dip. di Matematica
Università degli Studi di Pisa
Pisa
Italy

Donato Fortunato
Dip. di Matematica
Università degli Studi di Bari "Aldo Moro"
Bari
Italy

ISSN 1439-7382   ISSN 2196-9922 (electronic)
ISBN 978-3-319-36122-2   ISBN 978-3-319-06914-2 (eBook)
DOI 10.1007/978-3-319-06914-2
Springer Cham Heidelberg New York Dordrecht London

Printed on acid-free paper

Springer is part of Springer Science+Business Media (www.springer.com)

To the memory of our parents

Preface

The aim of this book is twofold.

(i) We want to present the mathematical structure of some fundamental principles
of Physics and to show how these principles lead to the existence of a particular
kind of solitons called hylomorphic solitons. Moreover, we analyse some
qualitative properties of them in the light of these principles.

(ii) We want to present the mathematical tools which allow to give rigorous proofs
of the existence of hylomorphic solitons.

(i) The basic principles which lead to the existence of hylomorphic solitons are
the variational principle and the symmetry principle; these two principles imply
the conservation of the energy and of the charge; the interplay between energy and
charge is the basic ingredient for the existence of this type of solitons. Large part
of this book is devoted to a qualitative discussion of these principles and to their
interpretation in relation to the various theorems.

(ii) After the pioneering papers of Rosen [123] and Coleman [60], the existence
of this type of solitary waves has been investigated by variational and topological
methods. A lot of papers appeared on the nonlinear Schrödinger equation, the
nonlinear Klein-Gordon equation, the Klein-Gordon-Maxwell equations etc.

In this book the main results obtained in the last 35 years are summarized.
However, they are not presented in the way they appeared in the literature. Actually,
we have framed all the above field equations in a single abstract theory (Chap. 2);
then, we have proved some abstract theorems which allow to deduce most of the
results relative to the existence and stability of solitary waves (Chaps. 3–6). We
think that this theory is sufficiently general and flexible to be applied to many
other situations. For example, we have applied it to the existence of solitons in the
nonlinear beam equation (Chap. 7) and to the study of vortices (Chap. 8).

This book is organized as follows.

• In Chap. 1, we discuss the general physical principles on which the theory of
hylomorphic soliton is based.

- Chapter 2 is the crucial part of the present book; in fact we give a general definition of solitary wave, soliton and hylomorphic soliton. Moreover, we prove two minimization results (Theorems 38 and 42) in a very general framework and we use them to prove an abstract theorem on the existence of hylomorphic solitons (Theorem 33). This theorem will be applied to several equations in Chaps. 3–7.
- Chapter 3 is devoted to the nonlinear Schrödinger equation

$$i\frac{\partial \psi}{\partial t} = -\frac{1}{2}\Delta \psi + \frac{1}{2}W'(\psi).$$

We discuss the general features of this equation, we apply Theorem 33 to prove the existence of solitons and we end describing the dynamics of solitons.
- Chapter 4 is devoted to the nonlinear Klein-Gordon equation

$$\Box \psi + W'(\psi) = 0.$$

After a discussion on the general features, we apply Theorem 33 to prove the existence of solitons. We end the chapter describing some dynamical properties of the solitons in relation to the principles of special relativity.
- In Chap. 5 the Maxwell equations are deduced from the general principles of Chap. 1. If these equations are coupled with the nonlinear Klein-Gordon equation, we get the simplest gauge theory with "matter":

$$\Box u + W'(u) + \frac{\mathbf{j}^2 - \rho^2}{q^2 u^3} = 0$$

$$\nabla \cdot \mathbf{E} = \rho$$

$$\nabla \times \mathbf{H} - \frac{\partial \mathbf{E}}{\partial t} = \mathbf{j}$$

$$\nabla \times \mathbf{E} + \frac{\partial \mathbf{H}}{\partial t} = 0$$

$$\nabla \cdot \mathbf{H} = 0$$

Theorem 33 permits to prove the existence of solitons.
- Chapter 6 is devoted to the nonlinear Schrödinger-Maxwell equations, namely to the nonlinear Schrödinger equations coupled with Maxwell equations. These equations represent another case to which Theorem 33 can be applied.
- Chapter 7 is devoted to the nonlinear beam equation

$$\frac{\partial^2 u}{\partial t^2} + \frac{\partial^4 u}{\partial x^4} + W'(u) = 0.$$

In the last years, this equation has been studied by McKenna, W. Walter [110,111] and others as a model for suspended bridges. Among the other things, they discovered by numerical simulations the existence of solitary waves. In this chapter we prove that these solitary waves can be considered hylomorphic solitons provided that the charge is replaced by the momentum.

- In Chap. 8, we define the notion of vortex for the equations considered in Chaps. 3–6 and we use Theorems 38 and 42 to find sufficient conditions for their existence.
- In the Appendix, we state some preliminary results used in the book.

Contents

Notation

- $\mathbf{e}_1, \mathbf{e}_2, ..., \mathbf{e}_N$ denotes the usual orthonormal basis in $\mathbb{R}^N$.
- $\mathbf{a} \times \mathbf{b}$ denotes the cross product between the vectors $\mathbf{a}$ and $\mathbf{b}$.
- $\text{Re}(z), \text{Im}(z)$ are the real and the imaginary part of z; $\bar{z}$ denotes the complex conjugate.
- $B_\rho(x_0) = B(x_0, \rho) = \{x \in \mathbb{R}^N : |x - x_0| \leq \rho\}$.
- $B_\rho^c(x_0) = \mathbb{R}^N \smallsetminus B_\rho(x_0)$.
- $|\cdot|$ is the Euclidean norm both of a vector or of a matrix, namely $|A| = \sqrt{Tr(AA^*)}$.
- ∇, $\nabla\cdot$ and $\nabla\times$ denote, respectively, the gradient, the divergence and the curl operators.

Let $\Omega \subset \mathbb{R}^N$ be an open set, then:

- $\mathfrak{F}(\Omega, V)$ denotes a generic linear space of functions $\Omega \to V$.
- $\mathcal{C}(\Omega)$ denotes the set of real continuous functions defined on a set Ω.
- $\mathcal{C}^k(\Omega)$ denotes the set of functions defined on an $\Omega \subset \mathbb{R}^N$ which have continuous derivatives up to the order k.
- $\mathcal{D}(\Omega)$ denotes the set of the infinitely differentiable functions with compact support in Ω; $\mathcal{D}'(\Omega)$ denotes the topological dual of $\mathcal{D}(\Omega)$, namely the set of distributions on Ω.
- $L^p(\Omega)$, $L^\infty(\Omega)$ denotes the usual Lebesgue spaces.
- $H^1(\Omega)$ is the usual Sobolev space defined as the set of functions $u \in L^2(\Omega)$ such that $\nabla u \in L^2(\Omega)$.
- $H_0^1(\Omega)$ is the closure of $\mathcal{D}(\Omega)$ in $H^1(\Omega)$.
- $H_r^1(\mathbb{R}^N)$ is the space of functions in $H^1(\mathbb{R}^N)$ having radial symmetry.
- $H^{-1}(\Omega)$ is the topological dual of $H_0^1(\Omega)$.

- $\mathcal{D}^{1,2}(\mathbb{R}^N)$ or $\mathcal{D}^{1,2}$ is the closure of $\mathcal{D}(\mathbb{R}^N)$ with respect to the norm

$$\|u\| = \left(\int |\nabla u|^2 \, dx \right)^{\frac{1}{2}}.$$

- $\mathcal{D}_r^{1,2}(\mathbb{R}^N)$ is the space of functions in $\mathcal{D}^{1,2}(\mathbb{R}^N)$ having radial symmetry.
- $(T_\tau \mathbf{u})(x) := \mathbf{u}(x - \tau)$ denotes a representation of the group of translations.

Introduction

Roughly speaking, a solitary wave is a solution of a field equation whose energy travels as a localized packet and which preserves this localization in time. A *soliton* is a solitary wave which exhibits some strong form of stability so that it has a particle-like behavior. There are several books devoted to the study of solitons (see e.g. [73, 122, 124, 147]).

Usually people make the history of soliton to start in 1834 when John Scott Russell[132] described his *Wave of Translation*:

> I was observing the motion of a boat which was rapidly drawn along a narrow channel by a pair of horses, when the boat suddenly stopped – not so the mass of water in the channel which it had put in motion; it accumulated round the prow of the vessel in a state of violent agitation, then suddenly leaving it behind, rolled forward with great velocity, assuming the form of a large solitary elevation, a rounded, smooth and well-defined heap of water, which continued its course along the channel apparently without change of form or diminution of speed. I followed it on horseback, and overtook it still rolling on at a rate of some eight or nine miles an hour, preserving its original figure some thirty feet long and a foot to a foot and a half in height. Its height gradually diminished, and after a chase of one or two miles I lost it in the windings of the channel. Such, in the month of August 1834, was my first chance interview with that singular and beautiful phenomenon which I have called the Wave of Translation.

Today, we know (at least) three mechanisms which might produce solitary waves and solitons:

- Complete integrability: e.g. Kortewg-de Vries equation (KdV)

$$u_t + u_{xxx} + 6uu_x = 0.$$

- Topological constraints: e.g. the Sine-Gordon equation

$$u_{tt} - u_{xx} + \sin u = 0.$$

- Ratio energy/charge: e.g. the Gross-Pitaevskii equation:

$$i\psi_t = -\frac{1}{2}\psi_{xx} - |\psi|^2 \psi; \ \psi \in \mathbb{C}.$$

Let us briefly describe the main features of these tree types of soliton (see e.g. [73, 138, 146]). The Kortewg-de Vries equation describes the movement of water waves in shallow water and give rise to the *Wave of Translation* of Scott Russell. These kinds of solitons can be understood by viewing these equations as infinite dimensional integrable Hamiltonian systems. Their study leads to very fruitful approaches for "integrating" such systems, the inverse scattering transform and more general inverse spectral methods.

Also the Sine-Gordon equation is completely integrable, but its solitons are also topological. A topological soliton (also called by the physicists *topological defect*) is a soliton whose stability is due to topological constraints, rather than integrability of the field equations. The constraints split the function space of the states into several homotopy classes. Thus, the solutions can be classified into homotopy classes. There is no continuous transformation that maps a solution in one homotopy class to another solution belonging to a different homotopy class. The solutions are truly distinct, and they maintain their integrity, even in the face of strong perturbations.

While complete integrability is an exceptional feature, the topological constraints are preserved by a large class of perturbations and variants of the Sine-Gordon equation. For example, the equations $u_{tt} - u_{xx} + \sin^3 u = 0$ is not completely integrable, but still it has topological solitons. The strongest request for an equation to have topological solitons is the continuity of the solutions which occurs in low space dimensions. For example, if we consider the equation,

$$\psi_{tt} - \Delta\psi + W'(\psi) = 0, \ \ W \geq 0,$$

then the Derrick theorem shows that there are not solitons if $N \geq 2$ ([72], see also the discussion in Sect. 4.2.1). A possible strategy to get topological solitons in higher dimensions consists in replacing the Laplace operator in the above equation as suggested by Derrick. This strategy has been adopted in [1,23,25,27]. Topological solitons appear also in many other situations; we refer to [73, 109, 124, 147] for the bibliography on this subject.

This book is devoted to the third type of solitons which will be called *hylomorphic solitons*. The Gross-Pitaevskii equation presents this type of solitons (see Theorem 52) even if it is also completely integrable. In order to have this kind of solitons, it is not necessary to have infinite integrals of motion as in the case of completely integrable systems, but only two, the energy E and the charge C, provided that a suitable relation holds between them. This relation, roughly speaking, can be expressed as follows:

a hylomorphic soliton is a state which realizes the minimum of the energy for a fixed charge.

A mathematical formalization of this definition will be given in Sect. 2.1.3.

The most general equations for which it is possible to have hylomorphic solitons need to have the following features:

A-1. *The equations are variational, namely they are the Euler-Lagrange equations relative to a Lagrangian density $\mathcal{L}$.*

A-2. *The equations are invariant for time translations, namely $\mathcal{L}$ does not depend explicitly on t.*

A-3. *The equations are invariant for a gauge action, namely $\mathcal{L}$ does not depend explicitly on the phase of the field Ψ which is supposed to be complex valued (or at least to have some complex valued component).*

By Noether's theorem (see Sect. 1.3.1) **A-1** and **A-2** guarantee the conservation of energy, while **A-1** and **A-3** guarantee the conservation of another constant of motion which we call *hylenic charge*.

The class of hylomorphic solitons include the Q-balls (see [60]), which are spherically symmetric solutions of the nonlinear Klein-Gordon equation and which have been first studied in the pioneering papers [60, 123]. Q-balls play an important role in the study of the origin of the matter that fills the universe (see [74]) and in the study of bosonic particles (see [99, 104]), when there is an attraction between the particles. The existence of Q-balls has been recently analysed in [14, 15, 21, 34].

Chapter 1
The General Principles

This chapter concerns the very general principles which are at the base of the existence of hylomorphic solitons such as the Variational Principle, the Invariance Principle, Noether's theorem, the Hamilton-Jacobi theory. A recent historical and epistemological analysis of these principles can be found in [44].

1.1 The Variational Principle

The fundamental equations of Physics are the Euler-Lagrange equations of a suitable functional. This fact is quite surprising. There is no logical reason for this. It is just an empirical fact: all the fundamental equations which have been discovered until now derive from a variational principle (see also [24]).

For example, the equations of motion of k particles whose positions at time t are given by $x_j(t)$, $x_j \in \mathbf{R}^3$, $j = 1, \ldots, k$ are obtained as the Euler Lagrange equations relative to the following functional

$$\mathcal{S} = \int \left(\sum_j \frac{m_j}{2} |\dot{x}_j|^2 - V(t, x_1, \ldots, x_k) \right) dt \qquad (1.1)$$

where m_j is the mass of the j-th particle and V is the potential energy of the system.

More generally, the equations of motion of a finite dimensional system whose generalized coordinates are $q_j(t)$, $j = 1, \ldots, k$ are obtained as the Euler Lagrange equations relative to the following functional

$$\mathcal{S} = \int \mathcal{L}(t, q_1, \ldots, q_k, \dot{q}_1, \ldots, \dot{q}_k) \, dt$$

where $\mathcal{L}$ is the Lagrangian of the system.

© Springer International Publishing Switzerland 2014
V. Benci, D. Fortunato, *Variational Methods in Nonlinear Field Equations*,
Springer Monographs in Mathematics, DOI 10.1007/978-3-319-06914-2_1

Also the Dynamics of fields can be determined by the Variational Principle. From a mathematical point of view a field is a function

$$u : \mathbf{R}^{N+1} \to V, \quad u = (u_1, \ldots, u_k).$$

where $\mathbf{R}^{N+1}$ is the space-time continuum and $V \cong \mathbf{R}^k$ is called the internal parameters space. Of course, in physical problems, the space dimension N is $1, 2$ or 3. The space and time coordinates will be denoted by $x = (x_1, \ldots, x_N)$ and t respectively. The function $u(t, x)$ describes the *internal* state of the ether (or vacuum) at the point x and time t.

Assumption **A-1** in the introduction states that the field equations are obtained by the variation of the action functional defined as follows:

$$S = \int \int \mathcal{L}(t, x, u, \partial_t u, \nabla u) \, dx \, dt. \tag{1.2}$$

The function $\mathcal{L}$ is called Lagrangian density function but in the following, as usual, we will call it just Lagrangian function.

If u is a scalar function, the variation of (1.2) gives the following equation:

$$\sum_{i=0}^{N} \frac{\partial}{\partial x_i} \left(\frac{\partial \mathcal{L}}{\partial u_{x_i}} \right) - \frac{\partial \mathcal{L}}{\partial u} = 0. \tag{1.3}$$

If $u = (u_1, \ldots, u_k)$, the Euler-Lagrange equations take the same form provided that we use the convention that

$$\frac{\partial \mathcal{L}}{\partial u_{x_i}} = \left(\frac{\partial \mathcal{L}}{\partial u_{1,x_i}}, \ldots, \frac{\partial \mathcal{L}}{\partial u_{k,x_i}} \right), \quad \frac{\partial \mathcal{L}}{\partial u} = \left(\frac{\partial \mathcal{L}}{\partial u_1}, \ldots, \frac{\partial \mathcal{L}}{\partial u_k} \right). \tag{1.4}$$

So, if u has k components ($k > 1$) then Eq. (1.3) is equivalent to the k equations:

$$\sum_{i=0}^{N} \frac{\partial}{\partial x_i} \left(\frac{\partial \mathcal{L}}{\partial u_{\ell,x_i}} \right) - \frac{\partial \mathcal{L}}{\partial u_\ell} = 0, \quad \ell = 1, \ldots, k. \tag{1.5}$$

1.2   The Invariance Principle

Given a Lie group G, a representation of G is a homomorphism

$$T : G \to Hom(\mathfrak{X})$$

where $Hom(\mathfrak{X})$ is the group of homeomorphism on a space $\mathfrak{X}$.

If $\mathcal{X}$ is a Hilbert space, $GL(\mathcal{X})$ is the group of linear, invertible operators on $\mathcal{X}$ and

$$T : G \to GL(\mathcal{X})$$

then T is called linear representation.

A functional J is called invariant under a representation T_g of a Lie group if

$$\forall u \in \mathcal{X}, \ J(T_g u) = J(u). \tag{1.6}$$

Now, let us consider the variational equation

$$\begin{cases} u \in \mathcal{X} \\ F(u) = 0 \end{cases}$$

where $F(u) = dJ(u)$. If J is invariant, given any solution u, we have that also $T_g u$ is a solution.

We need to be careful in the definition of invariance. In fact, if u belongs to some function space $\mathcal{X} = \mathfrak{F}(\Omega, V)$, (where $\Omega \subset \mathbb{R}^{N+1}$ and V is a finite dimensional vector space), it might happen that $T_g u \notin \mathfrak{F}(\Omega, V)$ and so (1.6) does not make sense. For example, if

$$T_h u = u(x - h), \ x \in \mathbb{R}^{N+1}, h \in \mathbb{R}^{N+1}$$

we have that $T_h u \in \mathfrak{F}(\Omega', V)$ where $\Omega' = \Omega + h$.

Thus we are led to give the following definition:

Definition 1. Given the Lagrangian $\mathcal{L}\left(x, u, \frac{\partial u}{\partial x}\right)$, (where $\frac{\partial u}{\partial x} = \left(\frac{\partial u}{\partial x_0}, \ldots, \frac{\partial u}{\partial x_N}\right)$), we say that the action

$$S(u, \Omega) = \int_\Omega \mathcal{L}\left(x, u, \frac{\partial u}{\partial x}\right) dx,$$

is invariant under the transformation

$$(x, u) \to (x', u')$$

if

$$\int_{\Omega'} \mathcal{L}\left(x', u', \frac{\partial u'}{\partial x'}\right) dx' = \int_\Omega \mathcal{L}\left(x, u, \frac{\partial u}{\partial x}\right) dx \tag{1.7}$$

where $x' = T_g x$, $u' = u'(x') = (T_g u)(T_g x)$ and $\Omega' = T_g \Omega = \{x' \in \mathbb{R}^{N+1} : x \in \Omega\}$.

If we consider (x', u') as functions of (x, u), (1.7) becomes

$$\int_\Omega \mathcal{L}\left(x'(x, u), u'(x, u), \frac{\partial u'}{\partial x'}(x, u)\right) \left|\det\left(\frac{\partial x'}{\partial x}\right)\right| dx = \int_\Omega \mathcal{L}\left(x, u, \frac{\partial u}{\partial x}\right) dx$$
(1.8)

where $\det\left(\frac{\partial x'}{\partial x}\right)$ is the Jacobian determinant of the transformation $(x, u) \to (x', u')$.

Remark 2. In most applications and in all the applications of this book $T_{g(\lambda)}$ is a unitary representation (in a suitable Sobolev space); then $\left|\det\left(\frac{\partial x'}{\partial x}\right)\right| \equiv 1$, and hence, (1.8) holds provided that

$$\mathcal{L}\left(x'(x, u), u'(x, u), \frac{\partial u'}{\partial x'}(x, u)\right) = \mathcal{L}\left(x, u, \frac{\partial u}{\partial x}\right).$$
(1.9)

This identity is easier to check. For example, the Lagrangian

$$\mathcal{L} = \frac{1}{2}\left(\frac{\partial u}{\partial t}\right)^2 - \frac{1}{2}|\nabla u|^2$$

does not depend explicitly on space and time, and hence (1.9) holds for the transformations

$$x'(x, u) = x + \tau_k, \ k = 0, \ldots, 3$$

$$u'(x, u) = u$$

$$\frac{\partial u'}{\partial x'}(x, u) = \frac{\partial u}{\partial x}$$

where $x = (t, x_1, x_2, x_3)$ and $\tau_k = \left(\delta_0^k, \delta_1^k, \delta_2^k, \delta_3^k\right)$ is a space/time translation.

1.2.1   The Poincaré Invariance

The fundamental equations of Physics are invariant for the Poincaré group: it is the *basic* principle on which the special theory of relativity is founded.

The Poincaré group $\mathfrak{P}$ is a generalization of the isometry group $\mathfrak{E}$. The isometry group $\mathfrak{E}$ in $\mathbb{R}^N$ is the group of transformations which preserve the quadratic form

$$|x|^2 := \sum_{i=1}^{N} x_i^2$$

i.e. the Euclidean norm and hence the Euclidean distance

$$d_E(x, y) = \sqrt{\sum_{i=1}^{N} |x_i - y_i|^2};$$

namely, $g \in \mathfrak{E}$, if and only if

$$d_E(gx, gy) = d_E(x, y).$$

In Euclidean geometry, the isometry group is also called congruence group. Roughly speaking, the content of Euclidean geometry is the study of the properties of geometric objects which are preserved by the *congruence* group.

The Poincaré group $\mathfrak{P}$, by definition, is the transformation group in $\mathbb{R}^{N+1}$ which preserves the quadratic form

$$|x|_M^2 = -x_0^2 + \sum_{i=1}^{N} x_i^2$$

which is induced by the Minkowski bilinear form

$$\langle x, y \rangle_M = -x_0 y_0 + \sum_{i=1}^{N} x_i y_i.$$

In Physics, of course we have $N = 3$, and $\mathbb{R}^4$ equipped with the Minkowski bilinear form is called space-time. The generic point $(x_0, x_1, x_2, x_3) \in \mathbb{R}^4$ usually is called *event* and it is denoted by (t, x_1, x_2, x_3) or $(t, \mathbf{x})$.

The Minkowski vectors $v = (v_0, \ldots, v_3) \in \mathbb{R}^4$ are classified according to their *causal* nature as follows:

- A vector is called space-like if $\langle v, v \rangle_M > 0$.
- A vector is called light-like if $\langle v, v \rangle_M = 0$.
- A vector is called time-like if $\langle v, v \rangle_M < 0$.

The causal nature is not changed by a Poincaré transformation, and hence it is not a transitive group (as the isometry group): space and time-are mixed, but not … so much.

In the physical world, where we have $N = 3$, the Poincaré group is a ten parameters Lie group generated by the following one-parameter transformations:

- Space translations in the directions x_1, x_2, x_3:

$$\begin{aligned} x_1' &= x_1 + x_{10} \\ x_2' &= x_2 \\ x_3' &= x_3 \\ t' &= t \end{aligned} \quad ; \quad \begin{aligned} x_1' &= x_1 \\ x_2' &= x_2 + x_{20} \\ x_3' &= x_3 \\ t' &= t \end{aligned} \quad ; \quad \begin{aligned} x_1' &= x_1 \\ x_2' &= x_2 \\ x_3' &= x_3 + x_{30} \\ t' &= t \end{aligned} \quad .$$

This invariance guarantees that space is homogeneous, namely that the laws of physics are independent of space: if an experiment is performed here or there, it gives the same results.

- Space rotations:

$$
\begin{array}{lll}
x_1' = x_1 & x_1' = x_1 \cos\theta_2 - x_3 \sin\theta_2 & x_1' = x_1 \cos\theta_3 - x_2 \sin\theta_3 \\
x_2' = x_2 \cos\theta_1 - x_3 \sin\theta_1 & x_2' = x_2 & x_2' = x_1 \sin\theta_3 + x_2 \cos\theta_3 \\
x_3' = x_2 \sin\theta_1 + x_3 \cos\theta_1 & x_3' = x_1 \sin\theta_2 + x_3 \cos\theta_2 & x_3' = x_3 \\
t' = t & t' = t & t' = t
\end{array}
$$

This invariance guarantees that space is isotropic, namely that the laws of physics are independent of orientation.

- Time translations:

$$
\begin{aligned}
x_1' &= x_1 \\
x_2' &= x_2 \\
x_3' &= x_3 \\
t' &= t + t_0.
\end{aligned}
$$

This invariance guarantees that time is isotropic; namely that the laws of physics are independent of time: if an experiment is performed earlier or later, it gives the same results.

- Lorentz boosts:

$$
\begin{array}{lll}
x_1' = \gamma(x_1 - v_1 t) & x_1' = x_1 & x_1' = x_1 \\
x_2' = x_2 & x_2' = \gamma(x_2 - v_2 t) & x_2' = x_2 \\
x_3' = x_3 & x_3' = x_3 & x_3' = \gamma(x_3 - v_3 t) \\
t' = \gamma(t - v_1 x_1) & t' = \gamma(t - v_2 x_2) & t' = \gamma(t - v_3 x_3)
\end{array}
\tag{1.10}
$$

where

$$
\gamma = \frac{1}{\sqrt{1 - v^2}}
$$

with $v = v_i$, $i = 1, 2, 3$. This invariance is an empirical fact and, as it will be shown in Sect. 4.3, it implies the remarkable facts of the theory of relativity such as the space contraction, the time dilation and the equality between mass and energy.

The Lorentz group is the six parameters Lie group generated by the space rotations and the Lorentz boosts (plus the time inversion, $t \to -t$, and the parity inversion $(x_1, x_2, x_3) \to (-x_1, -x_2, -x_3)$). Clearly it is a subgroup of $GL(4)$.

The Poincaré group $\mathfrak{P}$ is the ten parameters Lie group generated by the Lorentz group and the space-time translations. Then it is a subgroup of the affine group in $\mathbb{R}^4$.

So the Poincaré group acts on a scalar field ψ by the following representation:

$$\text{for all } g \in \mathfrak{P} : \left(T_g\psi\right)(t,x) = \psi\left(t',x'\right), \quad \left(t',x'\right) = g\left(t,x\right).$$

The "simplest" equation invariant under the Poincaré group is the D'Alembert equation:

$$\Box\psi = 0 \tag{1.11}$$

where

$$\Box\psi = \frac{\partial^2\psi}{\partial t^2} - \Delta\psi \text{ and } \Delta\psi = \frac{\partial^2\psi}{\partial x_1^2} + \frac{\partial^2\psi}{\partial x_2^2} + \frac{\partial^2\psi}{\partial x_3^2}.$$

Actually, the D'Alembert equation is the simplest *variational* field equation invariant under the Poincaré group.

In fact it is obtained from the variation of the action

$$S_0 = -\frac{1}{2}\int \langle d\psi, d\psi\rangle_M \, dx\,dt = \frac{1}{2}\int \left[|\partial_t\psi|^2 - |\nabla\psi|^2\right] dx\,dt. \tag{1.12}$$

In this case, the Lagrangian function is given by

$$\mathcal{L}_0 = -\frac{1}{2}\langle d\psi, d\psi\rangle_M = \frac{1}{2}|\partial_t\psi|^2 - \frac{1}{2}|\nabla\psi|^2. \tag{1.13}$$

It is easy to check directly that if ψ is a solution of D'Alembert equation, then also $T_g\psi$ is a solution of it for every $g \in \mathfrak{P}$.

1.2.2 The Galileo Invariance

The Galileo group $\mathfrak{G}$, as the Poincaré group, is a transformation group on the space time $\mathbb{R}^4$. The Galileo group, by definition, is the set of transformations which preserve the time intervals and the Euclidean distance between simultaneous events. More precisely, if $g \in \mathfrak{G}$, given two points (t,x) and $(\bar{t},\bar{x})$, we have that

$$t' - \bar{t}' = t - \bar{t}$$

and

$$t = \bar{t} \Rightarrow t' = \bar{t}' \text{ and } d_E\left(x',\bar{x}'\right) = d_E\left(x,\bar{x}\right)$$

where $(t',x') = g\left(t,x\right)$, $(\bar{t}',\bar{x}') = g\left(\bar{t},\bar{x}\right)$ and $d_E(x,y)$ is the Euclidean distance.

Thus the Galileo group is a ten parameters Lie group generated by the space-time translations and the space rotations, but the Lorentz boosts (1.10) are replaced be the *Galilean transformations* namely by the transformations

$$
\begin{array}{lll}
x_1' = x_1 - v_1 t & x_1' = x_1 & x_1' = x_1 \\
x_2' = x_2 & x_2' = x_2 - v_2 t & x_2' = x_2 \\
x_3' = x_3 & x_3' = x_3 & x_3' = x_3 - v_3 t \\
t' = t & t' = t & t' = t
\end{array} \qquad (1.14)
$$

We refer to the three parameters group of Galilean transformations as the restricted Galileo group. The equations of classical mechanics are invariant under the Galileo group. Let us see a field equation which is invariant under a representation of the restricted Galileo group, namely the Schrödinger equation.

Theorem 3. *Given the Galileo transformation* $g_v : \mathbb{R}^4 \to \mathbb{R}^4$ *defined by*

$$
g_v (t, x) = (t, x - \mathbf{v}t), \quad \mathbf{v} = (v_1, v_2, v_3) \qquad (1.15)
$$

and the family of transformations $T_v : L_{loc}^2 (\mathbb{R}^4, \mathbb{C}) \to L_{loc}^2 (\mathbb{R}^4, \mathbb{C})$ *defined by*

$$
(T_v \psi)(t, x) = \psi (t, x - \mathbf{v}t) \, e^{i(\mathbf{v} \cdot x - \frac{1}{2} \mathbf{v}^2 t)}. \qquad (1.16)
$$

we have that:

(i) $\{T_v\}_{v \in \mathbb{R}^3}$ *is a group of linear transformations on* $L_{loc}^2 (\mathbb{R}^4, \mathbb{C})$ *(namely it is a linear representation of the group (1.14)).*

(ii) *The Schrödinger equation for a free particle*

$$
i \frac{\partial \psi}{\partial t} = -\frac{1}{2} \Delta \psi \qquad (1.17)
$$

is invariant for $\{T_v\}_{v \in \mathbb{R}^4}$.

Proof. (i) It is immediate to check that T_v is linear; in order to see that $\{T_v\}_{v \in \mathbb{R}^4}$ is a group, it is sufficient to prove that

$$
T_w \circ T_v = T_{w+v}.
$$

Given a generic $\psi \in L_{loc}^2 (\mathbb{R}^4, \mathbb{C})$, we have

$$
(T_w \circ T_v) \, \psi \, (t, x)
$$
$$
= T_w \left[\psi (t, x - \mathbf{v}t) \, e^{i(\mathbf{v} \cdot x - \frac{1}{2} \mathbf{v}^2 t)} \right]
$$
$$
= \psi (t, x - \mathbf{w}t - \mathbf{v}t) \, e^{i[\mathbf{v} \cdot (x - \mathbf{w}t) - \frac{1}{2} \mathbf{v}^2 t]} \cdot e^{i(\mathbf{w} \cdot x - \frac{1}{2} \mathbf{w}^2 t)}
$$
$$
= \psi (t, x - (\mathbf{v} + \mathbf{w}) t) \, e^{i[(\mathbf{v}+\mathbf{w})x - \frac{1}{2}(\mathbf{v}+\mathbf{w})^2 t]} = T_{w+v} \psi \, (t, x).
$$

(ii) We have to prove that

$$T_{\mathbf{v}} \circ \left(i\frac{\partial}{\partial t} + \frac{1}{2}\Delta \right) = \left(i\frac{\partial}{\partial t} + \frac{1}{2}\Delta \right) \circ T_{\mathbf{v}}.$$

We have that

$$\left[T_{\mathbf{v}} \circ \left(i\frac{\partial}{\partial t} + \frac{1}{2}\Delta \right) - \left(i\frac{\partial}{\partial t} + \frac{1}{2}\Delta \right) \circ T_{\mathbf{v}} \right] \psi\,(t, x)$$

$$= \left[T_{\mathbf{v}} \left(i\frac{\partial}{\partial t}\psi \right) - i\frac{\partial}{\partial t}\,(T_{\mathbf{v}}\psi) \right] + \frac{1}{2}\left[T_{\mathbf{v}}\,(\Delta\psi) - \Delta\,(T_{\mathbf{v}}\psi) \right].$$

Let us compute each term separately:

$$T_{\mathbf{v}} \left(i\frac{\partial}{\partial t}\psi \right) - i\frac{\partial}{\partial t}\,(T_{\mathbf{v}}\psi) = i\psi_t\,(t, x - \mathbf{v}t)\,e^{i\left(\mathbf{v}\cdot x - \frac{1}{2}\mathbf{v}^2 t\right)}$$

$$- i\frac{\partial}{\partial t}\left[\psi\,(t, x - \mathbf{v}t)\,e^{i\left(\mathbf{v}\cdot x - \frac{1}{2}\mathbf{v}^2 t\right)} \right]$$

$$= \left[i\mathbf{v}\cdot\nabla\psi\,(t, x - \mathbf{v}t) - \frac{1}{2}\mathbf{v}^2\psi\,(t, x - \mathbf{v}t) \right]$$

$$\cdot e^{i\left(\mathbf{v}\cdot x - \frac{1}{2}\mathbf{v}^2 t\right)}$$

$$T_{\mathbf{v}}\,(\Delta\psi) - \Delta\,(T_{\mathbf{v}}\psi) = \Delta\psi\,(t, x - \mathbf{v}t)\,e^{i\left(\mathbf{v}\cdot x - \frac{1}{2}\mathbf{v}^2 t\right)}$$

$$- \Delta\left[\psi\,(t, x - \mathbf{v}t)\,e^{i\left(\mathbf{v}\cdot x - \frac{1}{2}\mathbf{v}^2 t\right)} \right]$$

$$= \Delta\psi\,(t, x - \mathbf{v}t)\,e^{i\left(\mathbf{v}\cdot x - \frac{1}{2}\mathbf{v}^2 t\right)}$$

$$- \nabla \cdot \left[\nabla\psi\,(t, x - \mathbf{v}t)\,e^{i\left(\mathbf{v}\cdot x - \frac{1}{2}\mathbf{v}^2 t\right)} \right.$$

$$\left. + i\mathbf{v}\psi\,(t, x - \mathbf{v}t)\,e^{i\left(\mathbf{v}\cdot x - \frac{1}{2}\mathbf{v}^2 t\right)} \right]$$

$$= \left[2i\mathbf{v} \cdot \nabla\psi\,(t, x - \mathbf{v}t) - \mathbf{v}^2\psi\,(t, x - \mathbf{v}t) \right] e^{i\left(\mathbf{v}\cdot x - \frac{1}{2}\mathbf{v}^2 t\right)}.$$

Thus,

$$\left[T_{\mathbf{v}} \circ \left(i\frac{\partial}{\partial t} + \frac{1}{2}\Delta \right) - \left(i\frac{\partial}{\partial t} + \frac{1}{2}\Delta \right) \circ T_{\mathbf{v}} \right] \psi\,(t, x)$$

$$= \left[i\mathbf{v}\cdot\nabla\psi\,(t, x - \mathbf{v}t) - \frac{1}{2}\mathbf{v}^2\psi\,(t, x - \mathbf{v}t) \right] e^{i\left(\mathbf{v}\cdot x - \frac{1}{2}\mathbf{v}^2 t\right)}$$

$$-\frac{1}{2}\left[2i\mathbf{v}\cdot\nabla\psi\left(t,x-\mathbf{v}t\right)-\mathbf{v}^2\psi\left(t,x-\mathbf{v}t\right)\right]e^{i\left(\mathbf{v}\cdot x-\frac{1}{2}\mathbf{v}^2 t\right)}$$

$$= 0.$$

$\square$

1.2.3   The Gauge Invariance

Take a function

$$\psi : \mathbb{R}^4 \to V$$

and assume that on V acts the representation T_g of some group $(G, \circ)$. This action induces two possible actions on ψ:

- A global action: $\psi\left(x\right) \mapsto T_g\psi\left(x\right)$ where $g \in G$.
- A local action: $\psi\left(x\right) \mapsto T_{g(x)}\psi\left(x\right)$ where $g(x)$ is a smooth function with values in G.

In the second case, we have a representation of the infinite dimensional group

$$\mathfrak{G} = C^\infty\left(\mathbb{R}^4, G\right) \tag{1.18}$$

equipped with the group operation

$$(g \circ h)\left(x\right) = g(x) \circ h(x).$$

Definition 4. If a Lagrangian $\mathcal{L}$ satisfies the following condition,

$$\mathcal{L}(t, x, \psi, \nabla\psi, \partial_t\psi) = \mathcal{L}\left(t, x, T_g\psi, \nabla\left(T_g\psi\right), \partial_t\left(T_g\psi\right)\right),\ g \in G$$

we say that it is invariant under a global action of the group G, or under a trivial gauge action of the group G; if $\mathcal{L}$ satisfies the following condition,

$$\mathcal{L}(t, x, \psi, \nabla\psi, \partial_t\psi) = \mathcal{L}\left(t, x, T_{g(x)}\psi, \nabla\left(T_{g(x)}\psi\right), \partial_t\left(T_{g(x)}\psi\right)\right),\ g \in \mathfrak{G}$$

we say that it is invariant under a local action of the group G, or under a gauge action of the group $\mathfrak{G}$.

Let us consider two simple examples: the functional

$$\int \mathcal{L}\left(\nabla u\right)\,dx\ ,\ u \in \mathbb{R}$$

is invariant under a global action of the group $(\mathbb{R}, +)$. In fact, if we set $T_r u = u + r$, $r \in \mathbb{R}$, we have that

$$\mathcal{L}(\nabla u) = \mathcal{L}(\nabla(T_r u)).$$

Next, consider the functional

$$\int \mathcal{L}(d\alpha) \, dx$$

where α is a 1-form and d is the exterior derivative of α. In this case, $\mathcal{L}(d\alpha)$ is not only invariant for a trivial action of $(\mathbb{R}, +)$, but also for the local action

$$T_{g(x)}\alpha = \alpha + dg(x), \ g \in \mathfrak{G} := C^{\infty}(\mathbb{R}^4, \mathbb{R})$$

in fact

$$\mathcal{L}(d(\alpha + dg(x))) = \mathcal{L}(d\alpha).$$

The simplest gauge invariance can be obtained taking a complex valued scalar field

$$\psi : \mathbb{R}^4 \to \mathbb{C},$$

and considering the group $S^1 = \{e^{i\theta} : \theta \in \mathbb{R}\}$ and the following representation

$$\psi \mapsto e^{i\theta}\psi. \tag{1.19}$$

The Schrödinger equation and the Klein-Gordon equation are invariant for the global action (1.19). The Klein-Gordon-Maxwell equations are invariant for the local action (5.37)–(5.39). For a discussion of these aspects we refer to section 2.7 in [124], to section 1.4 in [147] and to [30].

1.3 Conservation Laws

The conservation laws are one of the most important consequences of the structure of the variational equations. A conservation law states that a certain quantity, uniquely determined by the state of the system, is conserved during the evolution of the system itself. Such a quantity, for historical reasons, is called first integral. For example the law of conservation of energy is one ot them.

The conservation laws are a consequence of the variational principle, or more precisely, they are a consequence of the variational principle and the symmetry of the Lagrangian. The deep connection between the variational principle, symmetry and conservation laws is one of the most profound and important facts of our physical-mathematical description of the world.

This connection provides a very important theorem due to Emmy Noether (see [117]) which will be discussed in the next section. Roughly speaking, this theorem can be stated in the following way:

If the Lagrangian of a physical system is invariant under a continuous transformation group which depends on n independent parameters, then this system has a first integral for each of these parameters.

1.3.1   Noether's Theorem

In this section we will give a proof of Noether's theorem stated in a suitable form for the applications considered in this paper. A classical proof can be found in [80].

Suppose that a Lagrangian is invariant under the action T_g of some Lie group G. We denote by $T_{g(\lambda)}$ ($\lambda \in \mathbb{R}$) the action of a one-parameter subgroup $\{g(\lambda)\}_{\lambda \in \mathbb{R}}$. Under the action of $T_{g(\lambda)}$, t, x and u transform in new variables which we will denote by

$$t'(\lambda, t, x)$$
$$x'(\lambda, t, x)$$
$$u'(\lambda, t', x').$$

For example, the time translations in $\mathbb{R}^2$, are defined as follows:

$$t'(\lambda, t, x) = t + \lambda \tag{1.20}$$
$$x'(\lambda, t, x) = x \tag{1.21}$$
$$u'(\lambda, t', x') = u'(\lambda, t + \lambda, x) = u(t, x). \tag{1.22}$$

The theorem is valid also if the variables t', x' and u' depend also from u and its derivatives. The proof is the same and we do not consider the dependence on u just to simplify the notation. We set

$$\rho = \left(\frac{\partial \mathcal{L}}{\partial u_t} \frac{\partial u'}{\partial \lambda} + \mathcal{L} \frac{\partial t'}{\partial \lambda} \right)_{\lambda=0} \tag{1.23}$$

and

$$\mathbf{J} = \sum_{i=1}^{N} \left(\frac{\partial \mathcal{L}}{\partial u_{x^i}} \frac{\partial u'}{\partial \lambda} + \mathcal{L} \frac{\partial x_i'}{\partial \lambda} \right)_{\lambda=0} \mathbf{e}_i \tag{1.24}$$

where we denote by u_{x^i}, $i = 1, \ldots, N$ the last N independent variables of $\mathcal{L}$.

Theorem 5 (Noether's theorem). *Let $\mathcal{L}$ be invariant under the action of a one parameter group $T_{g(\lambda)}$, and let u be a smooth solution of the Euler-Lagrange equation (1.3). Suppose that u decays sufficiently fast so that*

$$\rho(\cdot, t), \ \frac{\partial \rho}{\partial t}(\cdot, t) \ and \ \mathbf{J}(\cdot, t) \ are \ in \ L^1(\mathbb{R}^N) \tag{1.25}$$

where ρ and $\mathbf{J}$ are defined by (1.23) and (1.24). Then,

$$\mathcal{I}(u) = \int \rho(x, t)dx$$

is an integral of motion.

For example take the Lagrangian

$$\mathcal{L} = \frac{1}{2} \left(\frac{\partial u}{\partial t} \right)^2 - \frac{1}{2} \left(\frac{\partial u}{\partial x} \right)^2.$$

This Lagrangian is invariant for time translations; then by (1.20), we have that

$$\frac{\partial t'}{\partial \lambda} = 1$$

and by (1.22), we have that

$$u'(\lambda, t, x) = u(t - \lambda, x)$$

and hence

$$\frac{\partial u'}{\partial \lambda}(\lambda, t, x) = -\frac{\partial u}{\partial t}(t - \lambda, x).$$

Thus

$$\mathcal{I}(u) = -\int \left[\left(\frac{\partial u}{\partial t} \right)^2 - \frac{1}{2}\left(\frac{\partial u}{\partial t} \right)^2 + \frac{1}{2}\left(\frac{\partial u}{\partial x} \right)^2 \right] dx = -\frac{1}{2} \int \left[\left(\frac{\partial u}{\partial t} \right)^2 + \left(\frac{\partial u}{\partial x} \right)^2 \right] dx$$

In order to prove Noether's theorem some work is necessary.

Lemma 6. *Let $\rho : \mathbb{R}^{N+1} \rightarrow \mathbb{R}$ and $\mathbf{J} : \mathbb{R}^{N+1} \rightarrow \mathbb{R}^N$ be two smooth functions defined on the "space-time". Assume that they satisfy the continuity equation*

$$\frac{\partial \rho}{\partial t} + \nabla \cdot \mathbf{J} = 0 \tag{1.26}$$

and that for all t (1.25) holds. Then for all t

$$\frac{d}{dt} \int_{\mathbb{R}^N} \rho(x,t)dx = 0. \tag{1.27}$$

Proof. Let

$$B_R = \{x \in \mathbb{R}^3 : |x| < R\}, \ R > 0$$

then, integrating over B_R, we get

$$\int_{B_R} \frac{\partial \rho}{\partial t} dx = - \int_{B_R} \nabla \cdot \mathbf{J} dx = - \int_{\partial B_R} (\mathbf{J} \cdot \mathbf{n}) d\sigma \tag{1.28}$$

where $\mathbf{n}$ denotes the outward normal to the boundary ∂B_R of B_R. Then

$$\left| \int_{B_R} \frac{\partial \rho}{\partial t} dx \right| \leq \int_{\partial B_R} |\mathbf{J} \cdot \mathbf{n}| d\sigma. \tag{1.29}$$

Since $\frac{\partial \rho}{\partial t}(.,t)$ is in $L^1(\mathbb{R}^3)$, there exists $\lim_{R \to \infty} \left| \int_{B_R} \frac{\partial \rho}{\partial t} dx \right|$, and we have to prove that this limit is 0. Arguing by contradiction we assume that

$$\lim_{R \to \infty} \left| \int_{B_R} \frac{\partial \rho}{\partial t} dx \right| = \alpha > 0. \tag{1.30}$$

Then, by (1.29) and (1.30), the map φ defined by

$$\varphi(R) = \int_{\partial B_R} |\mathbf{J} \cdot \mathbf{n}| d\sigma$$

is not integrable in $(0, +\infty)$ and

$$\int_{\mathbb{R}^N} |\mathbf{J} \cdot \mathbf{n}| dx = \int_0^{+\infty} \varphi(R)dR = +\infty$$

which contradicts assumption (1.25).   □

In the next lemmas, we set $x_0 = t$ and $x = (x_0, \dots, x_N)$.

Lemma 7. *Let $\mathcal{J}(x'(\lambda, x))$ be the Jacobian of the transformation $x \mapsto x'(\lambda, x)$. Assume that the action relative to $\mathcal{L}$ is invariant under the action of a one parameter group $T_{g(\lambda)}$ (see Definition 1), then*

$$\frac{d}{d\lambda} \left[\mathcal{L} \left(x'(\lambda, x), u'(\lambda, x), \frac{\partial u'}{\partial x}(\lambda, x) \right) \mathcal{J}(x'(\lambda, x)) \right]_{\lambda=0} = 0. \tag{1.31}$$

Proof. Since the action is invariant, we have that, for every $\lambda \in \mathbb{R}$ and every $\Omega \subset \mathbb{R}^{N+1}$ (Ω bounded)

$$\int_{\Omega'} \mathcal{L}\left(x', u', \frac{\partial u'}{\partial x'}\right) dx' = \int_{\Omega} \mathcal{L}\left(x, u, \frac{\partial u}{\partial x}\right) dx.$$

Now, making the change of variables $x \mapsto x'(\lambda, x)$

$$\int_{\Omega} \mathcal{L}\left(x'(\lambda, x), u'(\lambda, x), \frac{\partial u'}{\partial x'}(\lambda, x)\right) \mathcal{J}(x'(\lambda, x)) dx = \int_{\Omega} \mathcal{L}\left(x, u, \frac{\partial u}{\partial x}\right) dx.$$

Then

$$\frac{d}{d\lambda} \int_{\Omega} \mathcal{L}\left(x'(\lambda, x), u'(\lambda, x), \frac{\partial u'}{\partial x'}(\lambda, x)\right) \mathcal{J}(x'(\lambda, x)) dx = 0$$

and, by the arbitrariness of Ω, we have that

$$\frac{d}{d\lambda}\left[\mathcal{L}\left(x'(\lambda, x), u'(\lambda, x), \frac{\partial u'}{\partial x'}(\lambda, x)\right) \mathcal{J}(x'(\lambda, x))\right] = 0. \tag{1.32}$$

Notice that $\frac{\partial u'}{\partial x'}(\lambda, x) = \left(\frac{\partial u'}{\partial x_0'}, \ldots, \frac{\partial u'}{\partial x_N'}\right)$. For $\lambda = 0$, we have that $\frac{\partial x_i'}{\partial x_j} = \delta_{ij}$ and so

$$\frac{\partial u'}{\partial x_i} = \sum_j \frac{\partial u'}{\partial x_j'} \frac{\partial x_j'}{\partial x_i} = \sum_j \frac{\partial u'}{\partial x_j'} \delta_{ij} = \frac{\partial u'}{\partial x_i'}.$$

Then

$$\left(\frac{\partial u'}{\partial x'}\right)_{\lambda=0} = \left(\frac{\partial u'}{\partial x_0'}, \ldots, \frac{\partial u'}{\partial x_N'}\right)_{\lambda=0} = \left(\frac{\partial u'}{\partial x_0}, \ldots, \frac{\partial u'}{\partial x_N}\right)_{\lambda=0} = \left(\frac{\partial u'}{\partial x}\right)_{\lambda=0}. \tag{1.33}$$

So (1.31) follows from (1.32) and (1.33). $\qquad\square$

Lemma 8. *Let u be a solution of (1.3). Then, under the same assumptions of Lemma 7, we have*

$$\sum_{i=0}^{N} \frac{\partial}{\partial x_i}\left(\frac{\partial \mathcal{L}}{\partial u_{x^i}} \frac{\partial u'}{\partial \lambda} + \mathcal{L}\frac{\partial x_i'}{\partial \lambda}\right) = 0. \tag{1.34}$$

Proof. Let us compute $\frac{d\mathcal{L}}{d\lambda}$:

$$\frac{d\mathcal{L}}{d\lambda} = \sum_{i=0}^{N} \frac{\partial \mathcal{L}}{\partial u_{x^i}} \frac{d}{d\lambda} \frac{\partial u'}{\partial x^i} + \frac{\partial \mathcal{L}}{\partial u} \frac{du'}{d\lambda} + \sum_{i=0}^{N} \frac{\partial \mathcal{L}}{\partial x_i} \frac{\partial x_i'}{\partial \lambda}$$

$$= \sum_{i=0}^{N} \frac{\partial \mathcal{L}}{\partial u_{x^i}} \frac{\partial^2 u'}{\partial x^i \partial \lambda} + \sum_{ij=0}^{N} \frac{\partial \mathcal{L}}{\partial u_{x^j}} \frac{\partial^2 u'}{\partial x^j \partial x^i} \frac{\partial x_j'}{\partial \lambda} + \frac{\partial \mathcal{L}}{\partial u} \frac{\partial u'}{\partial \lambda}$$

$$+ \sum_{j=0}^{N} \frac{\partial \mathcal{L}}{\partial u} \frac{\partial u'}{\partial x^j} \frac{\partial x_j'}{\partial \lambda} + \sum_{i=0}^{N} \frac{\partial \mathcal{L}}{\partial x_i} \frac{\partial x_i'}{\partial \lambda}.$$

Now, let us compute $\frac{d\mathcal{J}}{d\lambda}$:

$$\frac{d}{d\lambda} \mathcal{J} = \frac{d}{d\lambda} \det \left\{ \frac{\partial x_i'}{\partial x_k} \right\} = tr \left\{ \frac{d}{d\lambda} \frac{\partial x_i'}{\partial x_k} \right\} = tr \left\{ \frac{\partial^2 x_i'}{\partial \lambda \partial x_k} \right\} = \sum_{i=0}^{N} \frac{\partial^2 x_i'}{\partial \lambda \partial x_i}.$$

Since $\mathcal{J}|_{\lambda=0} = 1$, we have:

$$\frac{d}{d\lambda} (\mathcal{L}\mathcal{J})|_{\lambda=0} = \left(\frac{d\mathcal{L}}{d\lambda} \mathcal{J} + \mathcal{L} \cdot \frac{d\mathcal{J}}{d\lambda} \right)_{\lambda=0} = \frac{d\mathcal{L}}{d\lambda} + \mathcal{L} \cdot \frac{d\mathcal{J}}{d\lambda} = \frac{d\mathcal{L}}{d\lambda} + \mathcal{L} \cdot \sum_{i=0}^{N} \frac{\partial^2 x_i'}{\partial \lambda \partial x_i}$$

$$= \sum_{i=0}^{N} \frac{\partial \mathcal{L}}{\partial u_{x^i}} \frac{\partial^2 u'}{\partial x^i \partial \lambda} + \sum_{ij=0}^{N} \frac{\partial \mathcal{L}}{\partial u_{x^j}} \frac{\partial^2 u'}{\partial x^j \partial x^i} \frac{\partial x_j'}{\partial \lambda} \qquad (1.35)$$

$$+ \frac{\partial \mathcal{L}}{\partial u} \frac{\partial u'}{\partial \lambda} + \sum_{j=0}^{N} \frac{\partial \mathcal{L}}{\partial u} \frac{\partial u'}{\partial x^j} \frac{\partial x_j'}{\partial \lambda} + \sum_{i=0}^{N} \frac{\partial \mathcal{L}}{\partial x_i} \frac{\partial x_i'}{\partial \lambda} + \mathcal{L} \sum_{i=0}^{N} \frac{\partial^2 x_i'}{\partial \lambda \partial x_i}.$$

Also, we have

$$\sum_{i=0}^{N} \frac{\partial}{\partial x_i} \left(\mathcal{L} \frac{\partial x_i'}{\partial \lambda} \right) = \sum_{ij=0}^{N} \frac{\partial \mathcal{L}}{\partial u_{x^j}} \frac{\partial^2 u'}{\partial x^j \partial x^i} \frac{\partial x_i'}{\partial \lambda} + \sum_{i=0}^{N} \frac{\partial \mathcal{L}}{\partial u} \frac{\partial u'}{\partial x^i} \frac{\partial x_i'}{\partial \lambda} + \sum_{i=0}^{N} \frac{\partial \mathcal{L}}{\partial x_i} \frac{\partial x_i'}{\partial \lambda} + \mathcal{L} \frac{\partial^2 x_i'}{\partial x_i \partial \lambda}$$

and comparing the above expression with (1.35), we get

$$\frac{d}{d\lambda} (\mathcal{L}\mathcal{J})|_{\lambda=0} = \sum_{i=0}^{N} \frac{\partial \mathcal{L}}{\partial u_{x^i}} \frac{\partial^2 u'}{\partial x^i \partial \lambda} + \frac{\partial \mathcal{L}}{\partial u} \frac{\partial u'}{\partial \lambda} + \sum_{i=0}^{N} \frac{\partial}{\partial x_i} \left(\mathcal{L} \frac{\partial x_i'}{\partial \lambda} \right)$$

$$= \sum_{i=0}^{N} \frac{\partial}{\partial x_i} \left(\frac{\partial \mathcal{L}}{\partial u_{x^j}} \frac{\partial u'}{\partial \lambda} \right) - \sum_{i=0}^{N} \frac{\partial}{\partial x_i} \frac{\partial \mathcal{L}}{\partial u_{x^j}} \frac{\partial u'}{\partial \lambda} + \frac{\partial \mathcal{L}}{\partial u} \frac{\partial u'}{\partial \lambda}$$

$$= \sum_{i=0}^{N} \frac{\partial}{\partial x_i} \left(\frac{\partial \mathcal{L}}{\partial u_{x^j}} \frac{\partial u'}{\partial \lambda} + \mathcal{L} \frac{\partial x_i'}{\partial \lambda} \right) - \left(\sum_{i=0}^{N} \frac{\partial}{\partial x_i} \frac{\partial \mathcal{L}}{\partial u_{x^j}} - \frac{\partial \mathcal{L}}{\partial u} \right) \frac{\partial u'}{\partial \lambda}$$

and since u is solution of Eq. (1.3)

$$\frac{d}{d\lambda}(\mathcal{L}\mathcal{J})|_{\lambda=0} = \sum_{i=0}^{N} \frac{\partial}{\partial x_i} \left(\frac{\partial \mathcal{L}}{\partial u_{x^j}} \frac{\partial u'}{\partial \lambda} + \mathcal{L} \frac{\partial x_i'}{\partial \lambda} \right) \tag{1.36}$$

and finally, by Lemma 7 and (1.36), we have

$$\sum_{i=0}^{N} \frac{\partial}{\partial x_i} \left(\frac{\partial \mathcal{L}}{\partial u_{x^i}} \frac{\partial u'}{\partial \lambda} + \mathcal{L} \frac{\partial x_i'}{\partial \lambda} \right) = 0.$$

$\square$

Proof of Theorem 5. The Lagrangian $\mathcal{L}$ is invariant, then the relative action is invariant too. So assumptions of Lemma 8 are satisfied and (1.34) holds.

If we recall that $x_0 = t$, then Eq. (1.34) becomes the continuity equation

$$\frac{\partial \rho}{\partial t} + \nabla \cdot \mathbf{J} = 0 \tag{1.37}$$

where ρ and $\mathbf{J}$ are defined by (1.23) and (1.24). Finally, by using Lemma 6 we get the conclusion.

$\square$

1.3.2 Main Constants of Motion

Now, using Noether's theorem (Theorem 5), we can compute the main constants of motion. They are due to the homogeneity of time and to the homogeneity and isotropy of space which provide the invariance with respect to the time translations, space translations and space rotations. We consider the case in which $\mathcal{L}$ depends on a complex valued scalar function ψ.

- **Energy**. Energy, by definition, is the quantity which is preserved by the time invariance of the Lagrangian; it has the following form

$$E = \mathrm{Re} \int \left(\frac{\partial \mathcal{L}}{\partial \psi_t} \frac{\overline{\partial \psi}}{\partial t} - \mathcal{L} \right) dx \tag{1.38}$$

where ψ_t represents the independent variable of $\mathcal{L}$ corresponding to $\frac{\partial \psi}{\partial t}$ and $\frac{\partial \psi}{\partial t}$ is the partial derivative of ψ.

Proof. If we identify $\psi = u_1 + i u_2$, with (u_1, u_2), and using (1.4), we have that

$$\frac{\partial \mathcal{L}}{\partial \psi_t} = \frac{\partial \mathcal{L}}{\partial u_{1t}} + i \frac{\partial \mathcal{L}}{\partial u_{2t}}.$$

Hence, (1.23) becomes

$$\rho = \left(\frac{\partial \mathcal{L}}{\partial u_{1t}} \frac{\partial u_1'}{\partial \lambda} + \frac{\partial \mathcal{L}}{\partial u_{2t}} \frac{\partial u_2'}{\partial \lambda} + \mathcal{L} \frac{\partial t'}{\partial \lambda} \right)_{\lambda=0} ;$$

then, using the complex valued notation, we get

$$\rho = \text{Re} \left(\frac{\partial \mathcal{L}}{\partial \psi_t} \frac{\overline{\partial \psi'}}{\partial \lambda} + \mathcal{L} \frac{\partial t'}{\partial \lambda} \right)_{\lambda=0} . \qquad (1.39)$$

The group action is given by (1.20) and (1.22) and hence

$$\frac{\partial t'}{\partial \lambda} = 1$$

and

$$\frac{\partial \psi'}{\partial \lambda}(0, t, x) = -\frac{\partial \psi}{\partial t}(t, x) .$$

Then we have that

$$\rho = -\text{Re} \left(\frac{\partial \mathcal{L}}{\partial \psi_t} \frac{\overline{\partial \psi}}{\partial t} - \mathcal{L} \right) .$$

Notice that we have changed a sign to get the energy with the usual conventional sign.   □

- **Momentum.** Momentum, by definition, is the quantity which is preserved by virtue of the space invariance of the Lagrangian; the invariance for translations in the x_i direction gives the following invariant:

$$P_i = -\text{Re} \int \frac{\partial \mathcal{L}}{\partial \psi_t} \frac{\overline{\partial \psi}}{\partial x_i} \, dx \ i = 1, 2, 3.$$

The numbers P_i are the components of the vector

$$\mathbf{P} = -\text{Re} \int \frac{\partial \mathcal{L}}{\partial \psi_t} \overline{\nabla \psi} \, dx. \qquad (1.40)$$

Proof. The group action, for $i = 1$ is given by

$$t' = t$$

$$x' = x + \lambda \mathbf{e}_1,$$

$$\psi'(\lambda, t', x') = \psi'(\lambda, t, x + \lambda \mathbf{e}_1) = \psi(t, x)$$

and hence

$$\frac{\partial t'}{\partial \lambda} = 0.$$

We have that

$$\psi'(\lambda, t, x) = \psi(\lambda, t, x - \lambda \mathbf{e}_1)$$

and hence

$$\frac{\partial \psi'}{\partial \lambda}(\lambda, t, x) = -\frac{\partial \psi}{\partial x_1}(t, x - \lambda \mathbf{e}_1).$$

Setting $\psi = u_1 + iu_2$, and computing (1.23) considering $\mathcal{L}$ as function of $u = (u_1, u_2)$ we get

$$\rho = \text{Re}\left(\frac{\partial \mathcal{L}}{\partial \psi_t}\frac{\overline{\partial \psi'}}{\partial \lambda} + \mathcal{L}\frac{\partial t'}{\partial \lambda}\right)_{\lambda=0} = -\text{Re}\left(\frac{\partial \mathcal{L}}{\partial \psi_t}\frac{\overline{\partial \psi}}{\partial x_1}\right).$$

$\square$

- **Angular momentum.** By definition, the angular momentum $\mathbf{M} = (M_1, M_2, M_3)$ is the quantity which is preserved by virtue of the invariance under space rotations of the Lagrangian $\mathcal{L}$ with respect to the origin

$$\mathbf{M} = \text{Re}\int \frac{\partial \mathcal{L}}{\partial \psi_t}\overline{(\mathbf{x} \times \nabla \psi)}\, dx. \tag{1.41}$$

Proof. First, we compute the third component M_3. Setting $x = (x_1, x_2, x_3)$, the rotations around the x_3 axis are described by

$$x_1' = x_1 \cos \lambda - x_2 \sin \lambda$$
$$x_2' = x_1 \sin \lambda + x_2 \cos \lambda$$
$$x_3' = x_3$$
$$t' = t$$
$$\psi'(\lambda, t', x') = \psi(t, x).$$

By the last equation, we get that

$$\psi'(\lambda, t, x_1, x_2, x_3) = \psi(t, x_1 \cos \lambda + x_2 \sin \lambda, -x_1 \sin \lambda + x_2 \cos \lambda, x_3)$$

then, setting $\psi = u_1 + iu_2$,

$$u_i'(\lambda, t, x_1, x_2, x_3) = u_i(t, x_1 \cos \lambda + x_2 \sin \lambda, -x_1 \sin \lambda + x_2 \cos \lambda, x_3), \quad i = 1, 2;$$

so

$$\left(\frac{\partial u_1'}{\partial \lambda}\right)_{\lambda=0} = \frac{\partial u_1}{\partial x_1}x_2 - \frac{\partial u_1}{\partial x_2}x_1; \quad \left(\frac{\partial u_2'}{\partial \lambda}\right)_{\lambda=0} = \frac{\partial u_2}{\partial x_1}x_2 - \frac{\partial u_2}{\partial x_2}x_1.$$

We can compute (1.23):

$$\left(\frac{\partial \mathcal{L}}{\partial u_{1,t}}\frac{\partial u_1'}{\partial \lambda} + \frac{\partial \mathcal{L}}{\partial u_{2,t}}\frac{\partial u_2'}{\partial \lambda} + \mathcal{L}\frac{\partial t}{\partial \lambda}\right)_{\lambda=0} = \frac{\partial \mathcal{L}}{\partial u_{1,t}}\left(\frac{\partial u_1}{\partial x_1}x_2 - \frac{\partial u_1}{\partial x_2}x_1\right)$$
$$+ \frac{\partial \mathcal{L}}{\partial u_{2,t}}\left(\frac{\partial u_2}{\partial x_1}x_2 - \frac{\partial u_2}{\partial x_2}x_1\right)$$
$$= \mathrm{Re}\left[\frac{\partial \mathcal{L}}{\partial \psi_t}\overline{\left(\frac{\partial \psi}{\partial x_1}x_2 - \frac{\partial \psi}{\partial x_2}x_1\right)}\right].$$

Analogously, we have

$$M_1 = \mathrm{Re}\int \frac{\partial \mathcal{L}}{\partial \psi_t}\cdot\overline{\left(\frac{\partial \psi}{\partial x_2}x_3 - \frac{\partial \psi}{\partial x_3}x_2\right)}$$

$$M_2 = \mathrm{Re}\int \frac{\partial \mathcal{L}}{\partial \psi_t}\cdot\overline{\left(\frac{\partial \psi}{\partial x_3}x_1 - \frac{\partial \psi}{\partial x_1}x_3\right)}.$$

Then we get the conclusion.   □

- Hylenic charge. The hylenic charge, by definition, is the quantity which is preserved by virtue of the invariance of the Lagrangian $\mathcal{L}$ under the gauge action (1.19). The hylenic charge has the following expression

$$C = \mathrm{Im}\int \frac{\partial \mathcal{L}}{\partial \psi_t}\cdot\bar{\psi}\,dx \ . \tag{1.42}$$

Proof. The group action is given by

$$t' = t$$
$$x' = x$$
$$\psi'(\lambda, t', x') = e^{i\lambda}\psi(t, x)$$

and hence

$$\frac{\partial t'}{\partial \lambda} = 0; \quad \left(\frac{\partial \psi'}{\partial \lambda}\right)_{\lambda=0} = i\psi(t, x).$$

Then, by (1.39)

$$\rho = \mathrm{Re}\left(\frac{\partial \mathcal{L}}{\partial \psi_t}\frac{\overline{\partial \psi_\lambda}}{\partial \lambda} + \mathcal{L}\frac{\partial t}{\partial \lambda}\right)_{\lambda=0} = -\mathrm{Re}\left(\frac{\partial \mathcal{L}}{\partial \psi_t}i\overline{\psi}\right)$$

$$= \mathrm{Im}\left(\frac{\partial \mathcal{L}}{\partial \psi_t}\overline{\psi}\right).$$

$\square$

1.4 The Hamilton-Jacobi Theory

In order to understand the motion of hylomorphic solitons, it is necessary to know the basic notions of the Hamilton-Jacobi formulation of the laws of Mechanics. Then, in this section, we will briefly recall these notions.

The Lagrangian formulation of the laws of Mechanics assumes the existence of a function

$$\mathcal{L} = \mathcal{L}(t, q, \dot{q})$$

of the generalized coordinates of the system $q = (q_1, \ldots, q_k)$, of their derivatives $\dot{q} = (\dot{q}_1, \ldots, \dot{q}_k)$ and of time. The trajectories $q(t)$ such that $q(t_0) = x_0$ and $q(t_1) = x_1$ are the critical points of the action functional

$$S(q) = \int_{t_0}^{t_1} \mathcal{L}(t, q, \dot{q})\ dt \tag{1.43}$$

defined on the space

$$\mathcal{C}^1_{x_0, x_1}[t_0, t_1] = \left\{q \in \mathcal{C}^1[t_0, t_1] : q(t_0) = x_0 \text{ and } q(t_1) = x_1\right\}.$$

Thus a trajectory $q(t)$ satisfies the "Euler-Lagrange" equations:

$$\frac{d}{dt}\frac{\partial \mathcal{L}}{\partial \dot{q}_j} - \frac{\partial \mathcal{L}}{\partial q_j} = 0, \quad j = 1, \ldots, k. \tag{1.44}$$

However, this is not the only formulation of the laws of Mechanics. Another very important formulation can be obtained as a first order system provided that the matrix

$$\left[\frac{\partial^2 \mathcal{L}}{\partial \dot{q}_i \partial \dot{q}_j}\right] \tag{1.45}$$

is positive definite. In this case, we set

$$p_j = \frac{\partial \mathcal{L}}{\partial \dot{q}_j} (t, q, \dot{q}).$$ (1.46)

By (1.45), we have that the function

$$(\dot{q}_1, \ldots, \dot{q}_k) \mapsto \left(\frac{\partial \mathcal{L}}{\partial \dot{q}_1} (t, q, \dot{q}), \ldots, \frac{\partial \mathcal{L}}{\partial \dot{q}_k} (t, q, \dot{q}) \right)$$

is smoothly invertible and hence there exits a smooth function F such that (1.46) can be rewritten as follows:

$$\dot{q} = F (t, q, p).$$ (1.47)

Now, we can define the *Hamiltonian function* as follows:

$$\mathcal{H} (t, q, p) = [\langle p, \dot{q} \rangle - \mathcal{L} (q, \dot{q}, t)]_{\dot{q}=F(p)}$$

where

$$\langle p, \dot{q} \rangle = \sum_{j=1}^{k} p_j \dot{q}_j$$

denotes the pairing between the tangent space (the space of the $\dot{q}$'s) and the relative cotangent space.

Then, the action (1.43) can be rewritten as follows

$$\mathcal{S} (p, q) = \int_{t_0}^{t_1} [\langle p, \dot{q} \rangle - \mathcal{H} (t, q, p)] \, dt$$

and the relative "Euler-Lagrange" equations take the form

$$\dot{q} = \frac{\partial \mathcal{H}}{\partial p} (t, q, p)$$

$$\dot{p} = -\frac{\partial \mathcal{H}}{\partial q} (t, q, p).$$

This is the Hamiltonian formulation of the laws of Mechanics and the above equations are called Hamilton equations. A third equivalent formulation of the laws of Dynamics is given by the *Hamilton-Jacobi* theory which uses notions both of the Lagrangian formulation and the Hamiltonian one. It reduces the laws of Mechanics to a partial differential equation and to a first order ordinary differential equation. The Hamilton-Jacobi theory has been very useful to relate the laws of Optics to

Dynamics. For us, it is essential if we want to understand the motion of solitons regarded as material particles.

The starting point is the definition of a function $S = S(t, x)$ called action. We fix once for ever a point (t_0, x_0) and the point (t, x) will be considered variable. Moreover, we set

$$S(t, x) = \int_{t_0}^{t} \mathcal{L}(s, q_x, \dot{q}_x) \, ds \tag{1.48}$$

where q_x is a critical point of (1.43) on the space $C^1_{x_0, x}[t_0, t]$. In general, this point is not unique; however, if (1.45) holds, it is possible to prove the uniqueness of the minimum provided that (t, x) is sufficiently close to (t_0, x_0) $(t \neq t_0)$. Hence, there exists an open set Ω in which the function (1.48) is well defined. The function (1.48) is called *action* as the functional (1.43). However, even if they have formal similar definitions, they are quite different objects: S in (1.48) is a functions of $k + 1$ variables defined in an open set $\Omega \subset \mathbb{R}^{k+1}$ while S in (1.43) is a functional defined in the function space $C^1_{x_0, x}[t_0, t]$.

The Hamilton-Jacobi theory states that the function S defined by (1.48), satisfies in Ω the following partial differential equation

$$\partial_t S + \mathcal{H}(t, x, \nabla S) = 0 \tag{1.49}$$

which is called Hamilton-Jacobi equation.

Moreover, this result can be inverted in the sense stated by the following theorem:

Theorem 9. *Let S be a solution of Eq. (1.49) in Ω and let $q = (q_1, \ldots, q_k)$ be a solution of the following Cauchy problem:*

$$\frac{\partial \mathcal{L}}{\partial \dot{q}_j}(t, q, \dot{q}) = \frac{\partial S}{\partial q_j}(t, q_j), \quad j = 1, \ldots, k \tag{1.50}$$

$$q(\bar{t}) = \bar{x}$$

with $(\bar{t}, \bar{x}) \in \Omega$. Then, q satisfies Eq. (1.44) with initial conditions

$$q(\bar{t}) = \bar{x}$$
$$\dot{q}(\bar{t}) = F(\bar{t}, \bar{x}, \nabla S(\bar{t}, \bar{x}))$$

where F is given by (1.47).

The proof of this theorem can be found in any book of Classical Mechanics; for example in the beautiful book of Landau and Lifchitz [100].

Notice that the above Cauchy problem is well posed, at least for small times, since, by (1.47), Eq. (1.50) gets the form

$$\dot{q} = F(t, q, \nabla S(t, q))$$

where F is the smooth function given by (1.47).

Thus, we can say that the equations of motions (1.44) are equivalent to the set of equations,

$$\partial_t S + \mathcal{H}(t, x, \nabla S) = 0 \tag{1.51}$$

$$\dot{q} = F(t, q, \nabla S(t, q)). \tag{1.52}$$

If $\mathcal{L}$ does not depend on t, then $\mathcal{H}$ is a constant of motion (namely it is the energy of the system). In this case, by Eq. (1.49), $\partial_t S = -h$, namely it does not depend on time and it represents the energy of the system with the sign changed. In this case, Eq. (1.49) takes the form

$$\mathcal{H}(x, \nabla S) = h.$$

Let us see some examples:

• **Newtonian dynamics.**

$$\mathcal{L}(t, q, \dot{q}) = \frac{1}{2} m \dot{q}^2 - V(q).$$

Then

$$p = m \dot{q}$$

$$\mathcal{H} = \frac{1}{2m} p^2 + V(q)$$

and Eqs. (1.51) and (1.52) take the form

$$\partial_t S + \frac{1}{2m} |\nabla S|^2 + V(x) = 0 \tag{1.53}$$

$$\dot{q} = \frac{1}{m} \nabla S(t, q). \tag{1.54}$$

• **Relativistic dynamics.** The Lagrangian of a relativistic particle is given by:

$$\mathcal{L}(t, q, \dot{q}) = -m_0 \sqrt{1 - \dot{q}^2}$$

where m_0 is a parameter. We refer to Landau-Lifchitz [101] for a very elegant deduction of this Lagrangian from the Minkowski geometry of space-time.

We have

$$p = \frac{\partial \mathcal{L}}{\partial \dot{q}} = \frac{m_0}{\sqrt{1 - \dot{q}^2}} \dot{q} = \gamma m_0 \dot{q} \qquad (1.55)$$

with

$$\gamma = \frac{1}{\sqrt{1 - \dot{q}^2}}$$

and Eq. (1.47) becomes

$$\dot{q} = \frac{p}{\sqrt{m_0^2 + p^2}}. \qquad (1.56)$$

Since the mass of a particle is defined by the equation $m = p/\dot{q}$, we will get that the mass changes with velocity

$$m = \gamma m_0$$

and the interpretation of m_0 as rest mass. The Hamiltonian is:

$$\mathcal{H} = p\dot{q} + m_0\sqrt{1 - \dot{q}^2} = \frac{m_0}{\sqrt{1 - \dot{q}^2}} \dot{q}^2 + m_0\sqrt{1 - \dot{q}^2} = \frac{m_0}{\sqrt{1 - \dot{q}^2}} = \gamma m_0.$$

Since the Lagrangian is independent of time, the Hamiltonian represents the energy and this gives the Einstein equation:

$$E = \mathcal{H} = m = \gamma m_0. \qquad (1.57)$$

Now let us express $\mathcal{H}$ as function of p. Using Eq. (1.56) we get

$$\mathcal{H}(p, q) = \frac{m_0}{\sqrt{1 - \frac{p^2}{p^2 + m_0^2}}} = \sqrt{m_0^2 + p^2} \qquad (1.58)$$

and Eqs. (1.51) and (1.52) take the form

$$\partial_t S + \sqrt{m_0^2 + |\nabla S|^2} = 0 \qquad (1.59)$$

$$\dot{q} = \frac{\nabla S}{\sqrt{m_0^2 + |\nabla S|^2}}. \qquad (1.60)$$

Chapter 2
Solitary Waves and Solitons: Abstract Theory

In this chapter we construct a functional abstract framework which allows to define solitary waves, solitons and hylomorphic solitons (Sects. 2.1.1 and 2.1.3). Then, we will give some abstract existence theorems (Sect. 2.2). These theorems are based on two general minimization principles related to the concentration compactness techniques (see Sects. 2.2.3 and 2.2.4). These results are able to cover all the situations considered in the rest of this book and in most of the present literature on this subject. In the last two Sects. 2.3.1 and 2.3.2, we will discuss the meaning, the structure and possible interpretations of hylomorphic solitons.

2.1 Solitary Waves and Solitons

2.1.1 Definitions

Solitary waves and solitons are particular states of a dynamical system described by one or more partial differential equations. Thus, we assume that the states of this system are described by one or more fields which mathematically are represented by functions

$$\mathbf{u} : \mathbb{R}^N \to V$$

where V is a vector space with norm $|\cdot|_V$ which is called the internal parameters space. We assume the system to be deterministic; this means that it can be described as a dynamical system (X, γ) where X is the set of the states and $\gamma : \mathbb{R} \times X \to X$ is the time evolution map. If $\mathbf{u}_0(x) \in X$ denotes the initial state, the evolution of the system will be described by the function

$$\mathbf{u}(t, x) := \gamma_t \mathbf{u}_0(x). \tag{2.1}$$

© Springer International Publishing Switzerland 2014
V. Benci, D. Fortunato, *Variational Methods in Nonlinear Field Equations*,
Springer Monographs in Mathematics, DOI 10.1007/978-3-319-06914-2_2

We assume that the states of X have "finite energy" so that they decay at ∞ sufficiently fast and that

$$X \subset L^1_{loc}\left(\mathbb{R}^N, V\right). \tag{2.2}$$

Using this framework, we give the following definitions:

Definition 10. A dynamical system (X, γ) is called of FT type (field-theory-type) if X is a Hilbert space of functions satisfying (2.2).

For every $\tau \in \mathbb{R}^N$, and $\mathbf{u} \in X$, we set

$$(T_\tau \mathbf{u})(x) = \mathbf{u}(x - \tau). \tag{2.3}$$

Clearly, the group

$$\mathcal{T} = \left\{T_\tau \mid \tau \in \mathbb{R}^N\right\}; \tag{2.4}$$

is a representation of the group of translations.

Definition 11. A set $\Gamma \subset X$ is called compact up to space translations or $\mathcal{T}$-compact if for any sequence $\mathbf{u}_n(x) \in \Gamma$ there is a subsequence $\mathbf{u}_{n_k}$ and a sequence $\tau_k \in \mathbb{R}^N$ such that $\mathbf{u}_{n_k}(x - \tau_k)$ is convergent.

Now, we want to give an abstract definition of solitary wave. Roughly speaking a solitary wave is a field whose energy travels as a localized packet and which preserves this localization in time. For example, consider a solution of a field equation having the following form:

$$\mathbf{u}(t, x) = u(x - vt - x_0)e^{i(v \cdot x - \omega t)}; \ u \in L^2(\mathbb{R}^N). \tag{2.5}$$

The field (2.5) is a solitary wave depending on the constants $x_0, v \in \mathbb{R}^N$ and $\omega \in \mathbb{R}$. The evolution of a solitary wave is a translation plus a mild change of the internal parameters (in this case the phase).

This situation can be formalized by the following definition:

Definition 12. If $\mathbf{u} \in X$, we denote the closure of the orbit of $\mathbf{u}$ by

$$\mathcal{O}(\mathbf{u}) := \overline{\{\gamma_t \mathbf{u}(x) \mid t \in \mathbb{R}\}}.$$

A state $\mathbf{u} \in X$ is called solitary wave if

(i) $0 \notin \mathcal{O}(\mathbf{u})$.
(ii) $\mathcal{O}(\mathbf{u})$ is $\mathcal{T}$-compact.

Clearly, (2.5) describes a solitary wave according to the definition above. The standing waves, namely objects of the form

$$\gamma_t \mathbf{u} = \mathbf{u}(t, x) = u(x)e^{-i\omega t}, \ u \in L^2(\mathbb{R}^N), \ u \neq 0 \tag{2.6}$$

probably are the "simplest" solitary waves. In this case the orbit $\mathcal{O}(\mathbf{u})$ is compact.

Take $X = L^1(\mathbb{R}^N)$ and $\mathbf{u} \in X$; if $\gamma_t \mathbf{u} = \mathbf{u}(e^t x)$, $\mathbf{u}$ is not a solitary wave since $\|\gamma_t \mathbf{u}\|_X \to 0$ as $t \to +\infty$ and (i) is clearly violated. If $\gamma_t \mathbf{u} = e^t \mathbf{u}(e^t x)$, $\mathbf{u}$ is not a solitary wave since (ii) of Definition 12 does not hold. Also, according to our definition, a "couple" of solitary waves is not a solitary wave: for example the dynamics

$$\gamma_t \mathbf{u} = [u(x - vt) + u(x + vt)] e^{i(v \cdot x - \omega t)}, \ u \in L^2(\mathbb{R}^N)$$

does not give rise to a solitary wave since (ii) is violated.

Another characterization of solitary waves is the following:

Proposition 13. *A state $\mathbf{u} \in X$ is a solitary wave if and only if (i) holds and there exist a compact set $K \subset X$ and $x(t) \in \mathbb{R}^N$ such that*

$$\gamma_t \mathbf{u} = \mathbf{u}(t, x) = T_{x(t)} \mathbf{u} + \mathbf{w}(t, x) \tag{2.7}$$

where $\forall t \in \mathbb{R}, \mathbf{w}(t, \cdot) \in K$.

Proof. It is immediate to see that (2.7) implies (ii) of Definition 12. Let us prove the converse. Given a (non necessarily continuous) map $\theta : \mathcal{O}(\mathbf{u}) \to \mathbb{R}^N$ we set

$$K_\theta = \left\{ T_{\theta(z)} \mathbf{z} : \mathbf{z} \in \mathcal{O}(\mathbf{u}) \right\}.$$

Since $\mathcal{O}(\mathbf{u})$ is $\mathcal{T}$-compact, we can choose θ such that K_θ is compact. Since $\gamma_t \mathbf{u} \in \mathcal{O}(\mathbf{u})$, we have that $T_{\theta(\gamma_t \mathbf{u})} \gamma_t \mathbf{u} \in K_\theta$ and hence

$$\gamma_t \mathbf{u} \in T_{-\theta(\gamma_t \mathbf{u})} K_\theta;$$

so

$$\gamma_t \mathbf{u} = T_{-\theta(\gamma_t \mathbf{u})} \mathbf{u} + \mathbf{w}(t, x)$$

where

$$\mathbf{w}(t, x) := \gamma_t \mathbf{u} - T_{-\theta(\gamma_t \mathbf{u})} \mathbf{u} \in T_{-\theta(\gamma_t \mathbf{u})} K_\theta - T_{-\theta(\gamma_t \mathbf{u})} \mathbf{u}.$$

Thus (2.7) holds with $T_{x(t)} = -\theta(\gamma_t \mathbf{u})$ and $K = T_{-\theta(\gamma_t \mathbf{u})} K_\theta - T_{-\theta(\gamma_t \mathbf{u})} \mathbf{u}$. □

The *solitons* are solitary waves characterized by some form of stability. To define them at this level of abstractness, we need to recall some well known notions in the theory of dynamical systems.

Definition 14. A set $\Gamma \subset X$ is called *invariant* if $\forall \mathbf{u} \in \Gamma, \forall t \in \mathbb{R}, \gamma_t \mathbf{u} \in \Gamma$.

Definition 15. Let (X, d) be a metric space and let (X, γ) be a dynamical system. An invariant set $\Gamma \subset X$ is called stable, if $\forall \varepsilon > 0, \exists \delta > 0, \forall \mathbf{u} \in X$,

$$d(\mathbf{u}, \Gamma) \leq \delta,$$

implies that

$$\forall t \geq 0, \ d(\gamma_t \mathbf{u}, \Gamma) \leq \varepsilon.$$

Now we are ready to give the definition of soliton:

Definition 16. A state $\mathbf{u} \in X$ is called soliton if $\mathbf{u} \in \Gamma \subset X$ where

(i) Γ is an invariant, stable set.
(ii) Γ is $\mathcal{T}$-compact.
(iii) $0 \notin \Gamma$.

The set Γ is called soliton manifold.

The above definition needs some explanation. First of all notice that every $\mathbf{u} \in \Gamma$ is a soliton and that every soliton is a solitary wave. Now for simplicity, we assume that Γ is a manifold.[1] Then (ii) implies that Γ is finite dimensional. Since Γ is invariant, $\mathbf{u} \in \Gamma \Rightarrow \gamma_t \mathbf{u} \in \Gamma$ for every time. Thus, since Γ is finite dimensional, the evolution of $\mathbf{u}$ is described by a finite number of parameters. The dynamical system (Γ, γ) behaves as a point in a finite dimensional phase space. By the stability of Γ, a small perturbation of $\mathbf{u}$ remains close to Γ. However, in this case, its evolution depends on an infinite number of parameters. Thus, this system appears as a finite dimensional system with a small perturbation.

Example. We will illustrate the Definition 16 with an example. Consider the solitary wave $\mathbf{u}$ in (2.5) and the set

$$\Gamma_v = \left\{ u(x - x_0)e^{i(v \cdot x - \theta)} \in L^2(\mathbb{R}^N) : x_0 \in \mathbb{R}^N; \ \theta \in \mathbb{R} \right\}. \tag{2.8}$$

Clearly $\mathbf{u} \in \Gamma_v$. Now assume that Γ_v is stable. It is easy to see that Γ_v is a soliton manifold and (2.5) is a soliton according to our Definition 16, in fact:

(i) Is satisfied since Γ_v is invariant under the dynamics

$$\gamma_t \left[u(x - x_0)e^{i(v \cdot x - \theta)} \right] = u(x - vt - x_0)e^{i(v \cdot x - \omega t - \theta)}$$

and since Γ_v has been assumed to be stable.
(ii) Γ_v is $\mathcal{T}$-compact, actually it is isomorphic to $\mathbb{R}^N \times S^1$.
(iii) Is obviously satisfied.

Observe that, since Γ_v is stable, any perturbation u_ε of our soliton has the following structure:

$$u_\varepsilon(t, x) = u(x - vt - x_0(t))e^{i(v \cdot x - \theta(t))} + w(t, x)$$

---

[1] Actually, in many concrete models, this is the generic case; this is the reason why Γ is called *soliton manifold* even if it might happen that it is not a manifold.

where $x_0(t)$, $\theta(t)$ are suitable functions and $w(t, x)$ is a perturbation small in $L^2(\mathbb{R}^N)$.

The nonlinear Schrödinger equation admits solitons like (2.5) provided that the nonlinear term satisfies suitable conditions. We refer to Chap. 3 for a discussion and the proof of this point.

2.1.2 Solitons and Symmetry

In Chap. 1 we have seen the relevance of the symmetry when the dynamics is induced by a variational principle. In this section we will investigate other relevant consequences that symmetry induces on the dynamical system.

Definition 17. Given a dynamical system (X, γ) and a Lie group $H \subset Hom(X)$ acting on X, we say that (X, γ) is invariant under H if, for any $\mathbf{u} \in X$, $\forall h \in H$,

$$h\gamma_t \mathbf{u} = \gamma_t h\mathbf{u}. \tag{2.9}$$

Proposition 18. *If (X, γ) is invariant under a Lie group $H \subset Hom(X)$ and $\mathbf{u}$ is a soliton, then $\forall g \in H$, $g\mathbf{u}$ is a soliton.*

Proof. If Γ is the soliton manifold relative to $\mathbf{u}$, then it is immediate to check that

$$g\Gamma := \{g\mathbf{w} : \mathbf{w} \in \Gamma\}$$

satisfies (i)–(iii) of Definition 16 and hence $g\mathbf{u}$ is a soliton.   □

Now, we will assume that the dynamical system (X, γ) is induced by the Euler-Lagrange equations (1.3) where the lagrangian $\mathcal{L}$ is invariant under a representation $\{T_g\}_{g \in \mathfrak{G}}$ of a group $\mathfrak{G}$. Of course we will be particularly interested in the case in which $\mathfrak{G}$ is a subgroup of the Poincaré group or the Galileo group. Let us investigate the relations between $\{T_g\}_{g \in \mathfrak{G}}$ and H as defined in Definition 17.

First of all we notice that $\mathfrak{G}$ acts on the spacetime $\mathbb{R}^{N+1}$ and, sometimes also on the internal parameter space V. Hence $\mathfrak{G}$ induces an action

$$T_g : (t, x, \mathbf{u}) \to (t', x', \mathbf{u}')$$

on the set

$$\mathfrak{X} = \mathfrak{F}\left(\mathbb{R}^{N+1}, V\right)$$

of all the functions defined on $\mathbb{R}^{N+1}$ taking values in V.

We assume that an orbit $\gamma_t \mathbf{u}$ ($t \in \mathbb{R}$) can be identified with a function $\mathbf{u}(t, x) \in \mathfrak{X}$. The family of all the orbits

$$\mathfrak{G} := \{\gamma_t \mathbf{u} \mid t \in \mathbb{R}, \, \mathbf{u} \in X\}$$

can be identified with the family of solutions of the Euler-Lagrange equations (1.3). Since the Lagrangian is T_g-invariant, $\mathfrak{S}$ is a T_g-invariant subset of $\mathfrak{X}$. Moreover there is an isomorphism

$$L : X \to \mathfrak{S}$$

$$\mathbf{u} \mapsto \gamma_t \mathbf{u}.$$

We now set

$$\rho(g) = L^{-1} \circ T_g \circ L : X \to X, \ g \in \mathfrak{S} \tag{2.10}$$

$$\mathbf{u} \mapsto \left[T_g \left(\gamma_t \mathbf{u} \right) \right]_{t=0}.$$

It is easy to see that (X, γ) is invariant under the group $H = \rho(\mathfrak{S})$ according to Definition 17: in fact

$$\rho(g) \left(\gamma_\tau \mathbf{u} \right) = \left[T_g \left(\gamma_t \gamma_\tau \mathbf{u} \right) \right]_{t=0} = \left[T_g \left(\gamma_\tau \gamma_t \mathbf{u} \right) \right]_{t=0}$$

$$= \gamma_\tau \left[T_g \left(\gamma_t \mathbf{u} \right) \right]_{t=0} = \gamma_\tau \rho(g) \mathbf{u}.$$

By the above discussion and Proposition 18 we get

Corollary 19. *Using the above notation, if* $\mathbf{u}$ *is a soliton, then, for every* $g \in \mathfrak{S}$, $\rho(g)\mathbf{u}$ *is a soliton.*

Example. Take $\mathbf{u} = u(x) \in H^1 \left(\mathbb{R}^N \right)$ and let us assume that it is a soliton relative to an equation invariant under the action (1.16) of the Galileo group. To fix the idea, consider the nonlinear Schrödinger equation (NS), page 62; we assume that the dynamics on u is given by

$$u(t, x) = \gamma_t u(x) = u(x) e^{-i\omega_0 t}, \quad \omega_0 \in \mathbb{R}$$

(in Sect. 3.2.2, we will show that such an $u(x)$ exists). Then, if we take $\mathbf{v}$ defined by (1.15), we have

$$\rho \left(g_\mathbf{v} \right) u = \left[T_{g_\mathbf{v}} \left(\gamma_t u \right) \right]_{t=0}$$

$$= \left[T_{g_\mathbf{v}} \left(u(x) e^{-i\omega_0 t} \right) \right]_{t=0}$$

$$= \left[u \left(x - \mathbf{v}t \right) e^{-i\omega_0 t} e^{i \left(\mathbf{v} \cdot x - \frac{1}{2} \mathbf{v}^2 t \right)} \right]_{t=0}$$

$$= u(x) e^{i\mathbf{v} \cdot x}.$$

And hence, we have that

$$H = \left\{ \rho \left(g_\mathbf{v} \right) : \mathbf{v} \in \mathbb{R}^N \right\}; \ \rho \left(g_\mathbf{v} \right) u(x) = u(x) e^{i\mathbf{v} \cdot x}.$$

Then, by Proposition 18, we have a family of solitons

$$u(x)e^{i\mathbf{v}\cdot x}, \ \mathbf{v} \in \mathbb{R}^N$$

and their dynamics is given by

$$\gamma_t\left[u(x)e^{i\mathbf{v}\cdot x}\right] = u(x - \mathbf{v}t)e^{i(\mathbf{v}\cdot x - \omega t)} \text{ where } \omega = \frac{1}{2}\mathbf{v}^2 + \omega_0.$$

So, for each $\mathbf{v} \in \mathbb{R}^N$, we have a soliton which moves with velocity $\mathbf{v}$ (see Sect. 3.1.2), it has momentum proportional to $\mathbf{v}$ and energy proportional to ω; the energy consists of the kinetic energy $\frac{1}{2}\mathbf{v}^2$ and the internal energy ω_0. For more details, we refer to Sect. 3.1.2.

2.1.3 Hylomorphic Solitons and Minimizers

We now assume that the dynamical system (X, γ) has two constants of motion: the energy E and the hylenic charge C. At the level of abstractness of this section (and the next two), the name energy and hylenic charge are conventional, but in our applications, E and C will be the energy and the hylenic charge as defined in Sect. 1.3.2.

Definition 20. A solitary wave $\mathbf{u}_0 \in X$ is called **hylomorphic soliton** if it is a soliton according to Definition 16 and if the soliton manifold Γ has the following structure

$$\Gamma = \Gamma(e_0, c_0) = \{\mathbf{u} \in X \mid E(\mathbf{u}) = e_0, \ |C(\mathbf{u})| = c_0\} \tag{2.11}$$

where

$$e_0 = \min\{E(\mathbf{u}) \mid |C(\mathbf{u})| = c_0\}. \tag{2.12}$$

Notice that, by (2.12), we have that a hylomorphic soliton $\mathbf{u}_0$ satisfies the following nonlinear eigenvalue problem:

$$E'(\mathbf{u}_0) = \lambda C'(\mathbf{u}_0). \tag{2.13}$$

In the literature, the following notion is frequently used

Definition 21. A solution $\mathbf{u}_0 \neq 0$ of Eq. (2.13) is called **ground state solution** (with respect to the energy E and the set $\mathfrak{M}_{c_0}$) if it minimizes $E(\mathbf{u})$ on the set

$$\mathfrak{M}_{c_0} := \{\mathbf{u} \in X \mid |C(\mathbf{u})| = c_0\}.$$

In general, a ground state solution is not a soliton; in fact, according to Definition 16, it is necessary to check the following facts:

  (i) The set $\Gamma\left(e_0, c_0\right)$ is stable.
 (ii) The set $\Gamma\left(e_0, c_0\right)$ is $\mathcal{T}$-compact (i.e. compact up to translations).
(iii) $0 \notin \Gamma\left(e_0, c_0\right)$ since otherwise, some $\mathbf{u} \in \Gamma\left(e_0, c_0\right)$ is not even a solitary wave (see Definition 12,(i)).

In concrete cases, the point (i) is the most delicate to prove. If (i) does not hold, according to our definitions, $\mathbf{u}_0$ is a solitary wave but not a soliton.

Now, let us assume that the dynamical system (X, γ) is induced by the Euler-Lagrange equations (1.3) where the lagrangian $\mathcal{L}$ is invariant under a representation $\left\{T_g\right\}_{g \in \mathfrak{G}}$ of a subgroup $\mathfrak{G}$ of the Poincaré or Galileo group.

If $\mathbf{u}_0$ is a hylomorphic soliton and $\rho(g)$ is defined by (2.10), using Corollary 19, we have that $\mathbf{u} = \rho(g)\mathbf{u}_0$ is a soliton for every $g \in \mathfrak{G}$. So we get a family of solitons $\left\{\rho(g)\mathbf{u}_0\right\}_{g \in \mathfrak{G}}$.

We may ask if the solitons in $\left\{\rho(g)\mathbf{u}_0\right\}_{g \in \mathfrak{G}}$ are hylomorphic. In general, the answer is negative when $\mathfrak{G}$ is the Lorentz or the restricted Galilean group (see page 8).

In fact, in many concrete models, the "kinetic energy" is positive. If $\mathbf{u}_0$ is a "stationary" hylomorphic soliton, and $g_\mathbf{v}$ is the transformation (1.10) or (1.15), $\rho(g_\mathbf{v})\mathbf{u}_0$ has some positive kinetic energy and then $E\left(\rho(g_\mathbf{v})\mathbf{u}_0\right) > E\left(\mathbf{u}_0\right)$. So in general, $\rho(g_\mathbf{v})\mathbf{u}_0$ is not a hylomorphic soliton.

After this discussion, it makes sense to give the following definition:

Definition 22. A solitary wave $\mathbf{u} \in X$ is called **travelling hylomorphic soliton** if

$$\mathbf{u} = \rho(g_\mathbf{v})\mathbf{u}_0, \quad \mathbf{v} \neq 0,$$

where $\rho(g_\mathbf{v})$ is defined above and $\mathbf{u}_0$ is a hylomorphic soliton.

Example. Consider the example at page 32. If the solitary wave $u_0(x)$ is a hylomorphic soliton relative to a Lagrangian invariant under the Galileo action (1.16), then

$$\rho(g_\mathbf{v})u_0 = u_0(x)e^{i\mathbf{v}\cdot x}, \quad \mathbf{v} \neq 0$$

is a travelling soliton.

2.2 Existence Results of Hylomorphic Solitons

Let (X, γ), E and C as in Sect. 2.1.3. We have seen that the existence of hylomorphic soliton (see Definition 20) is related to the existence of minimizers of the energy E on sets of prescribed charge. In this section we will investigate the following minimization problem

$$\min_{\mathbf{u} \in \mathfrak{M}_c} E(\mathbf{u}) \quad \text{where} \quad \mathfrak{M}_c := \{\mathbf{u} \in X \mid |C(\mathbf{u})| = c\} \tag{2.14}$$

and under which conditions the set of minimizers

$$\Gamma(e, c) = \{\mathbf{u} \in X \mid E(\mathbf{u}) = e, \ |C(\mathbf{u})| = c\}; \ e = \min_{\mathbf{u} \in \mathfrak{M}_c} E(\mathbf{u})$$

is stable. The theory we develop is based on some results of the authors [39–41] related to the concentration-compactness principle [106, 107, 140] and their references.

2.2.1 The Abstract Framework

We will study problem (2.14) in an abstract framework; in order to do so, we need a few definitions:

Definition 23. Let G be a group acting on X. A subset $\Gamma \subset X$ is called G-invariant if

$$\forall \mathbf{u} \in \Gamma, \ \forall g \in G, \ g\mathbf{u} \in \Gamma.$$

In many concrete situations, G will be a subgroup of the translations group $\mathcal{T}$ such as $\mathbb{Z}^N$.

Definition 24. Let G be a group acting on X. A sequence $\mathbf{u}_n$ in X is called G-compact if there is a subsequence $\mathbf{u}_{n_k}$ and a sequence $g_k \in G$ such that $g_k \mathbf{u}_{n_k}$ is convergent. A subset $\Gamma \subset X$ is called G-compact if every sequence in Γ is G-compact.

If $G = \{Id\}$ or more in general it is a compact group, G-compactness implies compactness. If G is not compact such as the translation group $\mathcal{T}$, G-compactness is a weaker notion than compactness.

Definition 25. A G-invariant functional J on X is called G-compact if any minimizing sequence $\mathbf{u}_n$ is G-compact.

Clearly a G-compact functional has a G-compact set of minimizers.

Definition 26. We say that a functional F on X has the splitting property if given a sequence $\mathbf{u}_n = \mathbf{u} + \mathbf{w}_n \in X$ such that $\mathbf{w}_n$ converges weakly to 0, we have that

$$F(\mathbf{u}_n) = F(\mathbf{u}) + F(\mathbf{w}_n) + o(1).$$

Remark 27. Every quadratic form which is continuous and symmetric satisfies the splitting property; in fact, in this case, we have that $F(\mathbf{u}) := \langle L\mathbf{u}, \mathbf{u} \rangle$ for some continuous selfajoint operator L; then, given a sequence $\mathbf{u}_n = \mathbf{u} + \mathbf{w}_n$ with $\mathbf{w}_n \rightharpoonup 0$ weakly, we have that

$$F(\mathbf{u}_n) = \langle L\mathbf{u}, \mathbf{u} \rangle + \langle L\mathbf{w}_n, \mathbf{w}_n \rangle + 2 \langle L\mathbf{u}, \mathbf{w}_n \rangle$$
$$= F(\mathbf{u}) + F(\mathbf{w}_n) + o(1).$$

Definition 28. Let G a group of unitary operators acting on X. A sequence $\{\mathbf{u}_n\} \subset X$ is called G-**vanishing sequence** *if for* any sequence $\{g_n\} \subset G$ the sequence $\{g_n \mathbf{u}_n\}$ converges weakly to 0.

In the following, we will simply write "vanishing sequence" if there is no ambiguity relative to the group G. If $\mathbf{u}_n \to 0$ strongly, $\mathbf{u}_n$ is a vanishing sequence. However, if $\mathbf{u}_n \rightharpoonup 0$ weakly, it might happen that it is not a vanishing sequence; in this case there exist a subsequence $\mathbf{u}_{n_k}$ and a sequence $g_k \in G$ such that $g_k \mathbf{u}_{n_k}$ is weakly convergent to some $\bar{\mathbf{u}} \neq 0$. Let see an example; if $\mathbf{u}_0 \in X$ is a solitary wave and $t_n \to +\infty$, then the sequence $\gamma_{t_n} \mathbf{u}_0$ is not vanishing.

In the following E and C will denote two constants of the motion for the dynamical system (in the applications they will be the energy and the charge).

We set

$$\Lambda (\mathbf{u}) := \frac{E (\mathbf{u})}{|C (\mathbf{u})|}. \tag{2.15}$$

Since E and C are constants of motion, also Λ is a constant of motion; it will be called **hylenic ratio** and, as we will see it will play a central role in this theory.

The notions of vanishing sequence and of hylenic ratio allow to introduce the following important definition:

Definition 29. We say that the *hylomorphy condition holds* if

$$\inf_{\mathbf{u} \in X} \frac{E (\mathbf{u})}{|C (\mathbf{u})|} < \Lambda_0. \tag{2.16}$$

where

$$\Lambda_0 := \inf \{\liminf \Lambda(\mathbf{u}_n) \mid \mathbf{u}_n \text{ is a vanishing sequence}\} . \tag{2.17}$$

Moreover, we say that $\mathbf{u}_0 \in X$ satisfies the *hylomorphy condition* if

$$\frac{E (\mathbf{u}_0)}{|C (\mathbf{u}_0)|} < \Lambda_0. \tag{2.18}$$

Remark 30. Let $\{\mathbf{u}_n\} \subset X$ such that $\lim \Lambda (\mathbf{u}_n) = \lambda < \Lambda_0$. Then, by definition (2.17), $\mathbf{u}_n$ is not vanishing i.e. there exists a subsequence $\{\mathbf{u}_{n_k}\} \subset X$ and $\{g_k\} \subset G$ s.t $g_k \mathbf{u}_{n_k}$ does not converge weakly to 0. So, if $\mathbf{u}_n$ is also bounded, we have that, up to a subsequence, $g_k \mathbf{u}_{n_k}$ weakly converges to $\mathbf{u} \neq 0$.

In order to apply the existence theorems of the next Sect. 2.2.2, it is necessary to estimate Λ_0; the following propositions may help to do this.

Proposition 31. *Assume that there exists a seminorm $\|\cdot\|_{\sharp}$ on X such that*

$$(\mathbf{u}_n \text{ is a vanishing sequence}) \Rightarrow \left(\|\mathbf{u}_n\|_{\sharp} \to 0\right). \tag{2.19}$$

Then

$$\liminf_{\|\mathbf{u}\|_{\sharp} \to 0} \Lambda(\mathbf{u}) \le \Lambda_0 \le \liminf_{\|\mathbf{u}\| \to 0} \Lambda(\mathbf{u}). \tag{2.20}$$

Proof. By Definition 28 and by (2.19) we have

$$(\|\mathbf{u}_n\| \to 0) \Rightarrow (\mathbf{u}_n \text{ vanishing sequence}) \Rightarrow \left(\|\mathbf{u}_n\|_{\sharp} \to 0\right). \tag{2.21}$$

Then, by (2.17) and (2.21), we get (2.20). □

Proposition 32. *If E and C are twice differentiable in 0 and*

$$E(0) = C(0) = 0; \; E'(0) = C'(0) = 0; C''(0) \ne 0$$

then we have that

$$\Lambda_0 \le \inf_{\mathbf{u} \ne 0} \frac{E''(0)[\mathbf{u}, \mathbf{u}]}{|C''(0)[\mathbf{u}, \mathbf{u}]|}.$$

Proof. By the above proposition,

$$\Lambda_0 \le \liminf_{\|\mathbf{u}\| \to 0} \Lambda(\mathbf{u}) = \liminf_{\|\mathbf{u}\| \to 0} \frac{E(0) + E'(0)[\mathbf{u}] + E''(0)[\mathbf{u}, \mathbf{u}] + o(\|\mathbf{u}\|^2)}{\left|C(0) + C'(0)[\mathbf{u}] + C''(0)[\mathbf{u}, \mathbf{u}] + o(\|\mathbf{u}\|^2)\right|}$$

$$= \inf_{\mathbf{u} \ne 0} \frac{E''(0)[\mathbf{u}, \mathbf{u}]}{|C''(0)[\mathbf{u}, \mathbf{u}]|}.$$

□

Now, finally, we can give some abstract theorems relative to the existence of hylomorphic solitons.

2.2.2 Statement of the Abstract Existence Theorems

Let (X, γ) be a dynamical system and E, C be two functionals on X which are constants of the motion. Let G be a unitary group acting on X. Although if almost all the results we shall state hold in this generality, we shall assume in the following that X is of FT type (see Definition 10) and that G is a representation of the group of translations (see (2.4)).

We distinguish two cases: $E(\mathbf{u}) \geq 0$ (positive energy case) and $C(\mathbf{u}) \geq 0$ (positive charge case). First, we formulate the properties we require in the positive energy case:

*(EC-0) (**Values at 0**) E and C are C^1 functionals which map bounded sets into bounded sets and such that*

$$E(0) = C(0) = 0; \; E'(0) = C'(0) = 0.$$

*(EC-1)(**Invariance**) $E(\mathbf{u})$ and $C(\mathbf{u})$ are G-invariant.*

*(EC-2)(**Splitting property**)   E and C satisfy the splitting property (see Definition 26).*

*(EC-3)(**Coercivity**)   We assume that*

   (i) $\forall \mathbf{u} \neq 0, \; E(\mathbf{u}) > 0$.
   (ii) *If* $\|\mathbf{u}_n\| \to \infty$, *then* $E(\mathbf{u}_n) \to \infty$.
   (iii) *If* $E(\mathbf{u}_n) \to 0$, *then* $\|\mathbf{u}_n\| \to 0$.

If the energy is not positive, assumption *(EC-3)* needs to be replaced by the following one:

(EC-3)(**Coercivity**)   We assume that there exist a > 0 and s > 1 such that*

   (i) $\forall \mathbf{u} \neq 0, \; C(\mathbf{u}) > 0$.
   (ii) *If* $\|\mathbf{u}\| \to \infty$, *then* $E(\mathbf{u}) + aC(\mathbf{u})^s \to \infty$.
   (iii) *For any bounded sequence* $\mathbf{u}_n$ *in X such that* $E(\mathbf{u}_n) + aC(\mathbf{u}_n)^s \to 0$, *we have that* $\|\mathbf{u}_n\| \to 0$.

If (EC-3)(i) holds (positive energy case) we shall set

$$\delta_\infty = \sup \{\delta > 0 \mid \exists \mathbf{v} : \Lambda (\mathbf{v}) + \delta E(\mathbf{v}) < \Lambda_0 \} . \tag{2.22}$$

If (EC-3*)(i) holds (positive charge case) we shall set

$$\delta_\infty = \sup \{\delta > 0 \mid \exists \mathbf{v} : \Lambda (\mathbf{v}) + \delta \Phi(\mathbf{v}) < \Lambda_0 \} \tag{2.23}$$

where

$$\Phi(\mathbf{u}) = E(\mathbf{u}) + 2aC(\mathbf{u})^s. \tag{2.24}$$

Observe that, if the hylomorphy condition (2.16) is satisfied, in both case we have

$$\delta_\infty > 0.$$

Now we can state the main results:

Theorem 33. *Assume that E and C satisfy (EC-0)–(EC-2) and (EC-3) or (EC-3*). Moreover assume that the hylomorphy condition (2.16) is satisfied. Then there exist hylomorphic solitons (see Sect. 2.1.3).*

Theorem 34. *Let the assumptions of Theorem 33 hold. Moreover assume that*

$$\left\| E'(\mathbf{u}) \right\| + \left\| C'(\mathbf{u}) \right\| = 0 \Leftrightarrow \mathbf{u} = 0. \tag{2.25}$$

Then for every $\delta \in (0, \delta_\infty)$ (δ_∞ as in (2.22) if (EC-3) is satisfied or as in (2.23) if (EC-3*) is satisfied) there exists a hylomorphic soliton $\mathbf{u}_\delta$. Moreover, if $\delta_1 < \delta_2$, the corresponding solitons $\mathbf{u}_{\delta_1}, \mathbf{u}_{\delta_2}$ are distinct and we have that

(a) $\Lambda(\mathbf{u}_{\delta_1}) < \Lambda(\mathbf{u}_{\delta_2})$.
(b) $|C(\mathbf{u}_{\delta_1})| > |C(\mathbf{u}_{\delta_2})|$.

It is interesting to observe that by Theorems 33 and 34 we immediately get the following existence theorem for constrained minimizers:

Corollary 35. *Assume that E and C satisfy (EC-0)–(EC-2) and (EC-3) or (EC-3*). Moreover assume that the hylomorphy condition (2.16) is satisfied. Then, for a suitable $c > 0$, there exists the minimum*

$$\min_{\mathbf{u} \in \mathfrak{M}_c} E(\mathbf{u}) \ \text{where} \ \mathfrak{M}_c := \{ \mathbf{u} \in X \mid |C(\mathbf{u})| = c \}.$$

Moreover, if we assume also (2.25), there exists a family $\mathbf{u}_\delta$, $\delta \in (0, \delta_\infty)$ (δ_∞ as in (2.22) or as in (2.23) according to (EC-3) or (EC-3*) are satisfied) of distinct minimizers. Namely, if $\delta_1 < \delta_2$, we have that

$$\Lambda(\mathbf{u}_{\delta_1}) < \Lambda(\mathbf{u}_{\delta_2}) \ \text{and} \ |C(\mathbf{u}_{\delta_1})| > |C(\mathbf{u}_{\delta_2})|.$$

Remark 36. If assumption (EC-3) holds, by Lemma 40, besides (a) and (b) of Theorem 34, we have also that

$$(\delta_1 < \delta_2) \implies E(\mathbf{u}_{\delta_1}) > E(\mathbf{u}_{\delta_2}).$$

If (EC-3*) holds, by Lemma 44, besides (a) and (b) of Theorem 34 , we have also that

$$(\delta_1 < \delta_2) \implies E(\mathbf{u}_{\delta_1}) + aC(\mathbf{u}_{\delta_1})^s > E(\mathbf{u}_{\delta_2}) + aC(\mathbf{u}_{\delta_2})^s.$$

The proofs of the above results are in the remaining part of this section. In the Sects. 2.2.3 and 2.2.4 we prove the existence of minimizers, namely that $\Gamma(e, c) \neq \varnothing$ (see (2.11)) and in Sect. 2.2.5, we prove the stability of $\Gamma(e, c)$ namely that the minimizers are hylomorphic solitons.

2.2.3 A Minimization Result in the Positive Energy Case

We start with a technical lemma.

Lemma 37. *Let* $\mathbf{u}_n = \mathbf{u} + \mathbf{w}_n \in X$ *be a sequence such that* $C(\mathbf{u}_n)$ *does not converge to* 0 *and* $\mathbf{w}_n$ *converges weakly to* 0. *Then, up to a subsequence, we have*

$$\lim \Lambda \,(\mathbf{u} + \mathbf{w}_n) \geq \min \{\Lambda \,(\mathbf{u}), \lim \Lambda \,(\mathbf{w}_n)\} \tag{2.26}$$

(we use the convention that $\frac{a}{0} = +\infty$*).*

Proof. Given four real numbers A, B, a, b, (with $B, b > 0$), we have that

$$\frac{A + a}{B + b} \geq \min \left(\frac{A}{B}, \frac{a}{b} \right). \tag{2.27}$$

In fact, suppose that $\frac{A}{B} \geq \frac{a}{b}$; then

$$\frac{A + a}{B + b} = \frac{\frac{A}{B} B + \frac{a}{b} b}{B + b} \geq \frac{\frac{a}{b} B + \frac{a}{b} b}{B + b} = \frac{a}{b} \geq \min \left(\frac{A}{B}, \frac{a}{b} \right).$$

Notice that the equality holds if and only if

$$\frac{A}{B} = \frac{a}{b}. \tag{2.28}$$

Now suppose that $|C(\mathbf{u})| > 0$ and $|C(\mathbf{w}_n)| > 0$. By the splitting property, by (2.27) and since the energy E is positive, we have that

$$\Lambda \,(\mathbf{u} + \mathbf{w}_n) = \frac{E(\mathbf{u}_n)}{|C(\mathbf{u}_n)|} = \frac{E(\mathbf{u}) + E(\mathbf{w}_n) + o(1)}{|C(\mathbf{u}) + C(\mathbf{w}_n) + o(1)|}$$

$$\geq \frac{E(\mathbf{u}) + o(1) + E(\mathbf{w}_n)}{|C(\mathbf{u}) + o(1)| + |C(\mathbf{w}_n)|} \geq \min \left\{ \frac{E(\mathbf{u}) + o(1)}{|C(\mathbf{u}) + o(1)|}, \frac{E(\mathbf{w}_n)}{|C(\mathbf{w}_n)|} \right\}.$$

Then we get (2.26). If $|C(\mathbf{u})| = 0$ or $|C(\mathbf{w}_n)| = 0$, since $C(\mathbf{u}_n)$ does not converge to 0, we get (2.26) by the convention that $\frac{a}{0} = +\infty$. $\qquad\square$

The proofs of Theorems 33 and 34 under the assumption (EC-3) (positive energy case) or (EC-3*) (positive charge case) are similar, but the differences make easier to prove some lemmas separately. In this section we will consider the case in which (EC-3) holds.

For any $\delta > 0$, set

$$J_\delta(\mathbf{u}) = \Lambda \,(\mathbf{u}) + \delta E(\mathbf{u}). \tag{2.29}$$

Clearly, if δ_∞ is as in (2.22), we have

$$(\delta \in [0, \delta_\infty)) \implies (\exists \mathbf{v} : J_\delta(\mathbf{v}) < \Lambda_0). \tag{2.30}$$

Theorem 38. *Assume that E and C satisfy (EC-0)–(EC-3) and the hylomorphy condition (2.16). Then, for every $\delta \in (0, \delta_\infty)$ (δ_∞ defined in (2.22)), J_δ is G-compact and it has a minimizer $\mathbf{u}_\delta \neq 0$. Moreover $\mathbf{u}_\delta$ is a minimizer of E on $\mathfrak{M}_\delta := \{\mathbf{u} \in X \mid |C(\mathbf{u})| = c_\delta\}$ where $c_\delta = |C(\mathbf{u}_\delta)|$.*

Proof. Let $\mathbf{u}_n$ be a minimizing sequence of J_δ, $\delta \in (0, \delta_\infty)$($\delta_\infty$ defined in (2.22)). This sequence $\mathbf{u}_n$ is bounded in X. In fact, arguing by contradiction, assume that, up to a subsequence, $\|\mathbf{u}_n\| \to \infty$. Then by (EC-3)(ii), $E(\mathbf{u}_n) \to \infty$ and hence $J_\delta(\mathbf{u}_n) \to \infty$ which contradicts the fact that $\mathbf{u}_n$ is a minimizing sequence of J_δ.

We now set

$$j_\delta := \inf_{\mathbf{u} \in X} J_\delta(\mathbf{u}). \tag{2.31}$$

Since $\delta \in (0, \delta_\infty)$ by (2.30) we have that

$$j_\delta < \Lambda_0. \tag{2.32}$$

Moreover, since $E \geq 0$, we have

$$0 \leq \Lambda(\mathbf{u}_n) \leq J_\delta(\mathbf{u}_n)$$

and

$$J_\delta(\mathbf{u}_n) \to j_\delta < \Lambda_0.$$

Then, up to a subsequence, $\Lambda(\mathbf{u}_n) \to \lambda < \Lambda_0$. So, arguing as in Remark 30, we can take a subsequence $\mathbf{u}_{n_k}$ and a sequence $\{g_k\} \subset G$ such that $\mathbf{u}'_k := g_k \mathbf{u}_{n_k}$ is weakly convergent to some

$$\mathbf{u}_\delta \neq 0. \tag{2.33}$$

We can write

$$\mathbf{u}'_n = \mathbf{u}_\delta + \mathbf{w}_n$$

with $\mathbf{w}_n \rightharpoonup 0$ weakly. We want to prove that $\mathbf{w}_n \to 0$ strongly. To this end first we show that

$$C(\mathbf{u}_\delta + \mathbf{w}_n) \text{ does not converge to } 0. \tag{2.34}$$

Arguing by contradiction assume that

$$C(\mathbf{u}_\delta + \mathbf{w}_n) \to 0.$$

Then, since $\mathbf{u}'_n$ is a minimizing sequence for J_δ, we have

$$E(\mathbf{u}'_n) \to 0.$$

Then, by (EC-3)(iii), we have

$$\mathbf{u}'_n \to 0 \text{ strongly in } X$$

and then $\mathbf{u}_\delta = 0$, contradicting (2.33). So (2.34) holds.

So, by Lemma 37 and since E satisfies the splitting property (Definition 26), we have that

$$j_\delta = \lim J_\delta (\mathbf{u}_\delta + \mathbf{w}_n) = \lim [\Lambda (\mathbf{u}_\delta + \mathbf{w}_n) + \delta E (\mathbf{u}_\delta + \mathbf{w}_n)]$$

$$\geq \min \{\Lambda (\mathbf{u}_\delta), \lim \Lambda (\mathbf{w}_n)\} + \delta E (\mathbf{u}_\delta) + \delta \lim E (\mathbf{w}_n). \qquad (2.35)$$

Now there are two possibilities (up subsequences):

(1) $\min \{\Lambda (\mathbf{u}_\delta), \lim \Lambda (\mathbf{w}_n)\} = \lim \Lambda (\mathbf{w}_n).$

(2) $\min \{\Lambda (\mathbf{u}_\delta), \lim \Lambda (\mathbf{w}_n)\} = \Lambda (\mathbf{u}_\delta).$

We will show that the possibility (1) cannot occur. In fact, if it holds, we have by (2.35) and by (2.31) that

$$j_\delta \geq \lim \Lambda (\mathbf{w}_n) + \delta E (\mathbf{u}_\delta) + \delta \lim E (\mathbf{w}_n)$$

$$= \lim J_\delta (\mathbf{w}_n) + \delta E (\mathbf{u}_\delta)$$

$$\geq j_\delta + \delta E (\mathbf{u}_\delta)$$

and hence, we get that $E (\mathbf{u}_\delta) \leq 0$, contradicting (2.33). Then possibility (2) occurs and by (2.35) we have that

$$j_\delta \geq \Lambda (\mathbf{u}_\delta) + \delta E (\mathbf{u}_\delta) + \delta \lim E (\mathbf{w}_n)$$

$$= J_\delta (\mathbf{u}_\delta) + \delta \lim E (\mathbf{w}_n)$$

$$\geq j_\delta + \delta \lim E (\mathbf{w}_n).$$

Then, $\lim E (\mathbf{w}_n) \to 0$ and by (EC-3)(iii), $\mathbf{w}_n \to 0$ strongly and consequently $\mathbf{u}'_n = \mathbf{u}_\delta + \mathbf{w}_n \to \mathbf{u}_\delta$ strongly. So we conclude that $J_\delta (\mathbf{u}'_n) \to J_\delta (\mathbf{u}_\delta)$ and $\mathbf{u}_\delta$ minimizes J_δ.

Now set

$$e_\delta = E(\mathbf{u}_\delta)$$

$$c_\delta = |C(\mathbf{u}_\delta)|$$

$$\mathfrak{M}_\delta := \{\mathbf{u} \in X \mid |C(\mathbf{u})| = c_\delta\}.$$

Obviously $\mathbf{u}_\delta$ is a minimizer of J_δ on $\mathfrak{M}_\delta$. Moreover for any $\mathbf{u} \in \mathfrak{M}_\delta$

$$J_\delta(\mathbf{u}) = \frac{E(\mathbf{u})}{c_\delta} + \delta E(\mathbf{u}) = \left(\frac{1}{c_\delta} + \delta\right) E(\mathbf{u}).$$

Then $\mathbf{u}_\delta$ minimizes also E on $\mathfrak{M}_\delta$. □

In the following $\mathbf{u}_\delta$ will denote a minimizer of J_δ.

Lemma 39. *Let the assumptions of Theorem 38 be satisfied. If $\delta_1, \delta_2 \in (0, \delta_\infty)$ (δ_∞ as in (2.22)) $\delta_1 < \delta_2$, then the minimizers $\mathbf{u}_{\delta_1}$ of J_{δ_1} and $\mathbf{u}_{\delta_2}$ of J_{δ_2} satisfy the following inequalities:*

(a) $J_{\delta_1}(\mathbf{u}_{\delta_1}) < J_{\delta_2}(\mathbf{u}_{\delta_2})$.
(b) $E(\mathbf{u}_{\delta_1}) \geq E(\mathbf{u}_{\delta_2})$.
(c) $\Lambda(\mathbf{u}_{\delta_1}) \leq \Lambda(\mathbf{u}_{\delta_2})$.
(d) $|C(\mathbf{u}_{\delta_1})| \geq |C(\mathbf{u}_{\delta_2})|$.

Proof. First we prove inequality (a)

$$\begin{aligned}
J_{\delta_1}(\mathbf{u}_{\delta_1}) &= \Lambda(\mathbf{u}_{\delta_1}) + \delta_1 E(\mathbf{u}_{\delta_1}) \\
&\leq \Lambda(\mathbf{u}_{\delta_2}) + \delta_1 E(\mathbf{u}_{\delta_2}) \quad \text{(since $\mathbf{u}_{\delta_1}$ minimizes J_{δ_1})} \\
&< \Lambda(\mathbf{u}_{\delta_2}) + \delta_2 E(\mathbf{u}_{\delta_2}) \quad \text{(since E is positive)} \\
&= J_{\delta_2}(\mathbf{u}_{\delta_2}).
\end{aligned}$$

Now in order to prove inequalities (b) and (c) we set

$$\Lambda(\mathbf{u}_{\delta_1}) = \Lambda(\mathbf{u}_{\delta_2}) + a$$
$$E(\mathbf{u}_{\delta_1}) = E(\mathbf{u}_{\delta_2}) + b.$$

We need to prove that $b \geq 0$ and $a \leq 0$. We have

$$J_{\delta_2}(\mathbf{u}_{\delta_2}) \leq J_{\delta_2}(\mathbf{u}_{\delta_1}) \Rightarrow$$
$$\Lambda(\mathbf{u}_{\delta_2}) + \delta_2 E(\mathbf{u}_{\delta_2}) \leq \Lambda(\mathbf{u}_{\delta_1}) + \delta_2 E(\mathbf{u}_{\delta_1}) \Rightarrow$$
$$\Lambda(\mathbf{u}_{\delta_2}) + \delta_2 E(\mathbf{u}_{\delta_2}) \leq (\Lambda(\mathbf{u}_{\delta_2}) + a) + \delta_2(E(\mathbf{u}_{\delta_2}) + b) \Rightarrow$$
$$0 \leq a + \delta_2 b. \tag{2.36}$$

On the other hand,

$$J_{\delta_1}(\mathbf{u}_{\delta_2}) \geq J_{\delta_1}(\mathbf{u}_{\delta_1}) \Rightarrow$$
$$\Lambda(\mathbf{u}_{\delta_2}) + \delta_1 E(\mathbf{u}_{\delta_2}) \geq \Lambda(\mathbf{u}_{\delta_1}) + \delta_1 E(\mathbf{u}_{\delta_1}) \Rightarrow$$
$$\Lambda(\mathbf{u}_{\delta_2}) + \delta_1 E(\mathbf{u}_{\delta_2}) \geq (\Lambda(\mathbf{u}_{\delta_2}) + a) + \delta_1(E(\mathbf{u}_{\delta_2}) + b) \Rightarrow$$
$$0 \geq a + \delta_1 b. \tag{2.37}$$

From (2.36) and (2.37) we get

$$(\delta_2 - \delta_1)\, b \geq 0$$

and hence $b \geq 0$.

Moreover, using again (2.37) we have $-\delta_1 b \geq a$. So, since $b \geq 0$, we get $a \leq 0$. Finally, since $|C(\mathbf{u})| = \frac{E(\mathbf{u})}{\Lambda(\mathbf{u})}$, also inequality (d) follows.   □

Lemma 40. *Let the assumptions of Theorem 38 be satisfied and assume that also (2.25) is satisfied. If $\delta_1, \delta_2 \in (0, \delta_\infty)$, $\delta_1 < \delta_2$, then the minimizers $\mathbf{u}_{\delta_1}$ of J_{δ_1} and $\mathbf{u}_{\delta_2}$ of J_{δ_2} satisfy the following inequalities:*

(a) $E(\mathbf{u}_{\delta_1}) > E(\mathbf{u}_{\delta_2})$.
(b) $\Lambda(\mathbf{u}_{\delta_1}) < \Lambda(\mathbf{u}_{\delta_2})$.
(c) $|C(\mathbf{u}_{\delta_1})| > |C(\mathbf{u}_{\delta_2})|$.

Proof. Let $\delta_1, \delta_2 \in (0, \delta_\infty)$ and assume that $\delta_1 < \delta_2$.

(a) Since by Lemma 39 $E(\mathbf{u}_{\delta_1}) \geq E(\mathbf{u}_{\delta_2})$, it will be sufficient to prove that $E(\mathbf{u}_{\delta_1}) \neq E(\mathbf{u}_{\delta_2})$. We argue indirectly and assume that

$$E(\mathbf{u}_{\delta_1}) = E(\mathbf{u}_{\delta_2}). \tag{2.38}$$

By the previous lemma, we have that

$$\Lambda\,(\mathbf{u}_{\delta_1}) \leq \Lambda\,(\mathbf{u}_{\delta_2}). \tag{2.39}$$

Also, we have that

$$\Lambda\,(\mathbf{u}_{\delta_2}) + \delta_2 E\,(\mathbf{u}_{\delta_2}) \leq \Lambda\,(\mathbf{u}_{\delta_1}) + \delta_2 E(\mathbf{u}_{\delta_1}) \quad \text{(since } \mathbf{u}_{\delta_2} \text{ minimizes } J_{\delta_2})$$
$$= \Lambda\,(\mathbf{u}_{\delta_1}) + \delta_2 E\,(\mathbf{u}_{\delta_2}) \quad \text{(by (2.38))}$$

and so

$$\Lambda\,(\mathbf{u}_{\delta_2}) \leq \Lambda\,(\mathbf{u}_{\delta_1})$$

and by (2.39) we get

$$\Lambda\,(\mathbf{u}_{\delta_1}) = \Lambda\,(\mathbf{u}_{\delta_2}). \tag{2.40}$$

By (2.40) and (2.38) we deduce that $\mathbf{u}_{\delta_1}$ is a minimizer of J_{δ_2}; in fact

$$J_{\delta_2}\,(\mathbf{u}_{\delta_1}) = \Lambda\,(\mathbf{u}_{\delta_1}) + \delta_2 E\,(\mathbf{u}_{\delta_1})$$
$$= \Lambda\,(\mathbf{u}_{\delta_2}) + \delta_2 E\,(\mathbf{u}_{\delta_2}) = J_{\delta_2}\,(\mathbf{u}_{\delta_2}).$$

So, since $\mathbf{u}_{\delta_2}$ minimizes J_{δ_2}, also $\mathbf{u}_{\delta_1}$ minimizes J_{δ_2}. Then $\mathbf{u}_{\delta_2}$ and $\mathbf{u}_{\delta_1}$ are both minimizers of J_{δ_2} and we have that $J'_{\delta_2}(\mathbf{u}_{\delta_1}) = 0$ and $J'_{\delta_1}(\mathbf{u}_{\delta_1}) = 0$ which give

$$\Lambda'(\mathbf{u}_{\delta_1}) + \delta_2 E'(\mathbf{u}_{\delta_1}) = 0$$
$$\Lambda'(\mathbf{u}_{\delta_1}) + \delta_1 E'(\mathbf{u}_{\delta_1}) = 0.$$

The above equations imply that $E'(\mathbf{u}_{\delta_1}) = 0$ and $\Lambda'(\mathbf{u}_{\delta_1}) = 0$, and since $\Lambda(\mathbf{u}) = \frac{E(\mathbf{u})}{|C(\mathbf{u})|}$, assuming without loss of generality that $C(\mathbf{u}_{\delta_1}) > 0$, we get that

$$\frac{E'(\mathbf{u}_{\delta_1})\, C(\mathbf{u}_{\delta_1}) - E(\mathbf{u}_{\delta_1})\, C'(\mathbf{u}_{\delta_1})}{C(\mathbf{u}_{\delta_1})^2} = 0$$

and hence $C'(\mathbf{u}_{\delta_1}) = 0$. Then

$$\|E'(\mathbf{u}_{\delta_1})\| + \|C'(\mathbf{u}_{\delta_1})\| = 0$$

and by (2.25) $\mathbf{u}_{\delta_1} = 0$. This fact contradicts the conclusion of Theorem 38.

(b) We argue indirectly and assume that

$$\Lambda(\mathbf{u}_{\delta_1}) = \Lambda(\mathbf{u}_{\delta_2}). \tag{2.41}$$

By (a), we have that

$$E(\mathbf{u}_{\delta_1}) > E(\mathbf{u}_{\delta_2}). \tag{2.42}$$

Also, we have that

$$\Lambda(\mathbf{u}_{\delta_1}) + \delta_1 E(\mathbf{u}_{\delta_1}) \leq \Lambda(\mathbf{u}_{\delta_2}) + \delta_1 E(\mathbf{u}_{\delta_2}) \quad \text{(since } \mathbf{u}_{\delta_1} \text{ minimizes } J_{\delta_1}\text{)}$$
$$= \Lambda(\mathbf{u}_{\delta_1}) + \delta_1 E(\mathbf{u}_{\delta_2}) \quad \text{(by (2.41))}$$

and so

$$E(\mathbf{u}_{\delta_1}) \leq E(\mathbf{u}_{\delta_2}).$$

This inequality contradicts (2.42).

(c) Since

$$|C(\mathbf{u}_\delta)| = \frac{E(\mathbf{u}_\delta)}{\Lambda(\mathbf{u}_\delta)}$$

the conclusion follows from (a) and (b). □

2.2.4 A Minimization Result in the Positive Charge Case

Here we consider the case in which $C \geq 0$ (positive charge case) and we shall assume that (EC-3*) (see page 37) holds. In this case, the function $\Phi(\mathbf{u})$, defined in (2.24) will play the role played by the energy E when the energy is positive

Now, for any $\delta > 0$, we set

$$K_\delta(\mathbf{u}) = \Lambda(\mathbf{u}) + \delta\Phi(\mathbf{u}). \tag{2.43}$$

Lemma 41. *Assume that* E *and* C *satisfy (EC-3*). Then for any* $\delta \geq 0$ *we have*

$$K_\delta(\mathbf{u}) \geq \frac{\delta}{2}\Phi(\mathbf{u}) - M_\delta$$

where

$$M_\delta = -a \min_{t \geq 0}\left(\frac{\delta}{2}t^s - t^{s-1}\right).$$

Proof. By (EC-3*)(i) we have

$$\frac{E(\mathbf{u})}{C(\mathbf{u})} \geq -aC(\mathbf{u})^{s-1}.$$

Then

$$K_\delta(\mathbf{u}) = \frac{E(\mathbf{u})}{C(\mathbf{u})} + \delta\Phi(\mathbf{u}) \geq -aC(\mathbf{u})^{s-1} + \frac{\delta}{2}\left[E(\mathbf{u}) + 2aC(\mathbf{u})^s\right] + \frac{\delta}{2}\Phi(\mathbf{u})$$

$$\geq -aC(\mathbf{u})^{s-1} + \frac{\delta}{2}\left[-aC(\mathbf{u})^s + 2aC(\mathbf{u})^s\right] + \frac{\delta}{2}\Phi(\mathbf{u})$$

$$= -aC(\mathbf{u})^{s-1} + \frac{a\delta}{2}C(\mathbf{u})^s + \frac{\delta}{2}\Phi(\mathbf{u}) \geq \frac{\delta}{2}\Phi(\mathbf{u}) - M_\delta$$

where

$$M_\delta = -a\min_{t \geq 0}\left(\frac{\delta}{2}t^s - t^{s-1}\right).$$

$\square$

The following theorem holds:

Theorem 42. *Assume that* E *and* C *satisfy (EC-0)–(EC-3*) (see page 37) and the hylomorphy condition (2.16). Then, for every* $\delta \in (0, \delta_\infty)$ *(see (2.23)),* K_δ *is G-compact and it has a minimizer* $\mathbf{u}_\delta \neq 0$*. Moreover* $\mathbf{u}_\delta$ *is a minimizer of* E *on* $\mathfrak{M}_\delta := \{\mathbf{u} \in X \mid C(\mathbf{u}) = c_\delta\}$ *where* $c_\delta = C(\mathbf{u}_\delta)$*.*

Proof. Let $\delta \in (0, \delta_\infty)$, where δ_∞ is defined in (2.23), and set

$$k_\delta := \inf_{\mathbf{u} \in X} K_\delta(\mathbf{u}).$$

By Lemma 41 and since $\Phi(\mathbf{u}) \geq 0$, we have $k_\delta > -\infty$. Then, since $\delta \in (0, \delta_\infty)$, we have

$$-\infty < k_\delta < \Lambda_0. \tag{2.44}$$

Now let $\mathbf{u}_n$ be a minimizing sequence of K_δ. Let us prove that $\mathbf{u}_n$ is G-compact. To this end we shall first prove that

$$\mathbf{u}_n \text{ is bounded.}$$

Arguing by contradiction assume that, up to a subsequence, $\|\mathbf{u}_n\| \longrightarrow +\infty$. Then, by (EC-3*)(ii), we have

$$\Phi(\mathbf{u}_n) = E(\mathbf{u}_n) + 2aC(\mathbf{u}_n)^s \longrightarrow +\infty. \tag{2.45}$$

By Lemma 41 and (2.45) we get

$$K_\delta(\mathbf{u}_n) \longrightarrow +\infty.$$

This contradicts the fact that $\mathbf{u}_n$ is a minimizing sequence of K_δ and hence we conclude that $\mathbf{u}_n$ is bounded.

Let us prove that

$$\mathbf{u}_n \text{ is nonvanishing.}$$

By (2.44) and since $\mathbf{u}_n$ is a minimizing sequence for K_δ, for large n we have

$$\Lambda(\mathbf{u}_n) \leq K_\delta(\mathbf{u}_n) < \Lambda_0 - \eta, \quad \eta > 0. \tag{2.46}$$

Then, by definition of Λ_0 (see (2.17)), $\mathbf{u}_n$ is a nonvanishing sequence. Hence, by Definition 28, we can extract a subsequence $\mathbf{u}_{n_k}$ and we can take a sequence $\{g_k\} \subset G$ such that $\mathbf{u}_k' := g_k \mathbf{u}_{n_k}$ is weakly convergent to some

$$\mathbf{u}_\delta \neq 0. \tag{2.47}$$

We can write

$$\mathbf{u}_n' = \mathbf{u}_\delta + \mathbf{w}_n$$

with $\mathbf{w}_n \rightharpoonup 0$ weakly.

In order to show that K_δ is G-compact we need to prove that

$$\mathbf{w}_n \to 0 \text{ strongly in } X.$$

To this end we argue as in the positive energy case and prove first that

$$C(\mathbf{u}_\delta + \mathbf{w}_n) \to 0 \text{ does not converge to } 0.$$

Arguing by contradiction assume that

$$C(\mathbf{u}_\delta + \mathbf{w}_n) \to 0. \tag{2.48}$$

Then, since $\mathbf{u}'_n$ is a minimizing sequence for K_δ, we have

$$E(\mathbf{u}'_n) \to 0. \tag{2.49}$$

So by (2.48) and (2.49) we get

$$E(\mathbf{u}'_n) + aC(\mathbf{u}'_n)^s \to 0.$$

Then, by (EC-3*)(iii), we have

$$\mathbf{u}'_n \to 0 \text{ strongly in } X$$

and then $\mathbf{u}_\delta = 0$, contradicting (2.47).

So we can apply Lemma 37 and we have

$$\lim \Lambda\,(\mathbf{u}_\delta + \mathbf{w_n}) \geq \min\left\{\Lambda\,(\mathbf{u}_\delta)\,, \lim \Lambda\,(\mathbf{w_n})\right\}.$$

By the above inequality we get

$$k_\delta = \lim K_\delta\,(\mathbf{u}_\delta + \mathbf{w}_n) = \lim\left[\Lambda\,(\mathbf{u}_\delta + \mathbf{w}_n) + \delta\Phi\,(\mathbf{u}_\delta + \mathbf{w}_n)\right] = \tag{2.50}$$

$$\geq \min\left\{\Lambda\,(\mathbf{u}_\delta)\,, \lim \Lambda\,(\mathbf{w}_n)\right\} + \delta \lim \Phi(\mathbf{u}_\delta + \mathbf{w}_n). \tag{2.51}$$

Using again the splitting property of E and C and since $s \geq 1$, we have that

$$\lim \Phi(\mathbf{u}_\delta + \mathbf{w}_n) = \lim\left(E(\mathbf{u}_\delta + \mathbf{w}_n) + 2aC(\mathbf{u}_\delta + \mathbf{w}_n)^s\right)$$

$$= E(\mathbf{u}_\delta) + \lim E(\mathbf{w}_n) + 2a \lim\left(C(\mathbf{u}_\delta) + C(\mathbf{w}_n)\right)^s$$

$$\geq E(\mathbf{u}_\delta) + \lim E(\mathbf{w}_n) + 2a \lim\left(C(\mathbf{u}_\delta)^s + C(\mathbf{w}_n)^s\right)$$

$$= E(\mathbf{u}_\delta) + 2aC(\mathbf{u}_\delta)^s + \lim E(\mathbf{w}_n) + 2a \lim C(\mathbf{w}_n)^s$$

$$= \Phi(\mathbf{u}_\delta) + \lim \Phi(\mathbf{w}_n). \tag{2.52}$$

Then by (2.51) and by (2.52) we have

$$k_\delta \geq \min\{\Lambda(\mathbf{u}_\delta), \lim \Lambda(\mathbf{w}_n)\} + \delta\Phi(\mathbf{u}_\delta) + \delta \lim \Phi(\mathbf{w}_n). \tag{2.53}$$

Now there are two possibilities:

$$\text{(a) } \min\{\Lambda(\mathbf{u}_\delta), \lim \Lambda(\mathbf{w}_n)\} = \lim \Lambda(\mathbf{w}_n).$$

$$\text{(b) } \min\{\Lambda(\mathbf{u}_\delta), \lim \Lambda(\mathbf{w}_n)\} = \Lambda(\mathbf{u}_\delta).$$

We will show that the possibility (a) cannot occur. In fact, if it holds, we have by (2.53) that

$$k_\delta \geq \lim \Lambda(\mathbf{w}_n) + \delta\Phi(\mathbf{u}_\delta) + \delta \lim \Phi(\mathbf{w}_n)$$
$$= \lim K_\delta(\mathbf{w}_n) + \delta\Phi(\mathbf{u}_\delta)$$
$$\geq k_\delta + \delta\Phi(\mathbf{u}_\delta)$$

and hence, we get that $\Phi(\mathbf{u}_\delta) \leq 0$; this, by (EC-3*)(i), implies that $\mathbf{u}_\delta = 0$, contradicting (2.47).

Then the possibility (b) holds and, by (2.53), we have that

$$k_\delta \geq \Lambda(\mathbf{u}_\delta) + \delta\Phi(\mathbf{u}_\delta) + \delta \lim \Phi(\mathbf{w}_n)$$
$$= K_\delta(\mathbf{u}_\delta) + \delta \lim \Phi(\mathbf{w}_n)$$
$$\geq k_\delta + \delta \lim \Phi(\mathbf{w}_n).$$

Then, $\lim \Phi(\mathbf{w}_n) \to 0$ and by (EC-3*)(iii), $\mathbf{w}_n \to 0$ strongly. We conclude that K_δ is G-compact and $\mathbf{u}_\delta$ is a minimizer of K_δ. Then $\mathbf{u}_\delta$ minimizes also the functional

$$\frac{E(\mathbf{u})}{c_\delta} + \delta\left[E(\mathbf{u}) + 2ac_\delta^s\right] = \left(\frac{1}{c_\delta} + \delta\right) E(\mathbf{u}) + 2\delta a c_\delta^s$$

on the set $\mathfrak{M}_\delta = \{\mathbf{u} \in X \mid C(\mathbf{u}) = c_\delta\}$ and hence $\mathbf{u}_\delta$ minimizes also $E|_{\mathfrak{M}_\delta}$. $\square$

In the following $\mathbf{u}_\delta$ will denote a minimizer of K_δ.

Lemma 43. *Let the assumptions of Theorem 42 be satisfied. Let $\delta_1, \delta_2 \in (0, \delta_\infty)$ $\delta_1 < \delta_2$ and let $\mathbf{u}_{\delta_1}, \mathbf{u}_{\delta_2}$ be minimizers of $K_{\delta_1}, K_{\delta_2}$ respectively. Then the following inequalities hold:*

(a) $K_{\delta_1}(\mathbf{u}_{\delta_1}) < K_{\delta_2}(\mathbf{u}_{\delta_2})$.
(b) $\Phi(\mathbf{u}_{\delta_1}) \geq \Phi(\mathbf{u}_{\delta_2})$.
(c) $\Lambda(\mathbf{u}_{\delta_1}) \leq \Lambda(\mathbf{u}_{\delta_2})$.
(d) $C(\mathbf{u}_{\delta_1}) \geq C(\mathbf{u}_{\delta_2})$.

Proof. We prove first the inequality (a)

$$
\begin{aligned}
K_{\delta_1}(\mathbf{u}_{\delta_1}) &= \Lambda(\mathbf{u}_{\delta_1}) + \delta_1 \Phi(\mathbf{u}_{\delta_1}) \\
&\leq \Lambda(\mathbf{u}_{\delta_2}) + \delta_1 \Phi(\mathbf{u}_{\delta_2}) \quad \text{(since } \mathbf{u}_{\delta_1} \text{ minimizes } K_{\delta_1}) \\
&< \Lambda(\mathbf{u}_{\delta_2}) + \delta_2 \Phi(\mathbf{u}_{\delta_2}) \quad \text{(since } \Phi \text{ is positive and } \delta_1 < \delta_2) \\
&= K_{\delta_2}(\mathbf{u}_{\delta_2}).
\end{aligned}
$$

In order to prove inequalities (b) and (c) we set

$$
\Lambda(\mathbf{u}_{\delta_1}) = \Lambda(\mathbf{u}_{\delta_2}) + a
$$
$$
\Phi(\mathbf{u}_{\delta_1}) = \Phi(\mathbf{u}_{\delta_2}) + b.
$$

We need to prove that $b \geq 0$ and $a \leq 0$. We have

$$
K_{\delta_2}(\mathbf{u}_{\delta_2}) \leq K_{\delta_2}(\mathbf{u}_{\delta_1}) \Rightarrow
$$
$$
\Lambda(\mathbf{u}_{\delta_2}) + \delta_2 \Phi(\mathbf{u}_{\delta_2}) \leq \Lambda(\mathbf{u}_{\delta_1}) + \delta_2 \Phi(\mathbf{u}_{\delta_1}) \Rightarrow
$$
$$
\Lambda(\mathbf{u}_{\delta_2}) + \delta_2 \Phi(\mathbf{u}_{\delta_2}) \leq (\Lambda(\mathbf{u}_{\delta_2}) + a) + \delta_2 (\Phi(\mathbf{u}_{\delta_2}) + b) \Rightarrow
$$
$$
0 \leq a + \delta_2 b. \tag{2.54}
$$

On the other hand,

$$
K_{\delta_1}(\mathbf{u}_{\delta_2}) \geq K_{\delta_1}(\mathbf{u}_{\delta_1}) \Rightarrow
$$
$$
\Lambda(\mathbf{u}_{\delta_2}) + \delta_1 \Phi(\mathbf{u}_{\delta_2}) \geq \Lambda(\mathbf{u}_{\delta_1}) + \delta_1 \Phi(\mathbf{u}_{\delta_1}) \Rightarrow
$$
$$
\Lambda(\mathbf{u}_{\delta_2}) + \delta_1 \Phi(\mathbf{u}_{\delta_2}) \geq (\Lambda(\mathbf{u}_{\delta_2}) + a) + \delta_1 (\Phi(\mathbf{u}_{\delta_2}) + b) \Rightarrow
$$
$$
0 \geq a + \delta_1 b. \tag{2.55}
$$

From (2.54) and (2.55) we get

$$
(\delta_2 - \delta_1) b \geq 0
$$

and hence $b \geq 0$. So, by (2.55), $a \leq 0$.

Finally we prove inequality (d). Arguing by contradiction we assume that

$$
C(\mathbf{u}_{\delta_1}) < C(\mathbf{u}_{\delta_2}). \tag{2.56}
$$

Then

$$
aC(\mathbf{u}_{\delta_1})^s < aC(\mathbf{u}_{\delta_2})^s. \tag{2.57}
$$

By (c) and (2.56) we get

$$
C(\mathbf{u}_{\delta_1})\Lambda(\mathbf{u}_{\delta_1}) < C(\mathbf{u}_{\delta_2})\Lambda(\mathbf{u}_{\delta_2}). \tag{2.58}
$$

By (2.57) and (2.58) we get

$$\Phi(\mathbf{u}_{\delta_1}) < \Phi(\mathbf{u}_{\delta_2})$$

and this contradicts (b). □

Lemma 44. *Let the assumptions of Theorem 42 be satisfied and assume that also (2.25) is satisfied. Let $\delta_1, \delta_2 \in (0, \delta_\infty)$ $\delta_1 < \delta_2$ and let $\mathbf{u}_{\delta_1}, \mathbf{u}_{\delta_2}$ be non zero minimizers of $K_{\delta_1}, K_{\delta_2}$ respectively. Then the following inequalities hold:*

(a) $\Phi(\mathbf{u}_{\delta_1}) > \Phi(\mathbf{u}_{\delta_2})$.
(b) $\Lambda(\mathbf{u}_{\delta_1}) < \Lambda(\mathbf{u}_{\delta_2})$.
(c) $C(\mathbf{u}_{\delta_1}) > C(\mathbf{u}_{\delta_2})$.

Proof. Let $\delta_1, \delta_2 \in (0, \delta_\infty)$ $\delta_1 < \delta_2$. By Lemma 43 there exist $\mathbf{u}_{\delta_1}, \mathbf{u}_{\delta_2}$ non zero minimizers of $K_{\delta_1}, K_{\delta_2}$.

By Lemma 43 we know that $\Phi(\mathbf{u}_{\delta_1}) \geq \Phi(\mathbf{u}_{\delta_2})$, so in order to prove (a) we need only to show that $\Phi(\mathbf{u}_{\delta_1}) \neq \Phi(\mathbf{u}_{\delta_2})$. We argue indirectly and assume that

$$\Phi(\mathbf{u}_{\delta_1}) = \Phi(\mathbf{u}_{\delta_2}). \qquad (2.59)$$

By the previous lemma, we have that

$$\Lambda(\mathbf{u}_{\delta_1}) \leq \Lambda(\mathbf{u}_{\delta_2}). \qquad (2.60)$$

Also, we have that

$$\Lambda(\mathbf{u}_{\delta_2}) + \delta_2 \Phi(\mathbf{u}_{\delta_2}) \leq \Lambda(\mathbf{u}_{\delta_1}) + \delta_2 \Phi(\mathbf{u}_{\delta_1}) \quad \text{(since } \mathbf{u}_{\delta_2} \text{ minimizes } K_{\delta_2})$$
$$= \Lambda(\mathbf{u}_{\delta_1}) + \delta_2 \Phi(\mathbf{u}_{\delta_2}) \quad \text{(by (2.59))}$$

and so

$$\Lambda(\mathbf{u}_{\delta_2}) \leq \Lambda(\mathbf{u}_{\delta_1})$$

and by (2.60) we get

$$\Lambda(\mathbf{u}_{\delta_1}) = \Lambda(\mathbf{u}_{\delta_2}). \qquad (2.61)$$

Then, it follows that $\mathbf{u}_{\delta_1}$ is also a minimizer of K_{δ_2}; in fact, by (2.61) and (2.59)

$$K_{\delta_2}(\mathbf{u}_{\delta_1}) = \Lambda(\mathbf{u}_{\delta_1}) + \delta_2 \Phi(\mathbf{u}_{\delta_1})$$
$$= \Lambda(\mathbf{u}_{\delta_2}) + \delta_2 \Phi(\mathbf{u}_{\delta_2}) = K_{\delta_2}(\mathbf{u}_{\delta_2}).$$

Then, we have that $K'_{\delta_2}(\mathbf{u}_{\delta_1}) = 0$ as well as $K'_{\delta_1}(\mathbf{u}_{\delta_1}) = 0$ which explicitly give

$$\Lambda'(\mathbf{u}_{\delta_1}) + \delta_2 \Phi'(\mathbf{u}_{\delta_1}) = 0$$
$$\Lambda'(\mathbf{u}_{\delta_1}) + \delta_1 \Phi'(\mathbf{u}_{\delta_1}) = 0.$$

The above equations imply that

$$\Phi' (\mathbf{u}_{\delta_1}) = 0$$
$$\Lambda' (\mathbf{u}_{\delta_1}) = 0.$$

Since $\Lambda (\mathbf{u}) = \frac{E(\mathbf{u})}{C(\mathbf{u})}$ and $\Phi(\mathbf{u}) = E(\mathbf{u}) + 2aC(\mathbf{u})^s$, the above system of equations becomes

$$\frac{E' (\mathbf{u}_{\delta_1})}{C (\mathbf{u}_{\delta_1})} - \frac{E (\mathbf{u}_{\delta_1})}{C (\mathbf{u}_{\delta_1})^2} C' (\mathbf{u}_{\delta_1}) = 0$$
$$E'(\mathbf{u}_{\delta_1}) + 2as C(\mathbf{u}_{\delta_1})^{s-1} C' (\mathbf{u}_{\delta_1}) = 0. \tag{2.62}$$

Eliminating $E'(\mathbf{u}_{\delta_1})$, we get

$$(2as C(\mathbf{u}_{\delta_1})^s + E(\mathbf{u}_{\delta_1})) \frac{C' (\mathbf{u}_{\delta_1})}{C (\mathbf{u}_{\delta_1})^2} = 0$$

and, using (2.24), we get

$$\frac{\Phi (\mathbf{u}_{\delta_1}) + 2a(s-1) C(\mathbf{u}_{\delta_1})^s}{C (\mathbf{u}_{\delta_1})^2} C' (\mathbf{u}_{\delta_1}) = 0. \tag{2.63}$$

By assumption (EC-3*) (i) and since $s > 1$, we have

$$\frac{\Phi (\mathbf{u}_{\delta_1}) + 2a(s-1) C(\mathbf{u}_{\delta_1})^s}{C (\mathbf{u}_{\delta_1})^2} > 0, \tag{2.64}$$

then (2.63) and (2.64) imply that $C' (\mathbf{u}_{\delta_1}) = 0$, and hence, by (2.62), also $E' (\mathbf{u}_{\delta_1}) = 0$. Finally by (2.25) $\mathbf{u}_{\delta_1} = 0$, and we get a contradiction.

Also to prove (b) we argue indirectly and assume that

$$\Lambda(\mathbf{u}_{\delta_1}) \geq \Lambda(\mathbf{u}_{\delta_2}). \tag{2.65}$$

So by (c) in Lemma 43 we have

$$\Lambda(\mathbf{u}_{\delta_1}) = \Lambda(\mathbf{u}_{\delta_2}). \tag{2.66}$$

Then we have that

$$\Lambda (\mathbf{u}_{\delta_1}) + \delta_1 \Phi (\mathbf{u}_{\delta_1}) \leq \Lambda (\mathbf{u}_{\delta_2}) + \delta_1 \Phi(\mathbf{u}_{\delta_2}) \quad \text{(since } \mathbf{u}_{\delta_1} \text{ minimizes } K_{\delta_1})$$
$$= \Lambda (\mathbf{u}_{\delta_1}) + \delta_1 \Phi (\mathbf{u}_{\delta_2}) \quad \text{(by (2.66))}$$

and so

$$\Phi (\mathbf{u}_{\delta_1}) \leq \Phi (\mathbf{u}_{\delta_2})$$

and this contradicts inequality (a).

Let us finally prove the inequality (c).
Since

$$\Lambda\left(\mathbf{u}_{\delta_i}\right)C\left(\mathbf{u}_{\delta_i}\right) = E\left(\mathbf{u}_{\delta_i}\right), \ i = 1,2$$

we have

$$\Phi\left(\mathbf{u}_{\delta_i}\right) = \Lambda\left(\mathbf{u}_{\delta_i}\right)C\left(\mathbf{u}_{\delta_i}\right) + 2aC\left(\mathbf{u}_{\delta_i}\right)^s, \ i = 1,2$$

and the conclusion easily follows from inequalities (a) and (b).   □

2.2.5   The Stability Result

In order to prove Theorem 33 it is sufficient to show that the minimizers $\mathbf{u}_\delta$ in
Theorems 38 and 42 provide solitons, so we have to prove that the set

$$\Gamma\left(e_\delta, c_\delta\right) = \{\mathbf{u} \in X \mid |C(\mathbf{u})| = c_\delta, E(\mathbf{u}) = e_\delta\}, \ (e_\delta = E(\mathbf{u}_\delta), c_\delta = |C(\mathbf{u}_\delta)|)$$
$$(2.67)$$

is stable and G compact. To do this, we need the well known Liapunov theorem in
following form:

Theorem 45. *Let (X, γ) be dynamical system and let Γ be an invariant set. Assume
that there exists a differentiable function V (called a Liapunov function) defined on
a neighborhood of Γ such that*

(a) $V(\mathbf{u}) \geq 0$ and $V(\mathbf{u}) = 0 \Leftrightarrow u \in \Gamma$.
(b) $\partial_t V(\gamma_t(\mathbf{u})) \leq 0$.
(c) $V(\mathbf{u}_n) \to 0 \Leftrightarrow d(\mathbf{u}_n, \Gamma) \to 0$.

Then Γ is stable.

Proof. For completeness, we give a proof of this well known result. Arguing by
contradiction, assume that Γ, satisfying (a)–(c), is not stable. Then there exists $\varepsilon > 0$
and sequences $\mathbf{u}_n \in X$ and $t_n > 0$ such that

$$d(\mathbf{u}_n, \Gamma) \to 0 \text{ and } d(\gamma_{t_n}(\mathbf{u}_n), \Gamma) > \varepsilon. \tag{2.68}$$

Then we have

$$d(\mathbf{u}_n, \Gamma) \to 0 \Longrightarrow V(\mathbf{u}_n) \to 0 \Longrightarrow V(\gamma_{t_n}(\mathbf{u}_n)) \to 0 \Longrightarrow d(\gamma_{t_n}(\mathbf{u}_n), \Gamma) \to 0$$

where the first and the third implications are consequence of property (c). The
second implication follows from property (b). Clearly, this fact contradicts (2.68).

□

Lemma 46. *Let V be a G-compact functional on X. Assume that $V \geq 0$. and let $\Gamma = V^{-1}(0)$ be the set of minimizers of V. Then Γ is G-compact and V satisfies the point (c) of Theorem 45.*

Proof. The fact that Γ is G-compact, is a trivial consequence of the fact that Γ is the set of minimizers of a G-compact functional V. Now we prove (c) of Theorem 45.

First we show the implication $\Rightarrow$. Let $\mathbf{u}_n$ be a sequence such that $V(\mathbf{u}_n) \to 0$. By contradiction, we assume that $d(\mathbf{u}_n, \Gamma) \nrightarrow 0$, namely that there is a subsequence $\mathbf{u}'_n$ such that

$$d(\mathbf{u}'_n, \Gamma) \geq a > 0. \tag{2.69}$$

Since $V(\mathbf{u}_n) \to 0$ also $V(\mathbf{u}'_n) \to 0$, and, since V is G compact, there exists a sequence g_n in G such that, for a subsequence $\mathbf{u}''_n$, we have $g_n \mathbf{u}''_n \to \mathbf{u}_0$. Then

$$d(\mathbf{u}''_n, \Gamma) = d(g_n \mathbf{u}''_n, \Gamma) \leq d(g_n \mathbf{u}''_n, \mathbf{u}_0) \to 0$$

and this contradicts (2.69).

Now we prove the other implication $\Leftarrow$. Let $\mathbf{u}_n$ be a sequence such that $d(\mathbf{u}_n, \Gamma) \to 0$, then there exists $\mathbf{v}_n \in \Gamma$ s.t.

$$d(\mathbf{u}_n, \Gamma) \geq d(\mathbf{u}_n, \mathbf{v}_n) - \frac{1}{n}. \tag{2.70}$$

Since V is G-compact, also Γ is G-compact; so, for a sequence g_n, we have $g_n \mathbf{v}_n \to \bar{\mathbf{w}} \in \Gamma$. We get the conclusion if we show that $V(\mathbf{u}_n) \to 0$. We have by (2.70), that $d(\mathbf{u}_n, \mathbf{v}_n) \to 0$ and hence $d(g_n \mathbf{u}_n, g_n \mathbf{v}_n) \to 0$ and so, since $g_n \mathbf{v}_n \to \bar{\mathbf{w}}$, we have $g_n \mathbf{u}_n \to \bar{\mathbf{w}} \in \Gamma$. Therefore, by the continuity of V and since $\bar{\mathbf{w}} \in \Gamma$, we have $V(g_n \mathbf{u}_n) \to V(\bar{\mathbf{w}}) = 0$ and, since V is G-invariant, we can conclude that $V(\mathbf{u}_n) \to 0$. $\qquad\square$

Proof of Theorem 33. First we consider the case in which (EC-3) holds (positive energy case) and let $\delta \in (0, \delta_\infty)$ as in Theorem 38. So by Theorem 38 there exists a minimizer $\mathbf{u}_\delta$ of J_δ.

Moreover $\mathbf{u}_\delta$ minimizes E on $\mathfrak{M}_\delta := \{\mathbf{u} \in X \mid |C(\mathbf{u})| = c_\delta\}$ where $c_\delta = |C(\mathbf{u}_\delta)|$. So, in order to show that $\mathbf{u}_\delta$ is an hylomorphic soliton, we need to show that $\Gamma(e_\delta, c_\delta)$ (see (2.67)) is G-compact and stable.

We set

$$V(\mathbf{u}) = (E(\mathbf{u}) - e_\delta)^2 + (|C(\mathbf{u})| - c_\delta)^2.$$

Since $\Gamma(e_\delta, c_\delta) = V^{-1}(0)$, we are reduced to show that $V^{-1}(0)$ is G-compact and stable.

To this hand first we prove that V is G compact.

Let $\mathbf{w}_n$ be a minimizing sequence for V, then $V(\mathbf{w}_n) \to 0$ and consequently $E(\mathbf{w}_n) \to e_\delta$ and $|C(\mathbf{w}_n)| \to c_\delta$. Now, since

$$\inf J_\delta = \frac{e_\delta}{c_\delta} + \delta e_\delta,$$

we have that $\mathbf{w}_n$ is a minimizing sequence also for J_δ. Then, since by Theorem 38 J_δ is G-compact, we get

$$\mathbf{w}_n \text{ is } G\text{-compact.}$$

So we conclude that V is G-compact.

By Lemma 46, we deduce that $V^{-1}(0)$ is G compact and that V satisfies the property (c) in Theorem 45. Moreover, since E and C are constants of motion, V satisfies also the properties (a) and (b). So, by Theorem 45, we conclude that $V^{-1}(0)$ is stable.

Now consider the case in which (EC-3*) holds (positive charge case) and let $\delta \in (0, \delta_\infty)$ as in Theorem 42. Then, by Theorem 42, there exists a minimizer $\mathbf{u}_\delta$ of K_δ. Moreover $\mathbf{u}_\delta$ minimizes E on the set $\mathfrak{M}_\delta = \{\mathbf{u} \in X \mid C(\mathbf{u}) = c_\delta\}$ ($c_\delta = C(\mathbf{u}_\delta)$).

In order to show that $\mathbf{u}_\delta$ is a hylomorphic soliton we argue as in the positive energy case. So we are reduced to show that

$$V(\mathbf{u}) = (E(\mathbf{u}) - e_\delta)^2 + (C(\mathbf{u}) - c_\delta)^2, \ e_\delta = E(\mathbf{u}_\delta)$$

is G compact.

Let $\mathbf{w}_n$ be a minimizing sequence for V, then $V(\mathbf{w}_n) \to 0$ and consequently $E(\mathbf{w}_n) \to e_\delta$ and $C(\mathbf{w}_n) \to c_\delta$. Now, since

$$\inf K_\delta = \frac{e_\delta}{c_\delta} + \delta \left[e_\delta + 2ac_\delta^s \right],$$

we have that $\mathbf{w}_n$ is a minimizing sequence also for K_δ. Then, since by Theorem 42 K_δ is G-compact, we get

$$\mathbf{w}_n \text{ is } G\text{-compact.} \tag{2.71}$$

So V is G-compact and the conclusion follows by using the same arguments as in the positive energy case. $\qquad\square$

Proof of Theorem 34. By Theorem 33 for any $\delta \in (0, \delta_\infty)$ there exists a hylomorphic soliton $\mathbf{u}_\delta$. By using Lemma 40, if (EC-3) holds, or Lemma 44, if (EC-3*) holds, we get different solitons for different values of δ. Namely for $\delta_1 < \delta_2$ we have $\Lambda(\mathbf{u}_{\delta_1}) < \Lambda(\mathbf{u}_{\delta_2})$ and $|C(\mathbf{u}_{\delta_1})| > |C(\mathbf{u}_{\delta_2})|$. $\qquad\square$

2.3 The Structure of Hylomorphic Solitons

2.3.1 The Meaning of Hylenic Ratio

Let (X, γ) be a dynamical system of type FT. If $\mathbf{u} \in X$ is a finite energy field, usually it disperses as time goes on, namely

$$\lim_{t \to \infty} \|\gamma_t \mathbf{u}\|_{\star} = 0.$$

where

$$\|\mathbf{u}\|_{\star} = \sup_{x \in \mathbb{R}^N} \int_{B_1(x)} |\mathbf{u}|_V \, dx,$$

V is the internal parameter space (cf. page 28) and $B_1(x) = \{y \in \mathbb{R}^N : |x - y| < 1\}$. However, if the hylomorphy condition (2.16) is satisfied, this dispersion in general does not occur. In fact we have the following result:

Proposition 47. *Assume that X is compactly embedded into $L_{loc}^1(\mathbb{R}^N, V)$. Let $\mathbf{u}_0 \in X$ such that $\Lambda(\mathbf{u}_0) < \Lambda_0$, then*

$$\min \lim_{t \to \infty} \|\mathbf{u}(t)\|_{\star} > 0 \qquad (2.72)$$

where $\mathbf{u}(t) = \gamma_t \mathbf{u}$ and $\gamma_0 \mathbf{u} = \mathbf{u}_0$.

Proof. Let $t_n \to \infty$ be a sequence of times such that

$$\lim_{n \to \infty} \|\mathbf{u}(t_n)\|_{\star} = \min \lim_{t \to \infty} \|\mathbf{u}(t)\|_{\star}. \qquad (2.73)$$

Since Λ is a constant of motion

$$\Lambda(\mathbf{u}(t_n)) = \Lambda(\mathbf{u}_0) < \Lambda_0$$

then, by the definition of Λ_0, may be taking a subsequence, there is a sequence of translations T_{x_n} such that

$$T_{x_n} \mathbf{u}(t_n) = \mathbf{u}(t_n, x - x_n) = \bar{\mathbf{u}} + \mathbf{w}_n \qquad (2.74)$$

where $\bar{\mathbf{u}} \neq 0$ and $\mathbf{w}_n \rightharpoonup 0$ in X. Without loss of generality, we may assume that $\bar{\mathbf{u}} \neq 0$ in $B_1(0)$. Since X is compactly embedded into $L_{loc}^1(\mathbb{R}^N, V)$, we have that

$$\int_{B_1(0)} |\mathbf{w}_n|_V \, dx \to 0. \qquad (2.75)$$

By (2.74), we have that

$$|T_{x_n}\mathbf{u}(t_n)|_V \geq |\bar{\mathbf{u}}|_V - |\mathbf{w}_n|_V . \tag{2.76}$$

Then, using (2.76) and (2.75), we have that

$$\min\lim_{n\to\infty} \int_{B_1(0)} |T_{x_n}\mathbf{u}(t_n)|_V \, dx$$

$$\geq \lim_{n\to\infty} \left(\int_{B_1(0)} |\bar{\mathbf{u}}|_V \, dx - \int_{B_1(0)} |\mathbf{w}_n|_V \, dx \right)$$

$$= \int_{B_1(0)} |\bar{\mathbf{u}}|_V \, dx > 0.$$

Then

$$\min\lim_{n\to\infty} \int_{B_1(0)} |T_{x_n}\mathbf{u}(t_n)|_V \, dx > 0. \tag{2.77}$$

Finally, by (2.73) and (2.77), we get

$$\min\lim_{t\to\infty} \|\mathbf{u}(t)\|_\star = \lim_{n\to\infty} \|\mathbf{u}(t_n)\|_\star \geq \min\lim_{n\to\infty} \int_{B_1(x_n)} |\mathbf{u}(t_n)|_V \, dx$$

$$= \min\lim_{n\to\infty} \int_{B_1(0)} |T_{x_n}\mathbf{u}(t_n)|_V \, dx > 0.$$

The (2.72) is proved. □

Thus the hylomorphy condition prevents the dispersion. As we have seen in the preceding section, (2.16) is also a fundamental assumption in proving the existence of hylomorphic solitons.

Now, as it happens in Noether's theorem (Theorem 5), we assume E and C to be local quantities, namely, given $\mathbf{u} \in X$, there exist the density functions $\rho_{E,\mathbf{u}}(x)$ and $\rho_{C,\mathbf{u}}(x) \in L^1(\mathbb{R}^N)$ such that

$$E(\mathbf{u}) = \int \rho_{E,\mathbf{u}}(x) \, dx$$

$$C(\mathbf{u}) = \int \rho_{C,\mathbf{u}}(x) \, dx.$$

Energy and hylenic densities $\rho_{E,\mathbf{u}}$, $\rho_{C,\mathbf{u}}$ allow to define the density of *binding energy* as follows:

$$\beta(t,x) = \beta_\mathbf{u}(t,x) = [\rho_{E,\mathbf{u}}(t,x) - \Lambda_0 \cdot |\rho_{C,\mathbf{u}}(t,x)|]^- \tag{2.78}$$

where $[f]^-$ denotes the negative part of f.

If $\mathbf{u}$ satisfies the hylomorphy condition (2.18), we have that $E(\mathbf{u}) < \Lambda_0 |C(\mathbf{u})|$ and hence he have that $\beta_{\mathbf{u}}(t, x) \neq 0$ for some $x \in \mathbb{R}^N$.

The support of the binding energy density is called *bound matter region;* more precisely we have the following definition

Definition 48. Given any configuration $\mathbf{u}$, we define the **bound matter region** as follows

$$\Sigma(\mathbf{u}) = \overline{\{x : \beta_{\mathbf{u}}(t, x) \neq 0\}}.$$

If $\mathbf{u}_0$ is a soliton, the set $\Sigma(\mathbf{u}_0)$ is called **support of the soliton** at time t.

In the situation considered in this book, the solitons satisfy the hylomorphy condition. Thus we may think that a soliton $\mathbf{u}_0$ consists of bound matter localized in a precise region of the space, namely $\Sigma(\mathbf{u}_0)$. This fact gives the name to this type of soliton from the Greek words "*hyle*" = "matter" and "*morphe*" = "*form*".

2.3.2 The Swarm Interpretation of Hylomorphic Solitons

Clearly the physical interpretation of hylomorphic solitons depends on the model which we are considering. However we can always assume a *conventional inter-pretation* which we will call *swarm interpretation* since the soliton is regarded as a *swarm* of particles bound together. In each particular physical situation this interpretation, might have or might not have any physical meaning; in any case it represents a pictorial way of thinking of the mathematical phenomena which occur. This interpretation is consistent with the names and the definitions given in the previous section.

We assume that $\mathbf{u}$ is a field which describes a fluid consisting of particles; the particles density is given by the function $\rho_C(t, x) = \rho_{C,\mathbf{u}}(t, x)$ which, of course satisfies a continuity equation

$$\partial_t \rho_C + \nabla \cdot \mathbf{J}_C = 0 \tag{2.79}$$

where $\mathbf{J}_C$ is the flow of particles. Hence C is the total number of particles. Here the particles are not intended to be as in "particle theory" but rather as in fluidynamics, so that C does not need to be an integer number. Alternatively, if you like, you may think that C is not the number of particles but it is proportional to it. Also, in some equations, as for example in the Nonlinear Klein Gordon Equation (see Sect. 4.1), C can be negative; in this case, the existence of *antiparticles* is assumed

Thus, the hylenic ratio

$$\Lambda(\mathbf{u}) = \frac{E(\mathbf{u})}{|C(\mathbf{u})|}$$

represents the average energy of each particle (or antiparticle). The number Λ_0 defined in (2.17) is interpreted as the rest energy of each particle when they do not interact with each other. If $\Lambda\left(\mathbf{u}\right) > \Lambda_0$, then the average energy of each particle is bigger than the rest energy; if $\Lambda\left(\mathbf{u}\right) < \Lambda_0$, the opposite occurs and this fact means that particles act with each other with an attractive force.

If the particles were at rest and they were not acting on each other, their energy density would be

$$\Lambda_0 \cdot |\rho_C(t,x)|.$$

If $\rho_E(t,x)$ denotes the energy density and if

$$\rho_E(t,x) < \Lambda_0 \cdot |\rho_C(t,x)|;$$

then, in the point x at time t, the particles attract each other with a force which is stronger than the repulsive forces; this explains the name *density of binding energy* given to $\beta(t,x)$ in (2.78).

Thus a soliton relative to the state $\mathbf{u}$ can be considered as a "rigid" object occupying the region of space $\Sigma\left(\mathbf{u}\right)$ (cf. Definition 48); it consists of particles which stick to each other; the energy to destroy the soliton is given by

$$\int \beta_{\mathbf{u}}(t,x)dx = \int_{\Sigma(\mathbf{u})} (\Lambda_0 |\rho_C(t,x)| - \rho_E(t,x)) \, dx.$$

However out of $\Sigma\left(\mathbf{u}\right)$ the energy density is bigger than $\Lambda_0 \cdot |\rho_C(t,x)|$; thus the total energy necessary to reduce the soliton to isolated particles is given by

$$\Lambda_0 \cdot |C\left(\mathbf{u}\right)| - E\left(\mathbf{u}\right).$$

It can be shown that there are states $\mathbf{u}$ such that $\Sigma\left(\mathbf{u}\right) \neq \varnothing$ but they are not necessarily solitons. In these states, $\Sigma\left(\mathbf{u}\right)$ is a region where the particles stick with each other but they do not have reached a stable configurations; the shape of $\Sigma\left(\mathbf{u}\right)$ might changes with time. In many concrete situations, such states may evolve toward one or more solitons. In these cases we say that the solitons are asymptotically stable. The study of asymptotical stability is a quite involved problem. We refer to [50, 61, 98] and [62] and their references.

Chapter 3
The Nonlinear Schrödinger Equation

This chapter deals with the Nonlinear Schrödinger Equation (NS). After having analyzed the general features of NS, we apply the abstract theory of Chap. 2 and we prove the existence of hylomorphic solitons. In the last part of this chapter, we analyze the dynamics of such solitons.

3.1   General Features of NS

The Schrödinger equation for a particle which moves in a potential $V(x)$ is given by

$$i\frac{\partial \psi}{\partial t} = -\frac{1}{2}\Delta \psi + V(x)\psi$$

where $\psi : \mathbb{R} \times \mathbb{R}^N \to \mathbb{C}$ and $V : \mathbb{R}^N \to \mathbb{R}$.

We are interested to the nonlinear Schrödinger equation:

$$i\frac{\partial \psi}{\partial t} = -\frac{1}{2}\Delta \psi + \frac{1}{2}W'(\psi) + V(x)\psi \tag{3.1}$$

where $W : \mathbb{C} \to \mathbb{R}$ and

$$W'(\psi) = \frac{\partial W}{\partial \psi_1} + i\frac{\partial W}{\partial \psi_2}. \tag{3.2}$$

We assume that W depends only on $|\psi|$, namely

$$W(\psi) = F(|\psi|) \text{ and so } W'(\psi) = F'(|\psi|)\frac{\psi}{|\psi|}$$

for some smooth function $F : [0, \infty) \to \mathbb{R}$. In the following we shall identify, with some abuse of notation, W with F.

© Springer International Publishing Switzerland 2014
V. Benci, D. Fortunato, *Variational Methods in Nonlinear Field Equations*,
Springer Monographs in Mathematics, DOI 10.1007/978-3-319-06914-2_3

If $V(x) = 0$, then we get the equation

$$i \frac{\partial \psi}{\partial t} = -\frac{1}{2} \Delta \psi + \frac{1}{2} W'(\psi); \tag{NS}$$

this equation is variational and invariant for a representation of the Galileo group (cf. page 8).

First of all let us check that it is variational:

Proposition 49. *Equation (3.1) is the Euler-Lagrange equation relative to the Lagrangian density*

$$\mathcal{L} = \mathrm{Re} \left(i \partial_t \psi \overline{\psi} \right) - \frac{1}{2} |\nabla \psi|^2 - W(\psi) - V(x) |\psi|^2. \tag{3.3}$$

Proof. Set

$$\mathcal{S}(\psi) = \mathcal{S}_1(\psi) + \mathcal{S}_2(\psi)$$

$$\mathcal{S}_1(\psi) = \int \mathrm{Re} \left(i \partial_t \psi \overline{\psi} \right) dxdt; \quad \mathcal{S}_2(\psi) = -\int \left[\frac{1}{2} |\nabla \psi|^2 + W(\psi) + V(x) |\psi|^2 \right] dxdt$$

and set $\psi = u_1 + i u_2$. We have

$$\mathcal{S}_1(\psi) = \int \mathrm{Re} \left(i \partial_t \psi \overline{\psi} \right) dxdt$$

$$= \int \mathrm{Re} \left[(i \partial_t u_1 - \partial_t u_2)(u_1 - i u_2) \right] dxdt$$

$$= \int (\partial_t u_1 u_2 - \partial_t u_2 u_1) dxdt.$$

Then, if $\varphi = v_1 + i v_2$

$$d\mathcal{S}_1(\psi)[\varphi] = \int (\partial_t u_1 v_2 + \partial_t v_1 u_2 - \partial_t u_2 v_1 - \partial_t v_2 u_1) dxdt$$

$$= \int (2\partial_t u_1 v_2 - 2\partial_t u_2 v_1) dxdt$$

$$= \int \mathrm{Re} \left[2i (\partial_t u_1 + i \partial_t u_2)(v_1 - i v_2) \right]$$

$$= \int \mathrm{Re} \left(2i \partial_t \psi \overline{\varphi} \right).$$

Moreover

$$d\,\mathcal{S}_2(\psi)\,[\varphi] = -\int \left[\text{Re}\,\langle \nabla\psi, \nabla\varphi \rangle + \text{Re}\left(W'\,(\psi)\,\overline{\varphi}\right) + 2\,\text{Re}\left(V(x)\psi\overline{\varphi}\right)\right] dxdt$$

$$= -\int \left[\text{Re}\,(-\Delta\psi\overline{\varphi}) + \text{Re}\left(W'\,(\psi)\,\overline{\varphi}\right) + 2\,\text{Re}\left(V(x)\psi\overline{\varphi}\right)\right] dxdt$$

$$= -\int \text{Re}\left[\left(-\Delta\psi + W'\,(\psi) + 2V(x)\psi\right)\overline{\varphi}\right] dxdt.$$

Then

$$d\,\mathcal{S}(\psi)\,[\varphi] = \int \text{Re}\left[\left(2i\,\partial_t\psi + \Delta\psi - W'\,(\psi) - 2V(x)\psi\right)\overline{\varphi}\right] dxdt.$$

So, the stationary points of $\mathcal{S}$, satisfy the equation

$$2i\,\partial_t\psi + \Delta\psi - W'\,(\psi) - 2V(x)\psi = 0$$

which is equivalent to (3.1). $\square$

Sometimes it is useful to write ψ in polar form

$$\psi(t,x) = u(t,x)e^{iS(t,x)} \tag{3.4}$$

where $u(t,x) \in \mathbb{R}$ and $S(t,x) \in \mathbb{R}/(2\pi\mathbb{Z})$. Thus the state of the system ψ is uniquely defined by the pair of variables (u, S). Using these variables, the action $S = \int \mathcal{L}dxdt$ takes the form

$$\mathcal{S}(u,S) = -\int \left[\frac{1}{2}|\nabla u|^2 + W(u) + \left(\partial_t S + \frac{1}{2}|\nabla S|^2 + V(x)\right)u^2\right] dx$$

and Eq. (3.1) becomes:

$$-\frac{1}{2}\Delta u + \frac{1}{2}W'(u) + \left(\partial_t S + \frac{1}{2}|\nabla S|^2 + V(x)\right)u = 0 \tag{3.5}$$

$$\partial_t\left(u^2\right) + \nabla \cdot \left(u^2\nabla S\right) = 0. \tag{3.6}$$

3.1.1   Constants of Motion of NS

Noether's theorem states that any invariance for a one-parameter group of the Lagrangian implies the existence of a constant of motion, namely of a quantity which is preserved with time by the solutions (see Theorem 5). If $V = 0$ the

Lagrangian (3.3) is invariant under the Galileo group (page 8), then Eq. (3.1) has ten constants of motion:

- **Energy.** Energy, by definition, is the quantity which is preserved by virtue of the time invariance of the Lagrangian; it has the following form (see (1.38))

$$E = \text{Re} \int \left(\frac{\partial \mathcal{L}}{\partial \psi_t} \overline{\partial_t \psi} - \mathcal{L} \right) dx$$

where $\bar{z}$ denotes the complex conjugate of z. The Lagrangian $\mathcal{L}$ in (3.3) is invariant by time translations and the energy E of a state ψ is

$$E(\psi) = \int \left[\frac{1}{2} |\nabla \psi|^2 + W(\psi) + V(x) |\psi|^2 \right] dx. \tag{3.7}$$

Using (3.4) we get:

$$E(\psi) = \int \left(\frac{1}{2} |\nabla u|^2 + W(u) \right) dx + \int \left(\frac{1}{2} |\nabla S|^2 + V(x) \right) u^2 dx. \tag{3.8}$$

In the following we shall assume that $V = 0$ then

$$\mathcal{L} = \text{Re} \left(i \partial_t \psi \overline{\psi} \right) - \frac{1}{2} |\nabla \psi|^2 - W(\psi) \tag{3.9}$$

which in polar form can be written

$$\mathcal{L} = -\frac{1}{2} |\nabla u|^2 - W(u) - \left(\partial_t S + \frac{1}{2} |\nabla S|^2 \right) u^2. \tag{3.10}$$

Besides time translations, the Lagrangian in (3.9) is invariant also under space translations, space rotations and the Galileo transformations (see page 8).

In order to avoid unessential complications we shall consider only the case of three space dimensions ($N = 3$).

- **Momentum.** Momentum, by definition, is the constant of the motion related to the invariance under space translations of the Lagrangian; the invariance for translations in the x_i direction gives rise to the following constant of the motion (see (1.40))

$$P_i = -\text{Re} \int \frac{\partial \mathcal{L}}{\partial \psi_t} \overline{\partial_i \psi} \, dx.$$

The momentum $\mathbf{P} = (P_1, P_2, P_3)$ related to the Lagrangian $\mathcal{L}$ in (3.9), is

$$\mathbf{P} = \text{Im} \int \nabla \psi \overline{\psi} \, dx. \tag{3.11}$$

Using (3.4) we get:

$$\mathbf{P} = \int u^2 \nabla S \ dx. \tag{3.12}$$

- **Angular momentum.** The angular momentum, by definition, is the quantity which is preserved by virtue of the invariance under space rotations with respect to the origin. Its expression (see (1.41)) is given by

$$\mathbf{M} = \mathrm{Re} \int \frac{\partial \mathcal{L}}{\partial \psi_t} (\mathbf{x} \times \nabla \psi) \ dx.$$

So, taking the Lagrangian as in (3.9), we have

$$\mathbf{M} = \mathrm{Im} \int \mathbf{x} \times \nabla \psi \overline{\psi} \ dx. \tag{3.13}$$

Using (3.4) we get:

$$\mathbf{M} = \int \mathbf{x} \times \nabla S \, u^2 \ dx. \tag{3.14}$$

- **Barycenter velocity.** The quantity preserved by virtue of the invariance with respect to the Galileo transformations (1.16) is the following

$$\mathbf{K} = \int \mathbf{x} u^2 dx - t\mathbf{P}. \tag{3.15}$$

We shall prove only that the first component

$$K_1 = \int x_1 u^2 dx - t P_1$$

of (3.15) is a constant of the motion, the other cases can be treated in the same way.

Proof. Let us compute K_1 using Theorem 5; in this case the parameter λ is the first component v of the velocity $\mathbf{v} = (v, 0, 0)$ which appears in (1.14).

The Galileo transformations are

$$\mathbf{x}_v = \mathbf{x} - \mathbf{v}t$$

$$t_v = t$$

and recalling (1.16), we have that the representation T_{g_v} acts on a state $\psi = u e^{iS}$ of the system as follows

$$\left(T_{g_v} \psi \right)(t, x) = \psi \left(t, x - \mathbf{v}t \right) e^{i \left(\mathbf{v} \cdot x - \frac{1}{2} v^2 t \right)} = u \left(t, x - \mathbf{v}t \right) e^{iS(t, x - \mathbf{v}t)} e^{i \left(\mathbf{v} \cdot x - \frac{1}{2} v^2 t \right)}.$$

Namely

$$u_v(t, x) = u(t, x - \mathbf{v}t)$$

$$S_v(t, x) = S(t, x - \mathbf{v}t) + vx_1 - \frac{1}{2}v^2t.$$

Then, by (1.23) and taking $\mathcal{L}$ as in (3.10), we have:

$$\rho_{K_1} = \left[\frac{\partial \mathcal{L}}{\partial u_t}\frac{\partial u_v}{\partial v} + \frac{\partial \mathcal{L}}{\partial S_t}\frac{\partial S_v}{\partial v} + \mathcal{L}\frac{\partial t_v}{\partial v}\right]_{v=0} = \left[u_v^2 \cdot \frac{\partial S_v}{\partial v}\right]_{v=0}$$

$$= \left[u_v^2\left(-\frac{\partial S_v}{\partial x_1}t + x_1 - vt\right)\right]_{v=0} = \left(x_1 - t\frac{\partial S}{\partial x_1}\right)u^2.$$

Thus

$$K_1 = \int x_1 u^2 dx - t\int \frac{\partial S}{\partial x_1}u^2 dx = \int x_1 u^2 dx - tP_1.$$

$\square$

Thus the three components K_1, K_2, K_3 of $\mathbf{K}$ are the constants of motion related to the invariance of $\mathcal{L}$ with respect to the Galileo transformations. Let us interpret this fact in a more meaningful way. If we derive both sides of (3.15) with respect to t, we get

$$0 = \frac{d}{dt}\left(\int \mathbf{x}u^2 dx\right) - \mathbf{P}. \tag{3.16}$$

Now we define the *barycenter* (or *hylocenter*) as follows

$$\mathbf{q} := \frac{\int \mathbf{x}u^2 dx}{\int u^2 dx} = \frac{\int \mathbf{x}u^2 dx}{C(u)}, \tag{3.17}$$

where $C(u) = \int u^2 dx$ is a constant of the motion (see (3.19)).
Then by (3.16) and (3.17) we get

$$\dot{\mathbf{q}} = \frac{\mathbf{P}}{C(u)}. \tag{3.18}$$

So, since $\mathbf{P}$ and $C(u)$ are constants of the motion, then also the velocity $\dot{\mathbf{q}}$ of the barycenter is constant of the motion.

Then we can conclude that the Galileo invariance of the Lagrangian (3.10) implies that velocity $\dot{\mathbf{q}}$ of the barycenter is constant of the motion.

Besides the constants of the motion related to the invariance under the Galileo group, we have another constant of the motion.

- **Hylenic Charge.** The hylenic charge, by definition, is the constant of the motion related to the invariance of (3.3) under the gauge action (1.19). The hylenic charge has the following expression (see page 20)

$$C(\psi) = \text{Im} \int \frac{\partial \mathcal{L}}{\partial \psi_t} \overline{\psi} \, dx.$$

Then, for the Lagrangian $\mathcal{L}$ in (3.3), we have

$$C(\psi) = \int |\psi|^2 \, dx = \int u^2 dx. \tag{3.19}$$

3.1.2   Swarm Interpretation of NS

Before giving the swarm interpretation to Eq. (NS), we will write it with the usual physical constants m and $\hbar$:

$$i\hbar \frac{\partial \psi}{\partial t} = -\frac{\hbar^2}{2m} \Delta \psi + \frac{1}{2} W'(\psi) + V(x)\psi. \tag{3.20}$$

Here m has the dimension of *mass* and $\hbar$, the Planck constant, has the dimension of *action*.

The polar form of a state ψ is written as follows

$$\psi(t, x) = u(t, x) e^{iS(t,x)/\hbar}, \quad u(t, x) \in \mathbb{R} \tag{3.21}$$

and Eqs. (3.5) and (3.6) become

$$-\frac{\hbar^2}{2m} \Delta u + \frac{1}{2} W'(u) + \left(\partial_t S + \frac{1}{2m} |\nabla S|^2 + V(x) \right) u = 0 \tag{3.22}$$

$$\partial_t \left(u^2 \right) + \nabla \cdot \left(u^2 \frac{\nabla S}{m} \right) = 0. \tag{3.23}$$

The continuity equation (2.79) for (NS) is given by (3.23). This equation allows us to interpret the matter field to be a fluid composed by particles whose density is given by

$$\rho_C = u^2$$

and which move in the velocity field

$$\mathbf{v} = \frac{\nabla S}{m}. \tag{3.24}$$

So Eq. (3.23) reads

$$\partial_t \rho_C + \nabla \cdot (\rho_C \mathbf{v}) = 0.$$

Now assume that

$$-\frac{\hbar^2}{2m}\Delta u + \frac{1}{2}W'(u) \ll \left(\partial_t S + \frac{1}{2m}|\nabla S|^2 + V(x)\right)u \tag{3.25}$$

where $a \ll b$ means that $|a|$ is "much smaller" than $|b|$. In this case, Eq. (3.22) can be "approximated" by

$$\partial_t S + \frac{1}{2m}|\nabla S|^2 + V(x) = 0. \tag{3.26}$$

This is the Hamilton-Jacobi equation of a particle of mass m in a potential field V (cf. (1.53)). The trajectory $\mathbf{q}(t)$ of each particle satisfies, by (3.24), the equation

$$\dot{\mathbf{q}} = \frac{\nabla S}{m} \tag{3.27}$$

(cf. (1.54)).

If we do not assume (3.25), Eq. (3.26) needs to be replaced by

$$\partial_t S + \frac{1}{2m}|\nabla S|^2 + V(x) + Q(u) = 0 \tag{3.28}$$

with

$$Q(u) = \frac{-(\hbar^2/m)\,\Delta u + W'(u)}{2u}. \tag{3.29}$$

The term $Q(u)$ can be regarded as a field describing a sort of interaction between particles.

Given a wave of the form (3.21), the local frequency and the local wave number are defined as follows:

$$\omega(t, x) = -\frac{\partial_t S(t, x)}{\hbar}$$

$$\mathbf{k}(t, x) = \frac{\nabla S(t, x)}{\hbar}.$$

The energy of each particle moving according to (3.28), by the Hamilton-Jacobi theory (see Sect. 1.4), is given by

$$E = -\partial_t S$$

and its momentum is given by

$$\mathbf{p} = \nabla S;$$

thus we have that

$$E = \hbar\omega$$

$$\mathbf{p} = \hbar\mathbf{k};$$

these two equations are the De Broglie relation (see [71]). It is interesting to observe that they have been deduced by the swarm interpretation of the Schrödinger equation.

We recall again that the swarm interpretation is just a useful pictorial way to look at NS. In physical models, in general, there are different interpretations. Here we will mention very shortly some of them.

In the traditional model of quantum mechanics for one particle, we have $W = 0$ and ρ_C is interpreted as a probability density of the position of this particle.

One of the most important applications of NS is in the *Bose-Einstein condensate*. In this case $W(\psi) = -\frac{1}{4}|\psi|^4$ and (NS) takes the name of *Gross-Pitaevskii equation*. (see e.g. [138] and its references). Here $|\psi|^2$ is interpreted as the particle density which in this case are bosons (as, for example, atoms).

3.2 Existence Results for NS

3.2.1 Existence of Solitary Waves

The *standing waves*, are finite energy solutions having the following form (see (2.6))

$$\psi_0(t, x) = u(x)e^{-i\omega t}, u \in \mathbb{R}. \tag{3.30}$$

In particular, substituting (3.30) in Eq. (NS), we get

$$-\Delta u + W'(u) = 2\omega u \tag{3.31}$$

where, as usual, we have identified W with the function $F : [0, \infty) \to \mathbb{R}$ such that $W(\psi) = F(|\psi|)$.

Clearly a standing wave gives rise to a solitary wave (see Definition 12).

For simplicity, the existence of solitary waves for (NS) will be studied in the case $N \geq 3$ (N denotes the dimension of the space) under quite general assumptions on W. In the next subsection, where the existence of solitons will be studied, the case $N \geq 1$ will be considered.

We make the following assumptions on W:

(WA-i) W is a C^2 function s.t. $W(0) = W'(0) = 0$.
(WA-ii) If we set

$$W(s) = \frac{1}{2}W''(0)s^2 + N(s), \tag{3.32}$$

then

$$\exists s_0 \in \mathbb{R}^+ : \ N(s_0) < 0. \tag{3.33}$$

(WA-iii)   There exists $p < 2^* = \frac{2N}{N-2}$, $N \geq 3$, s.t.

$$N'(s) \geq -cs^{p-1} \text{ for any } s \geq 1. \tag{3.34}$$

We will see that (WA-ii) implies the hylomorphy condition (2.16). Actually, N is the nonlinear term which, when it is negative, produces a attractive "force". As we will see this assumption is essentially needed to guarantee the hylomorphy condition.

Now we can apply the results of Sect. A.3 and get the following theorem:

Theorem 50. Let W satisfy (WA-i)–(WA-iii). Then Eq. (NS) has finite energy solitary waves of the form (3.30) for every frequency $\omega \in (E_1, E_0)$, where $E_0 = \frac{W''(0)}{2}$ and

$$E_1 = \inf \{ a \in \mathbb{R} : \exists s \in \mathbb{R}^+, \ as^2 > W(s) \}.$$

Notice that, by virtue of (3.33), $E_1 < E_0$. Also, it is possible that $E_1 = -\infty$; for example this happens if

$$W(s) = E_0 s^2 - \frac{1}{p} |s|^p, \ 2 < p < 2^*. \tag{3.35}$$

Proof. We want to apply Theorem 144 to Eq. (3.31); to this end we set

$$G(s) = W(s) - \omega s^2 = (E_0 - \omega)s^2 + N(s).$$

It is easy to verify that, for every $\omega \in (E_1, E_0)$, G satisfies the assumptions (G-i) and (G-iii) of Theorem 144. Now we prove that also (G-ii) is satisfied.

Clearly by (3.34) we have

$$G'(s) = 2(E_0 - \omega)s + N'(s) \geq (E_0 - \omega)s - cs^{p-1} \text{ for any } s \geq 1. \tag{3.36}$$

Since $N''(0) = 0$, we can take $\delta > 0$ so small that

$$G'(s) = 2(E_0 - \omega)s + N'(s) \geq (E_0 - \omega)s \text{ for } 0 < s < \delta. \tag{3.37}$$

Now take $d > 0$ sufficiently large so that we have

$$G'(s) \geq (E_0 - \omega)s - ds^{p-1} \text{ for } 1 \geq s \geq \delta. \tag{3.38}$$

Then by (3.36)–(3.38) we have that

$$G'(s) \geq c_1 s - c_2 s^{p-1} \text{ for all } s > 0$$

where $c_1 = E_0 - \omega$ and $c_2 = \max\{d, c\}$ □

If we are interested in standing waves, it makes sense to consider also the non autonomous case, namely Eq. (3.1). In this case, the ansatz (3.30) gives the equation

$$- \Delta u + W'(u) + 2V(x)u = 2\omega u \tag{3.39}$$

The finite energy solutions of this equation have been largely studied in the literature and we refer to [56] and the references therein. In this book, we will consider only the case in which V is a lattice potential (see (3.48)).

3.2.2 Existence of Solitons

Theorem 50 provides sufficient conditions for the existence of solitary waves. In order to prove the existence of hylomorphic solitons (cf. Definition 20), first of all it is necessary to assume that the Cauchy problem for (3.1) is well posed namely, that it has a unique global solution which depends continuously on the initial data. For example, this is the case if W' is a globally Liptschitz function; we refer to the books [90] and [53] or to [82, 83] for more general conditions.

Moreover, it is necessary to investigate under which assumptions the energy

$$E = \int \left(\frac{1}{2} |\nabla \psi|^2 + W(\psi) + V(x) |\psi|^2 \right) dx$$

$$= \int \left(\frac{1}{2} |\nabla u|^2 + \frac{1}{2} |\nabla S|^2 u^2 + W(u) + V(x) u^2 \right) dx \tag{3.40}$$

achieves the minimum on the manifold

$$\mathfrak{M}_\sigma = \left\{ \psi \in H^1(\mathbb{R}^N, \mathbb{C}) : \int |\psi|^2 \, dx = \sigma \right\}. \tag{3.41}$$

We make the following assumptions on W and V:

(WB-i) W is a C^2 function s.t.

$$W(0) = W'(0) = 0 \text{ and } W''(0) = 2E_0 > 0. \tag{3.42}$$

(WB-ii)   If we set

$$W(s) = E_0 s^2 + N(s), \tag{3.43}$$

then

$$\exists s_0 \in \mathbb{R}^+ \text{ such that } N(s_0) < -V_0 s_0^2 \qquad (3.44)$$

where

$$V_0 = \max V.$$

(WB-iii)   There exist q, r in $(2, 2^*)$, $2^* = \frac{2N}{N-2}$ if $N \geq 3$ and $2^* = +\infty$ if $N = 1, 2$, s.t.

$$|N'(s)| \leq c_1 s^{r-1} + c_2 s^{q-1}. \qquad (3.45)$$

(WB-iv)

$$N(s) \geq -c s^p, \ c \geq 0, \ 2 < p < 2 + \frac{4}{N} \text{ for } s \text{ large.} \qquad (3.46)$$

We assume that $V : \mathbb{R}^N \to \mathbb{R}$ is a function satisfying the following:

(V-i) V continuous and

$$V(x) \geq 0, \ x \in \mathbb{R}^N. \qquad (3.47)$$

(V-ii) V is a lattice potential, namely it satisfies the periodicity condition

$$V(x) = V(x + Az) \text{ for all } x \in \mathbb{R}^N \text{ and } z \in \mathbb{Z}^N \qquad (3.48)$$

where A is a $N \times N$ invertible matrix.

Now let us make some remarks on these assumptions and a comparison with the assumptions on page 69.

Remark 51. First observe that the conditions $W''(0) = 2E_0 > 0$ in (3.42) and $V(x) \geq 0$ in (3.47) are assumed for simplicity; in fact, if

$$W''(0) = a \leq 0$$

and V is bounded below, namely for some constant b we have

$$V(x) \geq b, \ x \in \mathbb{R}^N,$$

we can reduce the problem to the case (3.42) and (3.47). To do this, we replace $W(s)$ with

$$W_1(s) = W(s) + \frac{1}{2}(1 - a) s^2$$

and $V(x)$ with

$$V_1(x) = V(x) - b.$$

So W_1 and V_1 satisfy the assumptions (3.42) and (3.47). Now, if ψ_1 solves

$$i\frac{\partial\psi_1}{\partial t} = -\frac{1}{2}\Delta\psi_1 + \frac{1}{2}W_1'(\psi_1) + V_1(x)\psi_1, \tag{3.49}$$

it can be easily seen that $\psi = \psi_1(t,x)e^{i\frac{(a+2b-1)t}{2}}$ is a solution of (3.1).

(WB-ii) clearly reduces to (WA-ii) if $V(x) = 0$.

(WB-iii) is stronger than (WA-iii) and it will be used in the proof of the splitting property (see Lemma 53); we do not know if it can be weakened.

(WB-iv) guaranties that the energy is bounded from below provided that the charge is fixed.

There are many results on the existence of stable solutions for the nonlinear Schrödinger equation ([17, 50, 53, 54, 138, 144, 145] and its references). The following theorem concerns with the existence of solitons in a periodic potential $V(x)$ and in this case it states the existence of a one parameter family of hylomorphic solitons for (3.1). In particular, if $V(x) = 0$, this result reduces to a variant of the results contained in the quoted papers.

As usual, Λ will denote the hylenic ratio

$$\Lambda = \frac{E(\psi)}{C(\psi)}.$$

Theorem 52. *Let W and V satisfy assumptions (WB-i)–(WB-iv) and (V-i), (V-ii). Then there exists $\delta_\infty > 0$ such that the dynamical system described by the Schrödinger equation (3.1) has a family $\psi_\delta \in H^1(\mathbb{R}^N, \mathbb{C})$ ($\delta \in (0, \delta_\infty)$) of hylomorphic solitons (Definition 20). Moreover if $\delta_1 < \delta_2$ we have that*

(a) $\Lambda(\psi_{\delta_1}) < \Lambda(\psi_{\delta_2})$.
(b) $\|\psi_{\delta_1}\|_{L^2} > \|\psi_{\delta_2}\|_{L^2}$.

In order to prove Theorem 52 we will apply the abstract theory developed in Sect. 2.2.2. In this case we have:

- $X = H^1(\mathbb{R}^N, \mathbb{C})$.
- The state $\mathbf{u} \in X$ will be denoted by ψ.
- The energy is given by (3.40).
- The hylenic charge is given by $C(\psi) = \|\psi\|_{L^2(\mathbb{R}^N)}^2$.
- The group G is the following representation of $\mathbb{Z}^N$:

$$\text{for all } z \in \mathbb{Z}^N \text{ and } \psi \in H^1(\mathbb{R}^N, \mathbb{C}) : T_z\psi(x) = \psi(x + Az) \tag{3.50}$$

where A is the $N \times N$ invertible matrix in (3.48).

The proof of Theorem 52 is based on the abstract Theorem 34. We need to show that (WB-i)–(WB-iv) and (V-i), (V-ii) permit to prove that assumptions (EC-0)–(EC-2), (EC-3*), (2.16) and (2.25) of Theorem 34 are satisfied.

Assumptions (EC-0) and (EC-1) are trivially verified. In the next two subsections we shall prove that also the other assumptions of Theorem 34 are satisfied.

3.2.3 Splitting and Coercivity

In this section we shall prove that E and C satisfy assumptions (EC-2) (splitting) and (EC-3*) (coercivity) (see page 38)

Lemma 53. *Let assumption (WB-iii) (see (3.45)) be satisfied. Then E and C satisfy the splitting property (EC-2) (see Definition 26).*

Proof. We set

$$E(v) = A(v, v) + K(v)$$

where

$$A(v, v) = \int \left[\frac{1}{2} |\nabla v|^2 + (E_0 + V(x)) \, v^2 \right] dx \qquad (3.51)$$

and

$$K(v) = \int N(v) \, dx. \qquad (3.52)$$

The hylenic charge $C(v) = \int v^2$ and $A(v, v)$ in (3.51) are quadratic forms, then, by Remark 27, they satisfy the splitting property. So the energy $E(v)$ will satisfy (EC-2) if we show that $K(v)$ in (3.52) satisfies (EC-2).

For any measurable $A \subset \mathbb{R}^N$ and any $v \in H^1(\mathbb{R}^N, \mathbb{C})$, we set

$$K_A(v) = \int_A N(v) dx.$$

Now consider any sequence

$$\psi_n = \psi + w_n \in H^1(\mathbb{R}^N, \mathbb{C})$$

where w_n converges weakly to 0.

Choose $\varepsilon > 0$ and $R = R(\varepsilon) > 0$ such that

$$\left| K_{B_R^c}(\psi) \right| < \varepsilon \qquad (3.53)$$

where

$$B_R^c = \mathbb{R}^N - B_R \text{ and } B_R = \{x \in \mathbb{R}^N : |x| < R\}.$$

Since $w_n \rightharpoonup 0$ weakly in $H^1(\mathbb{R}^N, \mathbb{C})$, by usual compactness arguments, we have that

$$K_{B_R}(w_n) \to 0 \text{ and } K_{B_R}(\psi + w_n) \to K_{B_R}(\psi). \qquad (3.54)$$

Then, by (3.54) and (3.53), we have

$$\lim_{n \to \infty} |K(\psi + w_n) - K(\psi) - K(w_n)|$$

$$= \lim_{n \to \infty} \left| K_{B_R^c}(\psi + w_n) + K_{B_R}(\psi + w_n) - K_{B_R^c}(\psi) - K_{B_R}(\psi) - K_{B_R^c}(w_n) \right.$$

$$\left. - K_{B_R}(w_n) \right|$$

$$= \lim_{n \to \infty} \left| K_{B_R^c}(\psi + w_n) - K_{B_R^c}(\psi) - K_{B_R^c}(w_n) \right|$$

$$\leq \lim_{n \to \infty} \left| K_{B_R^c}(\psi + w_n) - K_{B_R^c}(w_n) \right| + \varepsilon. \qquad (3.55)$$

Now, by the intermediate value theorem, there exists $\zeta_n \in (0, 1)$ such that for $z_n = \zeta_n \psi + (1 - \zeta_n) w_n$, we have that

$$\left| K_{B_R^c}(\psi + w_n) - K_{B_R^c}(w_n) \right| = \left| \left\langle K'_{B_R^c}(z_n), \psi \right\rangle \right|$$

$$\leq \int_{B_R^c} |N'(z_n)\psi| \leq \text{(by (3.45))}$$

$$\leq \int_{B_R^c} c_1 |z_n|^{r-1} |\psi| + c_2 |z_n|^{q-1} |\psi|$$

$$\leq c_1 \|z_n\|_{L^r(B_R^c)}^{r-1} \|\psi\|_{L^r(B_R^c)} + c_2 \|z_n\|_{L^q(B_R^c)}^{q-1} \|\psi\|_{L^q(B_R^c)}$$

(if R is large enough)

$$\leq c_3 \left(\|z_n\|_{L^r(B_R^c)}^{r-1} + \|z_n\|_{L^q(B_R^c)}^{q-1} \right) \varepsilon.$$

So we have

$$\left| K_{B_R^c}(\psi + w_n) - K_{B_R^c}(w_n) \right| \leq c_3 \left(\|z_n\|_{L^r(B_R^c)}^{r-1} + \|z_n\|_{L^q(B_R^c)}^{q-1} \right) \varepsilon. \qquad (3.56)$$

Since z_n is bounded in $H^1(\mathbb{R}^N, \mathbb{C})$, the sequences $\|z_n\|_{L^r(B_R^c)}^{r-1}$ and $\|z_n\|_{L^q(B_R^c)}^{q-1}$ are bounded. Then, by (3.55) and (3.56), we easily get

$$\lim_{n\to\infty} |K(\psi + w_n) - K(\psi) - K(w_n)| \leq \varepsilon + M \cdot \varepsilon \tag{3.57}$$

where M is a suitable constant.

Since ε is arbitrary, from (3.57) we get

$$\lim_{n\to\infty} |K(\psi + w_n) - K(\psi) - K(w_n)| = 0.$$

$\square$

Lemma 54. *Let assumptions (WB-i) and (WB-iv) (see (3.42) and (3.46)) be satisfied. Then E and C satisfy the coercivity assumption (EC-3*) (see 38).*

Proof. By the Gagliardo-Nirenberg interpolation inequalities (see e.g. [116]) there exists $b > 0$ such that for any $\psi \in H^1(\mathbb{R}^N, \mathbb{C})$ we get

$$||\psi||^p_{L^p} \leq b||\psi||^{p-pN\left(\frac{1}{2}-\frac{1}{p}\right)}_{L^2} ||\nabla\psi||^{pN\left(\frac{1}{2}-\frac{1}{p}\right)}_{L^2}.$$

We set $l = pN\left(\frac{1}{2} - \frac{1}{p}\right)$. Since $2 < p < 2 + \frac{4}{N}$, then $l < 2$. So

$$||\psi||^p_{L^p} \leq b||\psi||^t_{L^2}||\nabla\psi||^l_{L^2} \tag{3.58}$$

where $t = p - pN\left(\frac{1}{2} - \frac{1}{p}\right) = p - l > 0$.

Then by Hölder inequality we have for $M > 0$

$$||\psi||^p_{L^p} \leq bM||\psi||^t_{L^2} \frac{1}{M}||\nabla\psi||^l_{L^2}$$

$$\leq \frac{1}{\gamma'}\left(bM||\psi||^t_{L^2}\right)^{\gamma'} + \frac{1}{\gamma}\left(\frac{1}{M}||\nabla\psi||^l_{L^2}\right)^{\gamma}$$

$$= \frac{(bM)^{\gamma'}}{\gamma'}||\psi||^{t\gamma'}_{L^2} + \frac{1}{\gamma M^\gamma}||\nabla\psi||^{l\gamma}_{L^2}.$$

Now chose $\gamma = \frac{2}{l}$ and $M = \left(\frac{2c}{\gamma}\right)^{1/\gamma}$, where c is the constant in assumption (**WB-iv**) (see Sect. 3.2.2), so that

$$||\psi||^p_{L^p} \leq \frac{(bM)^{\gamma'}}{\gamma'}||\psi||^{t\gamma'}_{L^2} + \frac{1}{2c}||\nabla\psi||^2_{L^2}.$$

Then

$$c||\psi||^p_{L^p} \leq a||\psi||^{2s}_{L^2} + \frac{1}{2}||\nabla\psi||^2_{L^2} \tag{3.59}$$

$$\lim_{n\to\infty} |K(\psi + w_n) - K(\psi) - K(w_n)| \le \varepsilon + M \cdot \varepsilon \qquad (3.57)$$

where M is a suitable constant.

Since ε is arbitrary, from (3.57) we get

$$\lim_{n\to\infty} |K(\psi + w_n) - K(\psi) - K(w_n)| = 0.$$

$\square$

Lemma 54. *Let assumptions (WB-i) and (WB-iv) (see (3.42) and (3.46)) be satisfied. Then E and C satisfy the coercivity assumption (EC-3*) (see 38).*

Proof. By the Gagliardo-Nirenberg interpolation inequalities (see e.g. [116]) there exists $b > 0$ such that for any $\psi \in H^1(\mathbb{R}^N, \mathbb{C})$ we get

$$||\psi||_{L^p}^p \le b||\psi||_{L^2}^{p - pN\left(\frac{1}{2} - \frac{1}{p}\right)} ||\nabla\psi||_{L^2}^{pN\left(\frac{1}{2} - \frac{1}{p}\right)}.$$

We set $l = pN\left(\frac{1}{2} - \frac{1}{p}\right)$. Since $2 < p < 2 + \frac{4}{N}$, then $l < 2$. So

$$||\psi||_{L^p}^p \le b||\psi||_{L^2}^t ||\nabla\psi||_{L^2}^l \qquad (3.58)$$

where $t = p - pN\left(\frac{1}{2} - \frac{1}{p}\right) = p - l > 0$.

Then by Hölder inequality we have for $M > 0$

$$\begin{aligned}
||\psi||_{L^p}^p &\le bM||\psi||_{L^2}^t \frac{1}{M}||\nabla\psi||_{L^2}^l \\
&\le \frac{1}{\gamma'}\left(bM||\psi||_{L^2}^t\right)^{\gamma'} + \frac{1}{\gamma}\left(\frac{1}{M}||\nabla\psi||_{L^2}^l\right)^{\gamma} \\
&= \frac{(bM)^{\gamma'}}{\gamma'}||\psi||_{L^2}^{t\gamma'} + \frac{1}{\gamma M^{\gamma}}||\nabla\psi||_{L^2}^{l\gamma}.
\end{aligned}$$

Now chose $\gamma = \frac{2}{l}$ and $M = \left(\frac{2c}{\gamma}\right)^{1/\gamma}$, where c is the constant in assumption (**WB-iv**) (see Sect. 3.2.2), so that

$$||\psi||_{L^p}^p \le \frac{(bM)^{\gamma'}}{\gamma'}||\psi||_{L^2}^{t\gamma'} + \frac{1}{2c}||\nabla\psi||_{L^2}^2.$$

Then

$$c||\psi||_{L^p}^p \le a||\psi||_{L^2}^{2s} + \frac{1}{2}||\nabla\psi||_{L^2}^2 \qquad (3.59)$$

where

$$B_R^c = \mathbb{R}^N - B_R \text{ and } B_R = \left\{ x \in \mathbb{R}^N : |x| < R \right\}.$$

Since $w_n \rightharpoonup 0$ weakly in $H^1(\mathbb{R}^N, \mathbb{C})$, by usual compactness arguments, we have that

$$K_{B_R}(w_n) \to 0 \text{ and } K_{B_R}(\psi + w_n) \to K_{B_R}(\psi). \qquad (3.54)$$

Then, by (3.54) and (3.53), we have

$$\lim_{n \to \infty} |K(\psi + w_n) - K(\psi) - K(w_n)|$$

$$= \lim_{n \to \infty} \left| K_{B_R^c}(\psi + w_n) + K_{B_R}(\psi + w_n) - K_{B_R^c}(\psi) - K_{B_R}(\psi) - K_{B_R^c}(w_n) \right.$$

$$\left. - K_{B_R}(w_n) \right|$$

$$= \lim_{n \to \infty} \left| K_{B_R^c}(\psi + w_n) - K_{B_R^c}(\psi) - K_{B_R^c}(w_n) \right|$$

$$\leq \lim_{n \to \infty} \left| K_{B_R^c}(\psi + w_n) - K_{B_R^c}(w_n) \right| + \varepsilon. \qquad (3.55)$$

Now, by the intermediate value theorem, there exists $\zeta_n \in (0, 1)$ such that for $z_n = \zeta_n \psi + (1 - \zeta_n) w_n$, we have that

$$\left| K_{B_R^c}(\psi + w_n) - K_{B_R^c}(w_n) \right| = \left| \left\langle K'_{B_R^c}(z_n), \psi \right\rangle \right|$$

$$\leq \int_{B_R^c} |N'(z_n)\psi| \leq \text{(by (3.45))}$$

$$\leq \int_{B_R^c} c_1 |z_n|^{r-1} |\psi| + c_2 |z_n|^{q-1} |\psi|$$

$$\leq c_1 \|z_n\|_{L^r(B_R^c)}^{r-1} \|\psi\|_{L^r(B_R^c)} + c_2 \|z_n\|_{L^q(B_R^c)}^{q-1} \|\psi\|_{L^q(B_R^c)}$$

(if R is large enough)

$$\leq c_3 \left(\|z_n\|_{L^r(B_R^c)}^{r-1} + \|z_n\|_{L^q(B_R^c)}^{q-1} \right) \varepsilon.$$

So we have

$$\left| K_{B_R^c}(\psi + w_n) - K_{B_R^c}(w_n) \right| \leq c_3 \left(\|z_n\|_{L^r(B_R^c)}^{r-1} + \|z_n\|_{L^q(B_R^c)}^{q-1} \right) \varepsilon. \qquad (3.56)$$

Since z_n is bounded in $H^1(\mathbb{R}^N, \mathbb{C})$, the sequences $\|z_n\|_{L^r(B_R^c)}^{r-1}$ and $\|z_n\|_{L^q(B_R^c)}^{q-1}$ are bounded. Then, by (3.55) and (3.56), we easily get

where

$$a = \frac{c\,(bM)^{\gamma'}}{\gamma'}; \quad s = \frac{t\gamma'}{2}.$$

So using (WB-iv) and (3.59)

$$E(\psi) + aC(\psi)^s = \frac{1}{2}||\nabla\psi||^2_{L^2} + \int V\,|\psi|^2 + \int W(\psi) + a||\psi||^{2s}_{L^2} \qquad (3.60)$$

$$\geq \frac{1}{2}||\nabla\psi||^2_{L^2} + E_0 \int |\psi|^2 + \int N(\psi) + a||\psi||^{2s}_{L^2}$$

$$\geq \frac{1}{2}||\nabla\psi||^2_{L^2} + E_0 \int |\psi|^2 - c \int |\psi|^p + a||\psi||^{2s}_{L^2} \qquad (3.61)$$

$$\geq E_0 \int |\psi|^2. \qquad (3.62)$$

Observe that, since $p > 2$, we have $s > 1$. So (EC-3*)(i) is satisfied. Now we prove that also (EC-3*)(ii) is satisfied.

Let ψ_n be a sequence in $H^1(\mathbb{R}^N, \mathbb{C})$ such that

$$\int |\psi_n|^2 + \int |\nabla\psi_n|^2 \to \infty. \qquad (3.63)$$

Now distinguish two cases:

- Assume first that $\int |\psi_n|^2$ is unbounded. Then by (3.62), we have (up to a subsequence)

$$E(\psi_n) + aC(\psi_n)^s \to \infty.$$

So in this case (EC-3*) (ii) is satisfied.
- Assume now that $\int |\psi_n|^2$ is bounded and set

$$d = \sup ||\psi_n||^t_{L^2}.$$

So by (3.58) we have

$$||\psi_n||^p_{L^p} \leq c_1 ||\nabla\psi_n||^l_{L^2} \text{ where } c_1 = bd. \qquad (3.64)$$

Since $\int |\psi_n|^2$ is bounded, by (3.63) we get

$$\int |\nabla\psi_n|^2 \to \infty. \qquad (3.65)$$

On the other hand by (3.61) we have

$$E(\psi_n) + aC(\psi_n)^s \geq \frac{1}{2}||\nabla\psi_n||_{L^2}^2 - c\int |\psi_n|^p \geq \text{ (by (3.64))}$$

$$\frac{1}{2}||\nabla\psi_n||_{L^2}^2 - c_2||\nabla\psi_n||_{L^2}^l \text{ where } c_2 = cc_1. \tag{3.66}$$

Since $l < 2$, by (3.65) and (3.66) we deduce that (EC-3*)(ii) holds.

Now let us prove (EC-3*)(iii). Let ψ_n be a bounded sequence in $H^1(\mathbb{R}^N, \mathbb{C})$ such that $E(\psi_n) + aC(\psi_n)^s \to 0$, then by (3.62) we have

$$\int |\psi_n|^2 \to 0. \tag{3.67}$$

Then, in order to show that ψ_n goes to 0 in $H^1(\mathbb{R}^N, \mathbb{C})$, it remains to prove that

$$||\nabla\psi_n||_{L^2} \to 0. \tag{3.68}$$

Since ψ_n is bounded in $H^1(\mathbb{R}^N, \mathbb{C})$, by (3.58) and (3.67) we get

$$\int |\psi_n|^p \to 0. \tag{3.69}$$

Since $E(\psi_n) + aC(\psi_n)^s \to 0$ and by assumption (**WB-iv**) (see Sect. 3.2.2), we have

$$0 = \lim(E(\psi_n) + aC(\psi_n)^s) \geq \limsup\left(\frac{1}{2}||\nabla\psi_n||_{L^2}^2 + D_n\right) \tag{3.70}$$

where

$$D_n = E_0\int |\psi_n|^2 - c\int |\psi_n|^p + a||\psi_n||_{L^2}^{2s}. \tag{3.71}$$

By (3.67) and (3.69) we get $D_n \to 0$. So by (3.70) we deduce (3.68). □

3.2.4 Analysis of the Hylenic Ratio

In this section we will verify that the hylomorphy condition (2.16) is satisfied. The following lemma, which is in the same spirit of some compactness results in [22, 105] and [48], plays a fundamental role in proving (2.16):

Lemma 55. *Let $X = H^1(\mathbb{R}^N, \mathbb{C})$ and $2 < t < 2^*$, where $2^* = \frac{2N}{N-2}$, if $N \geq 3$ and $2^* = +\infty$ if $N = 1, 2$. Then the norm $\|\psi\|_{L^t}$ satisfies the property (2.19), namely, if ψ_n is vanishing (see Definition 28), then $\|\psi_n\|_{L^t} \to 0$.*

Proof. We set for $j \in \mathbb{Z}^N$

$$Q_j = A(j + Q^0) = \{Aj + Aq : q \in Q^0\}$$

where Q^0 is now the cube defined as follows

$$Q^0 = \{(x_1, \ldots, x_n) \in \mathbb{R}^N : 0 \leq x_i < 1\}.$$

Now let $x \in \mathbb{R}^N$ and set $y = A^{-1}(x)$. Clearly there exist $q \in Q^0$ and $j \in \mathbb{Z}^N$ such that $y = j + q$. So

$$x = Ay = A(j + q) \in Q_j.$$

Then we conclude that

$$\mathbb{R}^N = \bigcup_j Q_j.$$

Now let $\{\psi_n\} \subset H^1(\mathbb{R}^N, \mathbb{C})$ be a vanishing sequence and prove that $\|\psi_n\|_{L^t} \to 0$. Arguing by contradiction, assume that, up to a subsequence, $\|\psi_n\|_{L^t} \geq a > 0$. Since ψ_n is vanishing, there exists $M > 0$ such that $\|\psi_n\|_{H^1}^2 \leq M$. Then, if L is the constant for the Sobolev embedding $H^1(Q_j) \subset L^t(Q_j)$, we have

$$0 < a^t \leq \int |\psi_n|^t = \sum_j \int_{Q_j} |\psi_n|^t = \sum_j \|\psi_n\|_{L^t(Q_j)}^{t-2} \|\psi_n\|_{L^t(Q_j)}^2$$

$$\leq \left(\sup_j \|\psi_n\|_{L^t(Q_j)}^{t-2} \right) \cdot \sum_j \|\psi_n\|_{L^t(Q_j)}^2$$

$$\leq L \left(\sup_j \|\psi_n\|_{L^t(Q_j)}^{t-2} \right) \cdot \sum_j \|\psi_n\|_{H^1(Q_j)}^2$$

$$= L \left(\sup_j \|\psi_n\|_{L^t(Q_j)}^{t-2} \right) \|\psi_n\|_{H^1}^2 \leq LM \left(\sup_j \|\psi_n\|_{L^t(Q_j)}^{t-2} \right).$$

Then

$$\left(\sup_j \|\psi_n\|_{L^t(Q_j)} \right) \geq \left(\frac{a^t}{LM} \right)^{1/(t-2)}.$$

Then, for any n, there exists $j_n \in \mathbb{Z}^N$ such that

$$\|\psi_n\|_{L^t(Q_{j_n})} \geq \alpha > 0. \tag{3.72}$$

Then, if we set $Q = AQ^0$, we easily have

$$\|T_{j_n}\psi_n\|_{L^t(Q)} = \|\psi_n\|_{L^t(Q_{j_n})} \geq \alpha > 0. \tag{3.73}$$

Since ψ_n is bounded, also $T_{j_n}\psi_n$ is bounded (in $H^1(\mathbb{R}^N, \mathbb{C})$). Then we have, up to a subsequence, that $T_{j_n}\psi_n \rightharpoonup \psi_0$ weakly in $H^1(\mathbb{R}^N, \mathbb{C})$ and hence strongly in $L^t(Q)$. By (3.73), $\psi_0 \neq 0$ and this contradicts the fact that ψ_n is vanishing. □

Clearly the hylenic ratio takes the following form:

$$\Lambda(\psi) = \frac{\int \left(\frac{1}{2}|\nabla\psi|^2 + W(\psi) + V(x)|\psi|^2\right) dx}{\int |\psi|^2 \, dx}.$$

Lemma 56. *If the assumptions of Theorem 52 are satisfied, then for $2 < t < 2^*$, we have*

$$\liminf_{\psi \in H^1, \|\psi\|_{L^t} \to 0} \Lambda(\psi) \geq E_0.$$

Proof. Clearly

$$\liminf_{\psi \in H^1, \|\psi\|_{L^t} \to 0} \Lambda(\psi) = \liminf_{\psi \in H^1, \|\psi\|_{L^t} = 1, \varepsilon \to 0} \frac{E(\varepsilon\psi)}{C(\varepsilon\psi)}$$

$$= \inf_{\psi \in H^1, \|\psi\|_{L^t} = 1,} \left(\frac{\int \left(\frac{1}{2}|\nabla\psi|^2 + (E_0 + V(x))|\psi|^2\right) dx}{\int |\psi|^2} \right)$$

$$+ \liminf_{\psi \in H^1, \|\psi\|_{L^t} = 1, \varepsilon \to 0} \frac{\int N(\varepsilon\psi)}{\varepsilon^2 \int |\psi|^2}.$$

$$\geq E_0 + \liminf_{\psi \in H^1, \|\psi\|_{L^t} = 1, \varepsilon \to 0} \frac{\int N(\varepsilon\psi)}{\varepsilon^2 \int |\psi|^2}.$$

So the proof of Lemma will be achieved if we show that

$$\liminf_{\psi \in H^1, \|\psi\|_{L^t} = 1, \varepsilon \to 0} \frac{\int N(\varepsilon\psi)}{\varepsilon^2 \int |\psi|^2} = 0. \tag{3.74}$$

By assumptions (3.45) and (3.46) we have

$$- cs^p \leq N(s) \leq \bar{c}(s^q + s^r) \tag{3.75}$$

where $c, \bar{c}$ are positive constants and q, r belong to the interval $(2, 2^*)$.

Then by (3.75) we have

$$- cA\varepsilon^{p-2} \leq \inf_{\|\psi\|_{L^t}=1} \frac{\int N(\varepsilon\psi)}{\varepsilon^2 \int |\psi|^2} \leq \bar{c}B(\varepsilon^{q-2} + \varepsilon^{r-2}) \tag{3.76}$$

where

$$A = \inf_{\psi \in H^1 \, \|\psi\|_{L^t}=1} \frac{\int |\psi|^p}{\int |\psi|^2}, \quad B = \inf_{\psi \in H^1 \, \|\psi\|_{L^t}=1} \frac{\int (|\psi|^q + |\psi|^r)}{\int |\psi|^2}.$$

By (3.76) we easily get (3.74). $\qquad \square$

Now we can give a lower bound to Λ_0 (see 2.17).

Corollary 57. *If the assumptions of Theorem 52 are satisfied, then*

$$E_0 \leq \Lambda_0.$$

Proof. By Proposition 31, Lemmas 55 and 56

$$\Lambda_0 \geq \liminf_{\|\psi\|_{L^t} \to 0} \Lambda(\psi) \geq E_0.$$

$\qquad \square$

Finally we can prove that the hylomorphy condition is satisfied.

Lemma 58. *If the assumptions of Theorem 52 are satisfied, then the hylomorphy condition (2.16) holds, namely*

$$\inf_{\psi \in H^1(\mathbb{R}^N, \mathbb{C})} \Lambda(\psi) < \Lambda_0.$$

Proof. Since $X = H^1(\mathbb{R}^N, \mathbb{C})$, we need only to construct a function $u \in H^1(\mathbb{R}^N) \subset H^1(\mathbb{R}^N, \mathbb{C})$ such that $\Lambda(u) < \Lambda_0$.

Such a function can be constructed as follows. Set

$$u_R = \begin{cases} s_0 & if \ |x| < R \\ 0 & if \ |x| > R+1 \\ \frac{|x|}{R}s_0 - (|x| - R)\frac{R+1}{R}s_0 & if \ R < |x| < R+1 \end{cases}.$$

Then

$$\int |\nabla u_R|^2 \, dx = O(R^{N-1}), \int |u_R|^2 \, dx = O(R^N),$$

so that

$$\frac{\int \left[\frac{1}{2} |\nabla u_R|^2 + (E_0 + V) u_R^2 \right] dx}{\int u_R^2} \le E_0 + V_0 + O\left(\frac{1}{R}\right). \tag{3.77}$$

Moreover

$$\int N(u_R) dx = N(s_0) m(B_R) + \int_{B_{R+1} \backslash B_R} N(u_R)$$

where $m(A)$ denotes the measure of A. So

$$\frac{\int N(u_R) dx}{\int u_R^2} \le \frac{N(s_0) m(B_R) + c_1 R^{N-1}}{\int u_R^2} \le (\text{ since } N(s_0) < 0) \tag{3.78}$$

$$\le \frac{N(s_0) m(B_R)}{s_0^2 m(B_{R+1})} + \frac{c_1 R^{N-1}}{s_0^2 m(B_R)} = \frac{N(s_0)}{s_0^2} \left(\frac{R}{R+1}\right)^N + \frac{c_2}{R}.$$

Then, by (3.77) and (3.78) we get

$$\Lambda(u_R) = \frac{\int \left(\frac{1}{2} |\nabla u_R|^2 + W(u_R) + V(x) u_R^2\right) dx}{\int u_R^2 dx} \tag{3.79}$$

$$= \frac{\int \left(\frac{1}{2} |\nabla u_R|^2 + (E_0 + V(x)) u_R^2\right) dx}{\int u_R^2 dx} + \frac{\int N(u_R) dx}{\int u_R^2 dx} \le$$

$$\le E_0 + V_0 + \frac{N(s_0)}{s_0^2} \left(\frac{R}{R+1}\right)^N + \frac{c_3}{R}. \tag{3.80}$$

Then by (3.44) we can easily deduce that for R large enough we have

$$\Lambda(u_R) < E_0. \tag{3.81}$$

Finally by (3.81) and Corollary 57 we get

$$\Lambda(u_R) < \Lambda_0.$$

$\square$

Proof of Theorem 52. Lemmas 53, 54 and 58 show that E, C satisfy assumptions (EC-2), (EC-3*) and (2.16) of Theorem 34. Moreover $C'(u) = 0$ if and only if $u = 0$, so also assumption (2.25) is satisfied. Then the conclusion follows by using Theorem 34. □

3.2.5 Symmetry, Travelling Solitary Waves and Solitons in NS

We start this section with the following

Proposition 59. *Let $\psi_0 \in H^1(\mathbb{R}^N, \mathbb{C})$ be a hylomorphic soliton relative to (3.1). Then there exist constants ω and c such that ψ_0 is a ground state solution (see Definition 21) of the equation*

$$-\frac{1}{2}\Delta\psi + \frac{1}{2}W'(\psi) + V(x)\psi = \omega\psi \tag{3.82}$$

with respect to the energy (3.7) and the set

$$\mathfrak{M}_c = \{\psi \in H^1(\mathbb{R}^N, \mathbb{C}) : C(\psi) = c\},$$

where $C(\psi) = \int |\psi|^2$ denotes the hylenic charge. Moreover there exist a real function $u_0(x) \in H^1$ and a constant $\theta_0 \in \mathbb{R}$ such that

$$\psi_0(x) = u_0(x)e^{-i\theta_0} \tag{3.83}$$

and

$$\psi(t, x) := u_0(x)e^{-i\theta_0 - i\omega t} \tag{3.84}$$

solves (3.1).

Proof. ψ_0 is a hylomorphic soliton, then there exists a constant c such that it is a minimizer of the energy E defined in (3.7) on $\mathfrak{M}_c$. Then we get

$$E'(\psi_0) = \omega C'(\psi_0) \tag{3.85}$$

where ω is a Lagrange multiplier. Clearly (3.85) gives (3.82). Thus ψ_0 is a ground state solution of (3.82).

Equation (3.82) implies that $\psi = \psi_0 e^{-i\omega t}$ solves (3.1). It remains to show that there exist a real function $u_0(x)$ and a real constant θ_0 such that

$$\psi_0(x) = u_0(x)e^{-i\theta_0}. \tag{3.86}$$

Now we set

$$\psi(x) = u(x)e^{-i\theta(x)}, \; u(x) \in \mathbb{R}$$

$$\psi_0(x) = u_0(x)e^{-i\theta_0(x)}, u_0(x) \in \mathbb{R}. \tag{3.87}$$

So we have

$$E(\psi) = \int \left[\frac{1}{2}|\nabla\psi|^2 + W(\psi) + V(x)|\psi|^2 \right] dx$$

$$= \int \left[\frac{1}{2}|\nabla u|^2 + \frac{1}{2}u^2|\nabla\theta|^2 + W(u) + V(x)u^2 \right] dx$$

and

$$E(u_0) = \int \left[\frac{1}{2}|\nabla u_0|^2 + W(u_0) + V(x)u_0^2 \right] dx$$

$$\leq \int \left[\frac{1}{2}|\nabla u_0|^2 + \frac{1}{2}u_0^2|\nabla\theta_0|^2 + W(u_0) + V(x)u_0^2 \right] dx = E(\psi_0).$$

So, since ψ_0 is a minimizer of $E(\psi)$ on $\mathfrak{M}_c$ and $u_0 \in \mathfrak{M}_c$, we easily get that θ_0 is constant.

So by (3.87) we get

$$\psi_0(x) = u_0(x)e^{-i\theta_0}$$

where $u_0(x)$ is real and θ_0 constant. Then (3.86) holds.   □

Remark 60. Clearly if (3.83) is a ground state solution of (3.82), also $u_0(x)$ satisfies (3.82); we may assume that $u_0(x)$ does not change sign since $E(|u_0|) = E(u_0)$ and $C(u_0) = C(|u_0|)$. Then, by our assumptions, the well known result of Gidas, Ni and Nirenberg [81] can be applied. Thus $u_0(x)$ is rotationally invariant around a point x_0; it is monotonically decreasing with $r = |x - x_0|$ and finally, we have that every real valued ground state solution of (3.82) is strictly positive.

Now we will exploit the other symmetries of Eq. (NS) to produce other solutions. Assume that

$$\psi_0(t, x) = u(x)e^{-i\omega t}, \; u \in \mathbb{R} \tag{3.88}$$

is a solution of (NS). First of all, since (NS) is invariant for translations, for any $x_0 \in \mathbb{R}^N$, the function $\psi_{x_0}(t, x) = u(x - x_0)e^{-i\omega t}$ is a standing wave concentrated around the point x_0. The space rotations do not produce other solutions since $u(x)$ is rotationally invariant by Remark 60.

Since the Lagrangian related to (NS) (see (3.3) with $V = 0$) is invariant for the Galileo group, we can obtain other solutions: we can produce travelling waves just applying the transformation (1.16)–(3.88)

$$\psi_{x_0,\mathbf{v}}(t, x) = u(x - x_0 - \mathbf{v}t)e^{i(\mathbf{v}\cdot x - Et)}, \quad E = \frac{1}{2}\mathbf{v}^2 + \omega$$

$\psi_{x_0,\mathbf{v}}(t, x)$ is a solitary wave concentrated in the point $x_0 + \mathbf{v}t$, and hence it travels with velocity $\mathbf{v}$.

Finally, other solutions can be produced by the invariance (1.19); for $\theta \in [0, 2\pi)$, we have the solutions

$$\psi_{x_0,\mathbf{v},\theta}(t, x) = u(x - x_0 - \mathbf{v}t)e^{i(\mathbf{v}\cdot x - Et + \theta)}. \tag{3.89}$$

The invariance by time translations does not produce new solutions, since a time translation on $\psi_{x_0,\mathbf{v},\theta}$ produces a space and phase translation.

Concluding, for every frequency $\omega \in (E_1, E_0)$, we obtain a radially symmetric solution of the form (3.30); by the invariance of the equation, this solution produces a $2N + 1$ (N being the space dimension) parameters family of solutions given by (3.89). The parameters are $x_0, \mathbf{v}, \theta$.

If we take into account the usual physical constants m and $\hbar$, (3.89) takes the form

$$\psi_{x_0,\mathbf{v},\theta}(t, x) = u(x - x_0 - \mathbf{v}t)e^{i(\mathbf{p}\cdot x - Et)/\hbar + i\theta} \tag{3.90}$$

where

$$\mathbf{p} = m\mathbf{v}$$

$$E = \frac{1}{2}m\mathbf{v}^2 + \omega\hbar.$$

The meaning of these relations within the swarm interpretation has been given in Sect. 3.1.2: $\psi_{x_0,\mathbf{v},\theta}(t, x)$ is interpreted as a swarm of particles of mass m; $\mathbf{p}$ is the momentum of each particle and E is its energy: $\frac{1}{2}m\mathbf{v}^2$ is its kinetic energy and $\omega\hbar$ its potential energy E_0 plus its binding energy $\omega\hbar - E_0$. Since the binding energy must be negative, this explains why we must have $\omega\hbar < E_0$ in Theorem 50.

If the assumptions of Theorem 52 are satisfied and $V = 0$, (3.1) admits travelling hylomorphic solitons namely (Definition 22):

$$\rho(g_\mathbf{v})u = u(x - x_0)e^{i(\mathbf{v}\cdot x + \theta)}.$$

The evolution of these solitons is described by (3.89).

Some other properties of travelling solitons will be described in the next section and in particular in the Sect. 3.3.3.

3.3   Dynamics of Solitons in NS

In Sect. 2.1.1 we have given a notion of soliton which applies to suitable equations. If one of these equations is perturbed, the Definition 16 is not suitable; in fact any solution of the perturbed equation, in general, will not live in $\mathcal{T}$-compact invariant set. However it is possible that there are solutions that can be splitted as follows:

$$\psi(t, x) = \Psi(t, x) + \varphi(t, x) \tag{3.91}$$

where $\Psi(t, x)$ is "similar" to the solution of the unperturbed equation. In this case, we can say that our solution splits in a "wave" $\varphi(t, x)$ and a "soliton" $\Psi(t, x)$. When this splitting is possible, it is interesting to study the dynamics of the soliton.

In this section we will study the dynamics of solitons arising in the nonlinear Schrödinger equation. In particular, we are interested in the case in which the "size" of the soliton is small with respect to the other quantities. We get a small sized soliton choosing a suitable nonlinear term W_ε. So, we will write (3.1) in the following way:

$$i\frac{\partial \psi}{\partial t} = -\frac{1}{2m}\Delta\psi + \frac{1}{2}W_\varepsilon'(\psi) + V(x)\psi. \tag{3.92}$$

We consider solutions of our equation which can be written as (3.91) namely

$$\psi(t, x) = \Psi_\varepsilon(t, x) + \varphi(t, x) \tag{3.93}$$

where $\varphi(t, x)$ the wave and $\Psi_\varepsilon(t, x)$ is the soliton. We want to assimilate the soliton to a particle; so, it is appropriate for Ψ_ε to have a compact support. For the moment, we think of a soliton as a bump of energy concentrated in a ball centered at the point $q = q_\varepsilon(t)$ with radius $R_\varepsilon \to 0$ (for $\varepsilon \to 0$). The main purpose of this section is to show that for ε sufficiently small, our soliton behaves as a classical particle in a potential $V(x)$. More exactly, we prove that the decomposition (3.93) holds for all times and the bump follows a dynamics which approaches the dynamics of a pointwise particle moving under the action of the potential V (Theorem 61); in particular the position $q_\varepsilon(t)$ of the soliton approaches the position of the particle uniformly on bounded time intervals (Corollary 62).

The attention of the mathematical community on the dynamics of soliton of NSE began with the pioneering paper of Bronski and Jerrard [49]; then Fröhlich, Gustafson, Jonsson, and Sigal faced this problem using a different approach [77,78]. In the last years, several others works appeared following the first approach [91, 92, 133, 134, 137] or the second one [2–4, 13, 79, 87, 131]. Here we follow a third approach introduced in [42] and [43].

In most of the quoted papers, the following equation

$$i\varepsilon\frac{\partial \psi}{\partial t} = -\frac{\varepsilon^2}{2m}\Delta\psi + \frac{1}{2}W'(\psi) + V(x)\psi. \tag{3.94}$$

and not Eq. (3.92) is studied. Following [43], we study Eq. (3.92) which gives a different problem. Actually, in Eq. (3.92), the parameter ε appears in the nonlinear term and it will be chosen in such a way that $|\psi|^2$ approaches the delta-measure as $\varepsilon \to 0$. Thus Eq. (3.92) describes the dynamics of a soliton when its support is small with respect to the other relevant elements (namely $V(x)$, the initial conditions and its L^2-size).

3.3.1 Rescaling the Soliton

In this section, we focus on the "rescaling" properties of a soliton solution of Eq. (3.92) without the potential term V. We consider the following Cauchy problem relative to the NS:

$$i \frac{\partial \psi}{\partial t} = -\frac{1}{2m} \Delta \psi + \frac{1}{2} W_\varepsilon' (\psi) \tag{3.95}$$

$$\psi (0, x) = U_\varepsilon (x - \bar{q}) \, e^{i \bar{p} \cdot x} \tag{3.96}$$

where, with some abuse of notation, we have set

$$W_\varepsilon (\psi) = \frac{1}{\varepsilon^{N+2}} W \left(\varepsilon^{N/2} |\psi| \right); \quad W_\varepsilon' (\psi) = \frac{1}{\varepsilon^{N/2+2}} W' \left(\varepsilon^{N/2} |\psi| \right) \frac{\psi}{|\psi|}, \tag{3.97}$$

and $W : \mathbb{R}^+ \to \mathbb{R}$ is a real function which satisfies the usual existence assumptions with $E_0 = V(x) = 0$; see (WB-i)–(WB-iv) on page 71 and Remark 51. Let U denote a *ground state solution* of the equation

$$-\frac{1}{2m} \Delta U + \frac{1}{2} W'(U) = \omega_1 U \tag{3.98}$$

namely a function such that

$$J(u) = e$$

where

$$J(u) = \int \left(\frac{|\nabla u|^2}{2m} + W(u) \right) dx \tag{3.99}$$

and

$$e = \inf_{\substack{\|u\|_{L^2} = 1 \\ u \in H^1}} \int \left(\frac{|\nabla u|^2}{2m} + W(u) \right) dx. \tag{3.100}$$

The existence of such function is given by Proposition 59. It is well known that we can choose U radially symmetric and positive (see [81]).

Direct computations show that, by virtue of (3.97), the function

$$U_\varepsilon(x) = \frac{1}{\varepsilon^{N/2}} U\left(\frac{x}{\varepsilon}\right)$$

satisfies the equation

$$-\frac{1}{2m}\Delta U_\varepsilon + \frac{1}{2}W_\varepsilon'(U_\varepsilon) = \omega_\varepsilon U_\varepsilon \tag{3.101}$$

where

$$\omega_\varepsilon = \frac{\omega_1}{\varepsilon^2}.$$

Moreover $U_\varepsilon(x)$ is a ground state solution of (3.101). In many cases, the ground state solution U is unique up to translations and change of sign, but we do not need this assumption.

Notice that the choice of W_ε given by (3.97) implies that

$$\|U_\varepsilon\|_{L^2} = 1$$

for every $\varepsilon > 0$.

By direct computations or using the arguments of Sect. 3.2.5, the solution of (3.95) and (3.96) has the following form:

$$\psi_{q,\varepsilon}(t,x) = U_\varepsilon(x - \bar{q} - \bar{v}t)\, e^{i(\bar{p}\cdot x - \omega t)} \quad \text{where} \quad \bar{v} = \frac{\bar{p}}{m} \tag{3.102}$$

with

$$\omega = \omega_\varepsilon + \frac{1}{2}m\bar{v}^2.$$

Thus $\psi_{q,\varepsilon}(t,x)$ behaves as a particle of "radius" ε living in the point

$$q = \bar{q} + \bar{v}t. \tag{3.103}$$

Since $\|\psi_{q,\varepsilon}(t,\cdot)\|_{L^2} = 1$ for every $\varepsilon > 0$, if $\varepsilon \to 0$, we have that

$$\left|\psi_{q,\varepsilon}(t,x)\right|^2 \to \delta(x - \bar{q} - \bar{v}t) \quad \text{in } \mathcal{D}'(\mathbb{R}^N) \quad \forall t \in \mathbb{R},$$

where $\delta(x - x_0)$ denotes the Dirac measure concentrated in the point x_0. By (3.7), the energy $E_\varepsilon(\psi)$ of the configuration ψ is given by

$$E_\varepsilon(\psi) = \int \left[\frac{1}{2m} |\nabla\psi|^2 + W_\varepsilon(\psi) \right] dx,$$

so the energy of $\psi_{q,\varepsilon}$ is

$$E_\varepsilon\left(\psi_{q,\varepsilon}\right) = \int \left(\frac{|\nabla U_\varepsilon|^2}{2m} + W_\varepsilon(U_\varepsilon) \right) dx + \frac{1}{2} m\bar{v}^2. \tag{3.104}$$

Thus $\psi_{q,\varepsilon}(t,x)$ behaves as a particle of mass m: $\bar{p}$ can be interpreted as its momentum, $\frac{1}{2m}\bar{p}^2 = \frac{1}{2}m\bar{v}^2$ as its kinetic energy and

$$\int \frac{|\nabla U_\varepsilon|^2}{2m} + W_\varepsilon(U_\varepsilon)\, dx = \frac{e}{\varepsilon^2}$$

as the *internal* energy; here e is a constant defined as follows

$$e := \int \frac{|\nabla U_1|^2}{2m} + W(U_1)\, dx.$$

3.3.2 Statement of the Problem and Main Results

We will study the dynamics of the solitons in the presence of a potential $V(x)$ namely to investigate the Cauchy problem

$$\begin{cases} i\frac{\partial\psi}{\partial t} = -\frac{1}{2m}\Delta\psi + \frac{1}{2}W_\varepsilon'(\psi) + V(x)\psi \\[2mm] \psi\,(0,x) = \psi_{0,\varepsilon}(x) \end{cases} \tag{3.105}$$

where $\psi_{0,\varepsilon}$ satisfies the following assumptions

$$\psi_{0,\varepsilon}(x) = U_\varepsilon\left(x - \bar{q}\right) e^{i\,\bar{p}\cdot x} + \varphi_{0,\varepsilon}(x), \quad \varphi_{0,\varepsilon} \in H^1(\mathbb{R}^N); \tag{3.106}$$

$$\|\psi_{0,\varepsilon}\|_{L^2} = 1 \tag{3.107}$$

$$E_\varepsilon\left(\psi_{0,\varepsilon}\right) \le \frac{e}{\varepsilon^2} + M \tag{3.108}$$

with $M > 0$ independent of ε; here $E_\varepsilon(\psi)$ denotes the energy in the presence of the potential V :

$$E_\varepsilon(\psi) = \int \left[\frac{1}{2m} |\nabla\psi|^2 + W_\varepsilon(\psi) + V(x)\,|\psi|^2 \right] dx.$$

Clearly, if $\varphi_{0,\varepsilon}(x) = 0$, the above assumptions are satisfied; it is not difficult to realize that for any $\varepsilon > 0$, you can find a family of small functions $\varphi_{0,\varepsilon}(x)$ which satisfy (3.106)–(3.108).

We make the following assumptions which are slightly stronger than (WB-i)–(WB-iv) on page 71: $W : \mathbb{R}^+ \to \mathbb{R}$ is a C^2 function which satisfies the following assumptions:

$$W(0) = W'(0) = W''(0) = 0 \tag{3.109}$$

$$|W''(s)| \leq c_1|s|^{q-2} + c_2|s|^{p-2} \text{ for some } 2 < q \leq p < 2^* \tag{3.110}$$

$$W(s) \geq -c|s|^\nu, \ c \geq 0, \ 2 < \nu < 2 + \frac{4}{N} \text{ for } s \text{ large} \tag{3.111}$$

$$\exists s_0 \in \mathbb{R}^+ \text{ such that } W(s_0) < 0. \tag{3.112}$$

Moreover we make the following assumptions on $V : \mathbb{R}^N \to \mathbb{R}$. V is a C^2-function such that:

$$0 \leq V(x) \leq V_0; \tag{3.113}$$

$$|\nabla V(x)| \leq V_1; \tag{3.114}$$

$$|V''(x)| \leq V_2; \tag{3.115}$$

where V_0, V_1, V_2 are positive constants.

The main result describes the shape and the dynamics of the soliton $\Psi_\varepsilon (t, x)$:

Theorem 61. *Assume (3.109)–(3.115); then the solution of problem (3.105) has the following form*

$$\psi_\varepsilon(t, x) = \Psi_\varepsilon (t, x) + \varphi_\varepsilon(t, x) \tag{3.116}$$

where $\Psi_\varepsilon (t, x)$ is a function having support in a ball $B_{R_\varepsilon}(q_\varepsilon)$, with radius $R_\varepsilon \to 0$ and center $q_\varepsilon = q_\varepsilon(t)$. Moreover,

$$\||\Psi_\varepsilon (t, x)| - U_\varepsilon (x - q_\varepsilon(t))\|_{L^2} \to 0 \text{ as } \varepsilon \to 0 \tag{3.117}$$

uniformly in t, where $U_\varepsilon = \frac{1}{\varepsilon^{N/2}} U \left(\frac{x}{\varepsilon}\right)$ and U is a ground state solution of (3.98).
The dynamics is given by the following equations:

$$\begin{cases} \dot{q}_\varepsilon(t) = \frac{1}{m_\varepsilon(t)} \, p_\varepsilon(t) + K_\varepsilon(t) \\[2mm] \dot{p}_\varepsilon(t) = -\nabla V(q_\varepsilon(t)) + F_\varepsilon(q_\varepsilon) + H_\varepsilon(t) \end{cases} \tag{3.118}$$

with initial data

$$\begin{cases} q_\varepsilon(0) = \bar{q} + o(1) \\ p_\varepsilon(0) = \bar{p} + o(1) \end{cases} \tag{3.119}$$

where

*(a) $q_\varepsilon(t)$ is the **barycenter** of the soliton and it has the following form:*

$$q_\varepsilon(t) = \frac{\displaystyle\int x \ |\Psi_\varepsilon|^2 \, dx}{\displaystyle\int |\Psi_\varepsilon|^2 \, dx}.$$

*(b) $m_\varepsilon(t) = m \displaystyle\int |\Psi_\varepsilon(t,x)|^2 \, dx = m + o(1)$ can be interpreted as the **mass** of the soliton.*

*(c) $p_\varepsilon(t)$ is the **momentum** of the soliton and it has the following form:*

$$p_\varepsilon(t) = \mathrm{Im} \int \nabla \Psi_\varepsilon(t,x) \ \overline{\Psi_\varepsilon(t,x)} \, dx.$$

(d) $K_\varepsilon(t)$ and $H_\varepsilon(t)$ are errors due to the fact that the soliton is not a point and

$$\sup_{t \in \mathbb{R}} (|H_\varepsilon(t)| + |K_\varepsilon(t)|) \to 0 \quad as \ \varepsilon \to 0.$$

(e) $F_\varepsilon(q_\varepsilon)$ is the force due to the pressure of the wave φ_ε on the soliton and $F_\varepsilon \to 0$ in the space of distributions, more exactly we have that

$$\forall \tau_0, \tau_1, \ \left| \int_{\tau_0}^{\tau_1} F_\varepsilon(q_\varepsilon) \, dt \right| \leq c(\varepsilon) \, (1 + |\tau_1 - \tau_0|)$$

where $c(\varepsilon) \to 0$ as $\varepsilon \to 0$.

Corollary 62. *Let $\mathfrak{q}$ and $\mathfrak{p}$ be the solution of the following Cauchy problem:*

$$\begin{cases} \dot{\mathfrak{q}}(t) = \frac{1}{m} \, \mathfrak{p}(t) \\ \\ \dot{\mathfrak{p}}(t) = -\nabla V(\mathfrak{q}(t)) \end{cases} \tag{3.120}$$

with initial data

$$\begin{cases} \mathfrak{q}(0) = q_\varepsilon(0) \\ \mathfrak{p}(0) = p_\varepsilon(0) \end{cases} \tag{3.121}$$

where $q_\varepsilon(t)$ and $p_\varepsilon(t)$ are as in Theorem 61. Then, as $\varepsilon \to 0$

$$(q_\varepsilon(t), p_\varepsilon(t)) \to (q(t), p(t))$$

uniformly on compact sets.

Let us discuss the set of our assumptions.

Remark 63. The conditions (3.109) and (3.113) are assumed for simplicity; in fact they can be weakened as follows

$$W(0) = W'(0) = 0, \quad W''(0) = E_0$$

and

$$E_1 \le V(x) \le V_\infty < +\infty.$$

In fact, in the general case, the solution of the Schrödinger equation is modified only by a phase factor (see also Remark 51).

Remark 64. By our assumptions, the problem (3.105) has a unique solution

$$\psi \in C^0(\mathbb{R}, H^2(\mathbb{R}^N)) \cap C^1(\mathbb{R}, L^2(\mathbb{R}^N)). \tag{3.122}$$

Let us recall a result on the global existence of solutions of the Cauchy problem (3.105) (see [53,82,90]). Assume (3.110)–(3.112) for W. Let $D(A)$ (resp. $D(A^{1/2})$) denote the domain of the self-adjoint operator A (resp. $A^{1/2}$) where

$$A = -\Delta + V : L^2(\mathbb{R}^N) \to L^2(\mathbb{R}^N).$$

If $V \ge 0$, $V \in C^2$ and $|V''| \in L^\infty$ and the initial data $\psi(0, x) \in D(A^{1/2})$ then there exists the global solution ψ of (3.105) and

$$\psi(t, x) \in C^0\left(\mathbb{R}, D(A^{1/2})\right) \cap C^1(\mathbb{R}, H^{-1}(\mathbb{R}^N)).$$

Furthermore, if $\psi(0, x) \in D(A)$ then

$$\psi(t, x) \in C^0(\mathbb{R}, D(A)) \cap C^1(\mathbb{R}, L^2(\mathbb{R}^N)).$$

In this case, since $D(A) \subset H^2(\mathbb{R}^N)$, (3.122) is satisfied.

Theorem 61 and Corollary 62 will be proved in Sect. 3.3.8. The next four sections will be devoted to the analysis of the dynamics of the soliton and will provide the tools necessary to the proof of Theorem 61 and Corollary 62.

3.3.3 Analysis of the Energy and Momentum of the Soliton

We can write the energy (3.8) as follows:

$$E_\varepsilon(\psi) = \int \left(\frac{1}{2m}|\nabla u|^2 + W_\varepsilon(u)\right) dx + \int \left(\frac{1}{2m}|\nabla S|^2 + V(x)\right) u^2 dx.$$
(3.123)

The energy has two components: the (which, sometimes, is also called *binding energy*)

$$J_\varepsilon(u) = \int \left(\frac{1}{2m}|\nabla u|^2 + W_\varepsilon(u)\right) dx$$
(3.124)

and the *dynamical energy*

$$G(u, S) = \int \left(\frac{1}{2m}|\nabla S|^2 + V(x)\right) u^2 dx$$
(3.125)

which is composed by the *kinetic energy* $\frac{1}{2m}\int |\nabla S|^2 u^2 dx$ and the *potential energy* $\int V(x) u^2 dx$. By our assumptions, the internal energy is bounded from below and the dynamical energy is positive.

The momentum (see (3.12)) is constant in time if the Lagrangian is space-translation invariant; this happens when V is constant. In general we have the following fact

Proposition 65. *The following equation holds*

$$\partial_t \left(u^2 \nabla S\right) = -u^2 \nabla V + \nabla \cdot T$$
(3.126)

where T is the stress tensor and it has the following form

$$T_{jk} = \sum_k \partial_{x_k} \left(-u^2 S_{x_k} S_{x_j} + u u_{x_k x_j} - u_{x_j} u_{x_k} + \frac{1}{2}\left[W_\varepsilon(u) - u W_\varepsilon'(u)\right] \delta_{kj}\right).$$

Proof. We make a direct computation:

$$\partial_t \left(u^2 \nabla S\right) = \partial_t \left(u^2\right) \nabla S + u^2 \nabla \left(\partial_t S\right).$$

Using Eqs. (3.5) and (3.6) we get

$$\partial_t \left(u^2 \nabla S\right) = -\nabla \cdot \left(u^2 \nabla S\right) \nabla S + u^2 \nabla \left(-\frac{|\nabla S|^2}{2} - V(x) + \frac{1}{2}\frac{\Delta u}{u} - \frac{1}{2}\frac{W_\varepsilon'(u)}{u}\right)$$

$$= -\nabla \cdot \left(u^2 \nabla S\right) \nabla S - u^2 \nabla \frac{|\nabla S|^2}{2} - u^2 \nabla V(x) + \frac{1}{2}u^2 \nabla \frac{\Delta u}{u} - \frac{1}{2}u^2 \nabla \frac{W_\varepsilon'(u)}{u}.$$

We compute each piece separately:

$$\left[-\nabla \cdot \left(u^2 \nabla S\right) \nabla S - u^2 \nabla \frac{|\nabla S|^2}{2}\right]_j$$

$$= -\sum_k \partial_{x_k} \left(u^2 S_{x_k}\right) S_{x_j} - u^2 \partial_{x_j} \frac{\sum_k S_{x_k}^2}{2}$$

$$= -\sum_k \partial_{x_k} \left(u^2 S_{x_k} S_{x_j}\right) + u^2 S_{x_k} S_{x_k x_j} - \sum_k u^2 S_{x_k} S_{x_k x_j}$$

$$= -\sum_k \partial_{x_k} \left(u^2 S_{x_k} S_{x_j}\right) = \sum_k \partial_{x_k} T_{kj}^{(1)}$$

where $T_{kj}^{(1)} = -u^2 S_{x_k} S_{x_j}$.

We have that:

$$\frac{1}{2} u^2 \nabla \frac{\Delta u}{u} = \frac{1}{2} \nabla \left(u^2 \frac{\Delta u}{u}\right) - \frac{1}{2} \frac{\Delta u}{u} \nabla u^2 = \frac{1}{2} \nabla \left(u \Delta u\right) - \Delta u \nabla u.$$

Then

$$\left[\frac{1}{2} u^2 \nabla \frac{\Delta u}{u}\right]_j = \frac{1}{2} \partial_{x_j} \left(u \sum_k u_{x_k x_k}\right) - u_{x_j} \sum_k u_{x_k x_k}$$

$$= \frac{1}{2} u \sum_k u_{x_k x_k x_j} - \frac{1}{2} u_{x_j} \sum_k u_{x_k x_k}$$

$$= \frac{1}{2} \sum_k \partial_{x_k} \left(u u_{x_k x_j}\right) - \frac{1}{2} \sum_k u_{x_k} u_{x_k x_j}$$

$$- \frac{1}{2} \sum_k \partial_{x_k} \left(u_{x_j} u_{x_k}\right) + \frac{1}{2} \sum_k u_{x_k} u_{x_k x_j}$$

$$= \frac{1}{2} \sum_k \partial_{x_k} \left(u u_{x_k x_j} - u_{x_j} u_{x_k}\right) = \sum_k \partial_{x_k} T_{kj}^{(2)}.$$

The last piece:

$$\frac{1}{2} u^2 \nabla \frac{W_\varepsilon'(u)}{u} = \frac{1}{2} \nabla \left(u^2 \frac{W_\varepsilon'(u)}{u}\right) - \frac{1}{2} \frac{W_\varepsilon'(u)}{u} \nabla u^2 = \frac{1}{2} \nabla \left(u W_\varepsilon'(u)\right) - \frac{1}{2} W_\varepsilon'(u) \nabla u$$

$$= \frac{1}{2} \nabla u W_\varepsilon'(u) - \frac{1}{2} u \nabla W_\varepsilon'(u) - \frac{1}{2} W_\varepsilon'(u) \nabla u = -\frac{1}{2} u \nabla W_\varepsilon'(u).$$

Then

$$\left[-\frac{1}{2}u\nabla W_\varepsilon'(u)\right]_j = -\frac{1}{2}u\partial_{x_j}W_\varepsilon'(u) = -\frac{1}{2}\partial_{x_j}\left(uW_\varepsilon'(u)\right) + \frac{1}{2}W_\varepsilon'(u)u_{x_j}$$

$$= -\frac{1}{2}\partial_{x_j}\left(uW_\varepsilon'(u)\right) + \frac{1}{2}\partial_{x_j}W_\varepsilon(u)$$

$$= \frac{1}{2}\partial_{x_j}\left[W_\varepsilon(u) - uW_\varepsilon'(u)\right] = \sum_k \partial_{x_k}T_{kj}^{(3)}$$

where $T_{kj}^{(3)} = \frac{1}{2}\left[W_\varepsilon(u) - uW_\varepsilon'(u)\right]\delta_{kj}$.

Concluding,

$$\partial_t\left(u^2\nabla S\right) = -u^2\nabla V(x) + \sum_k \partial_{x_k}(T_{kj}^{(1)} + T_{kj}^{(2)} + T_{kj}^{(3)})$$

$$= -u^2\nabla V(x) + \sum_k \partial_{x_k}T_{kj}\quad with\quad T_{kj} = T_{kj}^{(1)} + T_{kj}^{(2)} + T_{kj}^{(3)}.$$

□

If $V = const$, Eq. (3.126) is a continuity equation and the momentum

$$\mathbf{P}(\psi) = \int u^2\nabla S\,dx = \int \mathrm{Im}\left(\bar\psi\nabla\psi\right)dx \qquad (3.127)$$

is constant in time. Notice that, by Eq. (3.6), the density of momentum $u^2\nabla S$ is nothing else than the flow of hylenic charge.

Let us consider the soliton (3.102); in this case we have

$$E_\varepsilon\left(\psi_{q,\varepsilon}\right) = \frac{e}{\varepsilon^2} + \frac{1}{2m}\bar p^2$$

where e is defined by (3.100),

$$C(\psi_\varepsilon) = 1$$

and

$$\mathbf{P}(\psi_\varepsilon) = \bar p.$$

Now, let us see the rescaling properties of the internal energy and the L^2 norm of a function $u(x)$ having the form

$$u(x) := \frac{1}{\varepsilon^{N/2}}v\left(\frac{x}{\varepsilon}\right).$$

We have

$$\|u\|^2_{L^2} = \frac{1}{\varepsilon^N} \int v\left(\frac{x}{\varepsilon}\right)^2 dx = \int v\left(\xi\right)^2 d\xi = \|v\|^2_{L^2}$$

and

$$
\begin{aligned}
J_\varepsilon(u) &= \int \left[\frac{1}{2m}|\nabla u|^2 + W_\varepsilon(u)\right] dx \\
&= \int \left[\frac{1}{2m}|\nabla u|^2 + \frac{1}{\varepsilon^{N+2}} W(\varepsilon^{N/2} u)\right] dx \\
&= \int \left[\frac{1}{2m}\frac{1}{\varepsilon^N}\left|\nabla_x v\left(\frac{x}{\varepsilon}\right)\right|^2 + \frac{1}{\varepsilon^{N+2}} W\left(v\left(\frac{x}{\varepsilon}\right)\right)\right] dx \\
&= \int \left[\frac{1}{2m}\frac{1}{\varepsilon^{N+2}}\left|\nabla_\xi v\left(\xi\right)\right|^2 + \frac{1}{\varepsilon^{N+2}} W\left(v\left(\xi\right)\right)\right] \varepsilon^N d\xi \\
&= \frac{1}{\varepsilon^2} \int \frac{1}{2m}\left|\nabla_\xi v\left(\xi\right)\right|^2 + W\left(v\left(\xi\right)\right) d\xi = \frac{1}{\varepsilon^2} J_1(u).
\end{aligned}
$$

3.3.4   Definition of the Soliton

In this section we want to give a new definition of soliton and to describe a method to split a solution of Eq. (3.105) in a wave and a soliton. In fact, as we have discussed at the beginning of Sect. 3.3, Definition 16 is not suitable if we want to have a splitting as (3.93).

If our solution has the following form

$$\psi_\varepsilon(t, x) = u_\varepsilon e^{iS_\varepsilon} = [U_\varepsilon(x - \xi(t)) + w_\varepsilon(t, x)]\, e^{iS_\varepsilon(t,x)}$$

where w is sufficiently small, then a possible choice is to identify the soliton with $U(x - \xi(t))e^{iS(t,x)}$ and the wave with $w_\varepsilon(t, x)e^{iS(t,x)}$. However, we want to give a definition which localizes the soliton, namely to assume the soliton $\Psi_\varepsilon(t, x)$ to have compact support in space.

Roughly speaking, the soliton can be defined as the part of the field ψ_ε where some density function $\rho_\varepsilon(t, x)$ is sufficiently large (e.g., after a suitable normalization, $\rho_\varepsilon(t, x) \geq 1$).

For the moment we do not define $\rho_\varepsilon(t, x)$ explicitly. We just require that $\rho_\varepsilon(t, x)$ satisfies the following assumptions:

- $\rho_\varepsilon \in C^1(\mathbb{R}^{N+1})$ and $\rho_\varepsilon(t, x) \to 0$ as $|x| \to \infty$.
- ρ_ε satisfies the continuity equation

$$\partial_t \rho_\varepsilon + \nabla \cdot J_{\rho_\varepsilon} = 0 \qquad (3.128)$$

for some $J_{\rho_\varepsilon} \in C^1(\mathbb{R}^{N+1})$.

In order to fix the ideas you may think of $\rho_\varepsilon(t, x)$ as a smooth approximation of $u_\varepsilon(t, x)^2$. An explicit definition of $\rho_\varepsilon(t, x)$ is given in Sect. 3.3.7. However, in other problems, it might be more useful to make different choices of it such as the energy density. We have postponed the choice of ρ_ε since the results of this section are independent of this choice.

Next we set

$$\chi_\varepsilon(t, x) = \sqrt{\varphi(\rho_\varepsilon(t, x))}$$

where

$$\varphi(s) = \begin{cases} 0 & \text{if } s \le 1 \\ s - 1 & \text{if } 1 \le s \le 2. \\ 1 & \text{if } s \ge 2 \end{cases}$$

So we have that $\chi_\varepsilon(t, x) = 1$ where $\rho_\varepsilon(t, x) \ge 2$ and $\chi_\varepsilon(t, x) = 0$ where $\rho_\varepsilon(t, x) \le 1$: thus you may think of $\chi_\varepsilon(t, x)$ as a sort of approximation of the characteristic function of the region occupied by the soliton. Finally, we set

$$\Psi_\varepsilon(t, x) = \psi_\varepsilon(t, x)\chi_\varepsilon \qquad (3.129)$$

$$\varphi_\varepsilon(t, x) = \psi_\varepsilon(t, x)[1 - \chi_\varepsilon] \qquad (3.130)$$

$\Psi_\varepsilon(t, x)$ is the soliton and $\varphi_\varepsilon(t, x)$ is the wave; the region

$$\Sigma_{\varepsilon,t} = \{(t, x) \in \mathbb{R}^{N+1} | 1 < \rho(t, x) < 2\} \qquad (3.131)$$

$$= \{(t, x) \in \mathbb{R}^{N+1} | 0 < \chi_\varepsilon(t, x) < 1\} \qquad (3.132)$$

is the region where the soliton and the wave interact with each other; we will refer to it as the *halo* of the soliton.

3.3.5 The Equation of Dynamics of the Soliton

Definition 66. We define the following quantities relative to the soliton:

- The *barycenter*:

$$q_\varepsilon(t) = \frac{\int x \, |\Psi_\varepsilon|^2 \, dx}{\int |\Psi_\varepsilon|^2 \, dx}.$$

- The *momentum*:

$$p_\varepsilon(t) = \int \nabla S_\varepsilon |\Psi_\varepsilon|^2 \, dx.$$

- The *mass*:

$$m_\varepsilon(t) = m \int |\Psi_\varepsilon|^2 \, dx.$$

Remark 67. Notice that the mass of the soliton $m_\varepsilon(t)$ depends on t. The global mass is constant (namely m) but it is shared between the soliton and the wave whose mass is $\int u_\varepsilon^2 \left[1 - \chi_\varepsilon^2\right] dx$.

The next theorem shows the relation between $q_\varepsilon(t)$ and $p_\varepsilon(t)$ and their derivatives.

Theorem 68. *The following equations hold*

$$\dot{q}_\varepsilon = \frac{p_\varepsilon}{m_\varepsilon} + \frac{1}{m_\varepsilon} \int_{\Sigma_{\varepsilon,t}} (x - q_\varepsilon) \left[u_\varepsilon^2 \nabla S_\varepsilon \cdot \nabla \rho_\varepsilon - \nabla \cdot J_{\rho_\varepsilon}\right] dx \qquad (3.133)$$

$$\dot{p}_\varepsilon = -\int \nabla V \, |\Psi_\varepsilon|^2 \, dx - \int_{\Sigma_{\varepsilon,t}} \left[T \cdot \nabla \rho_\varepsilon + u_\varepsilon^2 \nabla S_\varepsilon \left(\nabla \cdot J_{\rho_\varepsilon}\right)\right] dx. \qquad (3.134)$$

Remark 69. The term $\int_{\Sigma_{\varepsilon,t}} T \cdot \nabla \rho_\varepsilon dx$ represents the pressure of the wave on the soliton; if $\varepsilon \to 0$ and $\partial \Sigma_{\varepsilon,t}$ is sufficiently regular then

$$\int_{\Sigma_{\varepsilon,t}} T \cdot \nabla \rho_\varepsilon dx = \int_{\sigma_\varepsilon} T \cdot \mathbf{n} \, d\sigma$$

where $\sigma_\varepsilon = \{x \mid \rho_\varepsilon(x) = 1\}$ and $\mathbf{n}$ is its outer normal.

Proof of Theorem 68. We calculate the first derivative of the barycenter.

$$\dot{q}_\varepsilon(t) = \frac{d}{dt} \left(\frac{\int x \, |\Psi_\varepsilon|^2 \, dx}{\int |\Psi_\varepsilon|^2 \, dx} \right)$$

$$= \frac{\frac{d}{dt} \int x \, |\Psi_\varepsilon|^2 \, dx}{\int |\Psi_\varepsilon|^2 \, dx} - \frac{\left(\int x \, |\Psi_\varepsilon|^2 \, dx\right) \left(\frac{d}{dt} \int |\Psi_\varepsilon|^2 \, dx\right)}{\left(\int |\Psi_\varepsilon|^2 \, dx\right)^2}$$

$$= \frac{\frac{d}{dt} \int x \, |\Psi_\varepsilon|^2 \, dx}{\int |\Psi_\varepsilon|^2 \, dx} - q_\varepsilon(t) \frac{\frac{d}{dt} \int |\Psi_\varepsilon|^2 \, dx}{\int |\Psi_\varepsilon|^2 \, dx} = \frac{\int (x - q_\varepsilon(t)) \frac{d}{dt} (|\Psi_\varepsilon|^2) dx}{\int |\Psi_\varepsilon|^2 \, dx}.$$

We have

$$\nabla \chi^2 = \nabla \varphi \left(\rho_\varepsilon(t, x) \right) = \varphi' \left(\rho_\varepsilon(t, x) \right) \nabla \rho_\varepsilon = \mathbb{I}_{\Sigma_{\varepsilon,t}} \nabla \rho_\varepsilon$$

and

$$\frac{d}{dt} \chi^2 = \frac{d}{dt} \varphi \left(\rho_\varepsilon(t, x) \right) = \varphi' \left(\rho_\varepsilon(t, x) \right) \partial_t \rho_\varepsilon$$

$$= \mathbb{I}_{\Sigma_{\varepsilon,t}} \partial_t \rho_\varepsilon = -\mathbb{I}_{\Sigma_{\varepsilon,t}} \nabla \cdot J_{\rho_\varepsilon}$$

where $\mathbb{I}_{\Sigma_{\varepsilon,t}}$ is the characteristic function of $\Sigma_{\varepsilon,t}$.

So, we have

$$
\dot{q}_\varepsilon(t) = \frac{\int (x - q_\varepsilon(t)) \frac{d}{dt} (\chi^2 u_\varepsilon^2) dx}{\int |\Psi_\varepsilon|^2 \, dx}
$$

$$
= \frac{\int (x - q_\varepsilon(t)) \left(\chi^2 \frac{d}{dt} u_\varepsilon^2 + u_\varepsilon^2 \frac{d}{dt} \chi^2 \right) dx}{\int |\Psi_\varepsilon|^2 \, dx} \qquad (3.135)
$$

$$
= \frac{\int_{\mathbb{R}^N} (x - q_\varepsilon(t)) \chi^2 \frac{d}{dt} u_\varepsilon^2 dx - \int_{\Sigma_{\varepsilon,t}} (x - q_\varepsilon(t)) \nabla \cdot J_{\rho_\varepsilon} dx}{\int_{\mathbb{R}^N} |\Psi_\varepsilon|^2 \, dx}.
$$

For the first term we use the continuity equation (3.6). We have

$$\int (x - q_\varepsilon(t)) \chi^2 \frac{d}{dt} u_\varepsilon^2 dx = \int (x - q_\varepsilon(t)) \nabla \cdot \left(u_\varepsilon^2 \frac{\nabla S_\varepsilon}{m} \right) \chi^2 dx$$

$$= \frac{1}{m} \int (u_\varepsilon^2 \nabla S_\varepsilon) \, \chi^2 dx + \frac{1}{m} \int (x - q_\varepsilon(t)) u_\varepsilon^2 \nabla S_\varepsilon \cdot \nabla \chi^2 dx$$

$$= \frac{p_\varepsilon(t)}{m} + \frac{1}{m} \int_{\Sigma_{\varepsilon,t}} (x - q_\varepsilon(t)) u_\varepsilon^2 \nabla S_\varepsilon \cdot \nabla \rho_\varepsilon dx.$$

Concluding, we get the first equation of motion:

$$\dot{q}_\varepsilon(t) = \frac{p_\varepsilon(t)}{m_\varepsilon} + \frac{\int_{\Sigma_{\varepsilon,t}} (x - q_\varepsilon(t)) \left[u_\varepsilon^2 \nabla S_\varepsilon \cdot \nabla \rho_\varepsilon - \nabla \cdot J_{\rho_\varepsilon} \right] dx}{m_\varepsilon}.$$

Next, we will get the second one. We have that

$$\dot{p}_\varepsilon = \int \left(\frac{\partial}{\partial t} u_\varepsilon^2 \nabla S_\varepsilon \right) \chi^2 dx + \int u_\varepsilon^2 \nabla S_\varepsilon \frac{\partial}{\partial t} \chi^2 dx. \qquad (3.136)$$

Now, using (3.126) we have that

$$\int \left(\frac{\partial}{\partial t} u_\varepsilon^2 \nabla S_\varepsilon \right) \chi^2 dx = - \int \nabla V |\Psi_\varepsilon|^2 \, dx + \int \nabla \cdot T \, \chi^2 dx$$

$$= - \int \nabla V |\Psi_\varepsilon|^2 \, dx - \int T \cdot \nabla \chi^2 dx$$

$$= - \int \nabla V |\Psi_\varepsilon|^2 \, dx - \int_{\Sigma_{\varepsilon,t}} T \cdot \nabla \rho_\varepsilon dx.$$

The second term of Eq. (3.136) takes the form:

$$\int u_\varepsilon^2 \nabla S_\varepsilon \frac{\partial}{\partial t} \chi^2 dx = - \int_{\Sigma_{\varepsilon,t}} u_\varepsilon^2 \nabla S_\varepsilon \left(\nabla \cdot J_{\rho_\varepsilon} \right) dx.$$

$\square$

It is possible to give a "pictorial" interpretation to Eqs. (3.133) and (3.134). We may assume that u_ε^2 represents the density of a fluid; so the soliton is a bump of fluid particles which stick together and the halo $\Sigma_{\varepsilon,t}$ can be regarded as the interface where the soliton and the wave might exchange particles, momentum and energy.

Hence,

- $m_\varepsilon(t)$ is the mass of the soliton.
- $\frac{\nabla S_\varepsilon}{m}$ is the velocity of the fluid particles and ∇S_ε is their momentum.

So each term of Eqs. (3.133) and (3.134) have the following interpretation

- $\frac{p_\varepsilon(t)}{m_\varepsilon}$ is the average velocity of each particle; in fact

$$\frac{p_\varepsilon(t)}{m_\varepsilon} = \frac{\int \nabla S_\varepsilon |\Psi_\varepsilon|^2 \, dx}{m_\varepsilon} = \frac{\int \frac{\nabla S_\varepsilon}{m} |\Psi_\varepsilon|^2 \, dx}{\int |\Psi_\varepsilon|^2 \, dx}.$$

- The "halo term" $\frac{1}{m_\varepsilon} \int_{\Sigma_{\varepsilon,t}} (x - q_\varepsilon) \left[u_\varepsilon^2 \nabla S_\varepsilon \cdot \nabla \rho_\varepsilon - \nabla \cdot J_{\rho_\varepsilon} \right] dx$ describes the change of the average velocity of the soliton due to the exchange of fluid particles.
- The term $- \int \nabla V |\Psi_\varepsilon|^2 \, dx$ describes the volume force acting on the soliton.
- The term $- \int_{\Sigma_{\varepsilon,t}} T \cdot \nabla \rho_\varepsilon dx$ describes the surface force exerted by the wave on the soliton.
- The term $- \int_{\Sigma_{\varepsilon,t}} u_\varepsilon^2 \nabla S_\varepsilon \left(\nabla \cdot J_{\rho_\varepsilon} \right) dx$ describes the change of the momentum of the soliton due to the exchange of fluid particles with the wave.

3.3.6 Analysis of the Concentration Point of the Soliton

If $\psi_\varepsilon(t, x)$ is a solution of the problem (3.105), we say that $\hat{q}_\varepsilon(t)$ is the *concentration point* of $\psi_\varepsilon(t, x)$ if it minimizes the following quantity

$$f(q) = \| |\psi_\varepsilon(t, x)| - U_\varepsilon(x - q) \|_{L^2}^2 .$$

It is easy to see that $f(q)$ has a minimizer; of course, it might happen that it is not unique; in this case we denote by $\hat{q}_\varepsilon(t)$ one of the minimizers of f at the time t.

Basically $\hat{q}_\varepsilon(t)$ is a good candidate for the position of our soliton, but it cannot satisfy an equation of type (3.118) since in general it is not uniquely defined and a fortiori is not differentiable. $\hat{q}_\varepsilon(t)$ could be uniquely defined if we make assumptions on the non degeneracy of the ground state, but we do not like to make such assumptions since they are very hard to be verified and in general they do not hold. Actually the position of the soliton is supposed to be $q_\varepsilon(t)$ given by Definition 66. However, as we will see, $\hat{q}_\varepsilon$ is useful to recover some estimates on q_ε. So, in this subsection we will analyze some properties of $\hat{q}_\varepsilon(t)$. We start with a variant of a result contained in [17], but first we recall some notation:

- $\mathfrak{M}_\sigma = \{u \in H^1(\mathbb{R}^N) : \int u^2 dx = \sigma\}$ (cf. (3.41)).
- $J^\sigma = \{u \in H^1(\mathbb{R}^N) : J_1(u) \le \sigma\}$ where $J_1(u)$ is the internal energy when $\varepsilon = 1$; see (3.124).

Lemma 70. *Given $u \in H^1$, we define (if it exists) $\hat{q} \in \mathbb{R}^N$ to be a minimizer of the function*

$$q \mapsto \| U(x - q) - u(x) \|_{L^2}^2.$$

For any η there exists a $\delta(\eta)$ such that, if $u \in J^{e + \delta(\eta)} \cap \mathfrak{M}_1$, $\hat{q}$ exists and it holds

$$\| U(x - \hat{q}) - u \|_{H^1} \le \eta \tag{3.137}$$

$$\int_{\mathbb{R}^N \setminus B(\hat{q}, \hat{R}_\eta)} u^2 dx \le \eta \tag{3.138}$$

where $\hat{R}_\eta = -C \log(\eta)$ and $U \in \Gamma$.

Proof. The proof of (3.137) can be found in [17]. If $U \in \Gamma$, again by [17] we know that, for R sufficiently large,

$$\int_{|x| > R} U^2(x) dx < \int_{|x| > R} C_1 e^{-C_2 |x|}.$$

Thus

$$\int_{|x| > R} U^2(x) dx = C_3 \int_R^\infty \rho^{N-1} e^{-C_2 \rho} d\rho = C_4 R^N e^{-C_2 R} \le e^{-C_5 R}$$

where the C_i's are suitable positive constants. We remark that R does not depend on U.

Now, it is sufficient to take $\hat{R}_\eta > -\frac{1}{C_5} \log(\eta)$ and by (3.137) we obtain (3.138). □

We define the set of admissible initial data as follows:

$$B_{\varepsilon,M} = \left\{ \psi(x) = U_\varepsilon (x - q_0) \, e^{ip_0 \cdot x} + \varphi(x) : E_\varepsilon (\psi) \le \frac{e}{\varepsilon^2} + M \text{ and } \|\psi\|_{L^2} = 1 \right\}.$$

Lemma 71. *For every $\eta > 0$, there exists $\varepsilon = \varepsilon(\eta) > 0$ such that*

$$\int_{\mathbb{R}^N \smallsetminus B(\hat{q}_\varepsilon, \varepsilon \hat{R}_\eta)} |\psi_\varepsilon(t, x)|^2 \, dx < \eta \tag{3.139}$$

where $\psi_\varepsilon(t, x)$ is a solution of problem (3.105), with initial data in $B_{\varepsilon,M}$ and $\hat{q}_\varepsilon$ is the concentration point of ψ_ε.

Proof. By the conservation law, the energy $E_\varepsilon(\psi_\varepsilon(t, x))$ is constant with respect to t. Then we have, by hypothesis on the initial datum

$$E_\varepsilon(\psi_\varepsilon(t, x)) = E_\varepsilon(\psi_\varepsilon(0, x)) \le \frac{e}{\varepsilon^2} + M.$$

Thus

$$J_\varepsilon(\psi_\varepsilon(t, x)) = E_\varepsilon(\psi(t, x)) - G(\psi(t, x))$$

$$= E_\varepsilon(\psi_\varepsilon(t, x)) - \int_{\mathbb{R}^N} \left[\frac{|\nabla S_\varepsilon(t, x)|^2}{2m} + V(x) \right] u_\varepsilon(t, x)^2 dx$$

$$\le E_\varepsilon(\psi_\varepsilon(t, x)) \le \frac{e}{\varepsilon^2} + M$$

because $V \ge 0$. By rescaling the above inequality, and setting $y = x/\varepsilon$ we get

$$J(|\varepsilon^{N/2} \psi_\varepsilon(t, \varepsilon y)|) \le e + \varepsilon^2 M.$$

We choose ε small such that $\varepsilon^2 M \le \delta(\eta)$. Then $\varepsilon^{N/2} \psi_\varepsilon(t, \varepsilon y) \in J^{e+\delta(\eta)} \cap \mathfrak{M}_1$, and so applying Lemma 70.

$$\int_{\mathbb{R}^N \smallsetminus B(\hat{q}, \hat{R}_\eta)} \varepsilon^N |\psi_\varepsilon(t, \varepsilon y)|^2 dy < \eta.$$

Now, making the change of variable $x = \varepsilon y$, we obtain the desired result. □

Lemma 72. *If $\psi_\varepsilon(t, x)$ is a solution of problem (3.105), with initial data in $B_{\varepsilon,M}$ and ε sufficiently small, then*

$$\int_{\mathbb{R}^N \smallsetminus B(\hat{q}_\varepsilon, \sqrt{\varepsilon})} |\psi_\varepsilon(t, x)|^2 \, dx = \eta(\varepsilon) \qquad (3.140)$$

where $\eta(\varepsilon) \to 0$ as $\varepsilon \to 0$.

Proof. First we prove that for every $\eta > 0$, there exists $\varepsilon_1(\eta) > 0$ such that, if $\psi_\varepsilon(0, x) \in B_{\varepsilon_1(\eta),M}$, we have

$$\int_{\mathbb{R}^N \smallsetminus B\left(\hat{q}_\varepsilon, \sqrt{\varepsilon_1(\eta)}\right)} |\psi_\varepsilon(t, x)|^2 \, dx < \eta.$$

Arguing as in the proof of Lemma 71, if $\varepsilon_1(\eta) \le \min\left[\sqrt{\frac{\delta(\eta)}{M}}, \frac{1}{\hat{R}_\eta^2}\right]$, we get (3.139). At this point, since $\varepsilon_1(\eta) \le \frac{1}{\hat{R}_\eta^2}$ we have that $\varepsilon_1(\eta) \hat{R}_\eta \le \sqrt{\varepsilon_1(\eta)}$.

Now set

$$\varepsilon(\eta) = \min_{\eta \le \zeta} \varepsilon_1(\zeta).$$

Clearly, $\varepsilon(\eta)$ is a non-increasing function (which might be discontinuous) and $\varepsilon(\eta) \to 0$ as $\eta \to 0$. Then it has a "pseudoinverse" function $\eta(\varepsilon)$ namely a function which is the inverse in the monotonicity points, which is discontinuous where $\varepsilon(\eta)$ is constant and constant where $\varepsilon(\eta)$ is discontinuous. Moreover $\eta(\varepsilon)$ as $\varepsilon \to 0$. $\square$

3.3.7 Definition of the Density ρ_ε

First of all we notice that, in Lemma 72, it is not restrictive to assume that

$$\eta = \eta(\varepsilon) \ge \varepsilon. \qquad (3.141)$$

Now we set

$$\rho_\varepsilon(t, x) = a(x) * u(t, x)^2$$

where, $a_\varepsilon(s) \in C^\infty$,

$$a_\varepsilon(s) = \begin{cases} 3 \; |s| \le \eta^{\frac{1}{8}}\left(1 - \eta^{\frac{1}{8}}\right) \\ 0 \; |s| \ge \eta^{\frac{1}{8}}\left(1 + \eta^{\frac{1}{8}}\right) \end{cases}$$

and

$$|\nabla a_\varepsilon(s)| \leq \eta^{-\frac{1}{4}}. \tag{3.142}$$

Lemma 73. *Take ψ_ε a solution of (3.105) with initial data in $B_{\varepsilon,M}$.*

$$\text{If } |x - \hat{q}_\varepsilon(t)| \leq \eta^{\frac{1}{8}}\left(1 - 2\eta^{\frac{1}{8}}\right) \text{ then } \rho_\varepsilon(t,x) \geq 3\left(1 - \eta\right)$$
$$\text{if } |x - \hat{q}_\varepsilon(t)| \geq \eta^{\frac{1}{8}}\left(1 + 2\eta^{\frac{1}{8}}\right) \quad \text{then } \rho_\varepsilon(t,x) \leq 3\eta$$

where $\eta = \eta(\varepsilon)$ as in Lemma 72. In particular we have that

$$\Sigma_{\varepsilon,t} \subset B\left(\hat{q}_\varepsilon(t), \eta^{\frac{1}{8}}\left(1 + 2\eta^{\frac{1}{8}}\right)\right) \setminus B\left(\hat{q}_\varepsilon(t), \eta^{\frac{1}{8}}\left(1 - 2\eta^{\frac{1}{8}}\right)\right) \tag{3.143}$$

where $\Sigma_{\varepsilon,t}$ is defined by (3.131).

Proof. If $|x - \hat{q}_\varepsilon| \leq \eta^{\frac{1}{8}}\left(1 - 2\eta^{\frac{1}{8}}\right)$, then

$$|x - \hat{q}_\varepsilon| + \sqrt{\varepsilon} \leq \eta^{\frac{1}{8}}\left(1 - 2\eta^{\frac{1}{8}}\right) + \sqrt{\eta} \leq \eta^{\frac{1}{8}}\left(1 - \eta^{\frac{1}{8}}\right)$$

and hence

$$B(\hat{q}_\varepsilon, \sqrt{\varepsilon}) \subset B\left(x, \eta^{\frac{1}{8}}\left(1 - \eta^{\frac{1}{8}}\right)\right).$$

Then, by using Lemma 72,

$$\rho_\varepsilon(t,x) = \int a_\varepsilon(y-x)u_\varepsilon(t,y)^2 dy \geq 3\int_{B(x,\ \eta^{1/8}-\eta^{1/4})} u_\varepsilon(t,y)^2 dy$$

$$\geq 3\int_{B(\hat{q}_\varepsilon,\ \varepsilon^{1/2})} u_\varepsilon(t,y)^2 dy \geq 3\left(1 - \eta\right).$$

If $|x - \hat{q}_\varepsilon(t)| \geq \eta^{\frac{1}{8}}\left(1 + 2\eta^{\frac{1}{8}}\right)$,

$$|x - \hat{q}_\varepsilon| - \sqrt{\varepsilon} \geq \eta^{\frac{1}{8}}\left(1 + 2\eta^{\frac{1}{8}}\right) - \sqrt{\eta} \geq \eta^{\frac{1}{8}}\left(1 + \eta^{\frac{1}{8}}\right)$$

and so

$$B\left(x, \eta^{\frac{1}{8}}\left(1 + \eta^{\frac{1}{8}}\right)\right) \subset \mathbb{R}^N \setminus B(\hat{q}_\varepsilon, \sqrt{\varepsilon}).$$

Then, using again Lemma 72,

$$\rho_\varepsilon(t,x) = \int a_\varepsilon(y-x)u_\varepsilon(t,y)^2 dy \le 3 \int_{B\left(x,\, \eta^{\frac{1}{8}}\left(1+\eta^{\frac{1}{8}}\right)\right)} u_\varepsilon(t,y)^2 dy$$

$$\le 3 \int_{\mathbb{R}^N \smallsetminus B(\hat{q}_\varepsilon,\, \sqrt{\varepsilon})} u_\varepsilon(t,y)^2 dy \le 3\eta.$$

□

Clearly, $\rho_\varepsilon = a_\varepsilon * u_\varepsilon^2 \in C^1(\mathbb{R}^{N+1})$ and, by (3.6), it satisfies the continuity equation (3.128) with

$$J_{\rho_\varepsilon} = a_\varepsilon * \left(u_\varepsilon^2 \nabla S_\varepsilon\right). \tag{3.144}$$

Therefore, the results of Sect. 3.3.4 hold. In particular, we have that the support of $\Psi_\varepsilon(t,x)$ is contained in $B\left(\hat{q}_\varepsilon, \eta^{\frac{1}{8}}\left(1 + 2\eta^{\frac{1}{8}}\right)\right)$ when η is sufficiently small (namely $\eta < 1/3$). Moreover, by (3.143), we see that the size of the halo is an infinitesimal of higher order with respect to the diameter of the soliton.

3.3.8   The Dynamics of the Soliton

Theorem 74. *The following equations hold*

$$\dot{q}_\varepsilon(t) = \frac{p_\varepsilon(t)}{m_\varepsilon(t)} + K_\varepsilon(t) \tag{3.145}$$

$$\dot{p}_\varepsilon = -\nabla V(q_\varepsilon(t)) + F_\varepsilon(q_\varepsilon) + H_\varepsilon(t) \tag{3.146}$$

where

$$\sup_{t \in \mathbb{R}} \left(|H_\varepsilon(t)| + |K_\varepsilon(t)|\right) \to 0 \ as \ \varepsilon \to 0 \tag{3.147}$$

and

$$F_\varepsilon(q_\varepsilon) = -\int_{\Sigma_{\varepsilon,t}} T \cdot \nabla \rho_\varepsilon dx. \tag{3.148}$$

Moreover we have that

$$\forall \tau_0, \tau_1, \ \left| \int_{\tau_0}^{\tau_1} F_\varepsilon(q_\varepsilon) \, dt \right| \le c(\varepsilon)(1 + |\tau_1 - \tau_0|) \tag{3.149}$$

where $c(\varepsilon) \to 0$ as $\varepsilon \to 0$.

Proof. We set

$$K_\varepsilon(t) = \frac{1}{m_\varepsilon} \int_{\Sigma_{\varepsilon,t}} (x - q_\varepsilon) \left[u_\varepsilon^2 \nabla S_\varepsilon \cdot \nabla \rho_\varepsilon - \nabla \cdot J_{\rho_\varepsilon} \right] dx,$$

$$H_{1,\varepsilon}(t) = \int_{\Sigma_{\varepsilon,t}} u_\varepsilon^2 \nabla S_\varepsilon \left(\nabla \cdot J_{\rho_\varepsilon} \right) dx,$$

$$H_{2,\varepsilon}(t) = \nabla V(q_\varepsilon(t)) - \int \nabla V(x) \, |\Psi_\varepsilon|^2 \, dx,$$

$$H_\varepsilon(t) = H_{1,\varepsilon}(t) + H_{2,\varepsilon}(t),$$

and hence, by Theorem 68, we need just to prove (3.147).

We estimate each individual term of K_ε. We have that

$$\sup_{x \in \Sigma_{\varepsilon,t}} |x - q_\varepsilon| \le 2 \left(\eta^{1/8} + 2\eta^{1/4} \right) \le 3\eta^{1/8} \qquad (3.150)$$

since $q_\varepsilon(t), x \in B(\hat{q}_\varepsilon, \eta^{1/8} + 2\eta^{1/4})$.

Also, by (3.142) and well known properties on convolutions,

$$\sup_{x \in \Sigma_{\varepsilon,t}} |\nabla \rho_\varepsilon| \le \sup_{x \in \mathbb{R}^N} \left| \nabla a_\varepsilon(x) * u_\varepsilon(t, x)^2 \right| \qquad (3.151)$$

$$\le \|\nabla a_\varepsilon\|_{L^\infty} \cdot \|u_\varepsilon\|_{L^2}^2 \le \frac{1}{\eta^{1/4}}.$$

If $\psi_\varepsilon(0, x) \in B_{\varepsilon,M}$, by (3.125), we have

$$G(\psi) = E_\varepsilon(\psi) - J_\varepsilon(\psi) \le \frac{e}{\varepsilon^2} + M - \frac{e}{\varepsilon^2} = M; \qquad (3.152)$$

so, by Lemma 72,

$$\int_{A_{\varepsilon,t}} u^2 |\nabla S_\varepsilon| \le \left(\int_{A_{\varepsilon,t}} u_\varepsilon^2 \right)^{\frac{1}{2}} \left(\int_{A_{\varepsilon,t}} u_\varepsilon^2 |\nabla S_\varepsilon|^2 \right)^{\frac{1}{2}}$$

$$\le \eta^{\frac{1}{2}} \left[2mG(\psi) \right]^{\frac{1}{2}} \le const.\eta^{\frac{1}{2}} \qquad (3.153)$$

where for simplicity, we have set

$$A_{\varepsilon,t} = \mathbb{R}^N \setminus B(\hat{q}_\varepsilon(t), \sqrt{\varepsilon}).$$

Since $\Sigma_{\varepsilon,t} \subset A_{\varepsilon,t}$,

$$\int_{\Sigma_{\varepsilon,t}} u^2 |\nabla S_\varepsilon| \, dx \le \eta^{\frac{1}{2}} \, [2mG(\psi)]^{\frac{1}{2}} \le const.\eta^{\frac{1}{2}}. \tag{3.154}$$

Finally, by (3.144)

$$\sup_{x \in \Sigma_{\varepsilon,t}} |\nabla \cdot J_{\rho_\varepsilon}| \le \sup_{x \in \mathbb{R}^N} |(\nabla \cdot a_\varepsilon) * (u_\varepsilon^2 \nabla S_\varepsilon)|$$

$$\le \|\nabla \cdot a_\varepsilon\|_{L^\infty} \cdot \int_{\mathbb{R}^N} u_\varepsilon^2 |\nabla S_\varepsilon|$$

$$\le \frac{1}{\eta^{1/4}} \left(\int_{\mathbb{R}^N} u_\varepsilon^2 \right)^{\frac{1}{2}} \cdot \left(\int_{\mathbb{R}^N} u_\varepsilon^2 |\nabla S_\varepsilon| \right)^{\frac{1}{2}}$$

$$\le \frac{[2mG(\psi)]^{\frac{1}{2}}}{\eta^{1/4}} = const.\eta^{-\frac{1}{4}}. \tag{3.155}$$

By (3.143),

$$|\Sigma_{\varepsilon,t}| \le \left| B\left(\hat{q}_\varepsilon(t), \eta^{\frac{1}{8}} \left(1 + 2\eta^{\frac{1}{8}} \right) \right) \right| - \left| B\left(\hat{q}_\varepsilon(t), \eta^{\frac{1}{8}} \left(1 - 2\eta^{\frac{1}{8}} \right) \right) \right|$$

$$= \omega_N \left[\eta^{\frac{1}{8}} \left(1 + 2\eta^{\frac{1}{8}} \right) \right]^N - \omega_N \left[\eta^{\frac{1}{8}} \left(1 - 2\eta^{\frac{1}{8}} \right) \right]^N$$

$$= \omega_N \eta^{\frac{N}{8}} \left[\left(1 + 2\eta^{\frac{1}{8}} \right)^N - \left(1 - 2\eta^{\frac{1}{8}} \right)^N \right]$$

$$\le \omega_N \eta^{\frac{N}{8}} \cdot 5N\eta^{\frac{1}{8}} \le const.\eta^{\frac{N+1}{8}}. \tag{3.156}$$

So, by (3.150)–(3.156)

$$|K_\varepsilon(t)| \le \int_{\Sigma_{\varepsilon,t}} |(x - q_\varepsilon) [u_\varepsilon^2 \nabla S_\varepsilon \cdot \nabla \rho_\varepsilon - \nabla \cdot J_{\rho_\varepsilon}]| \, dx$$

$$\le \sup_{x \in \Sigma_{\varepsilon,t}} |x - q_\varepsilon| \cdot \left[\int_{\Sigma_{\varepsilon,t}} |u_\varepsilon^2 \nabla S_\varepsilon \cdot \nabla \rho_\varepsilon| \, dx + \int_{\Sigma_{\varepsilon,t}} |(\nabla \cdot J_{\rho_\varepsilon})| \, dx \right]$$

$$\le \sup_{x \in \Sigma_{\varepsilon,t}} |x - q_\varepsilon| \cdot \left[\sup_{x \in \Sigma_{\varepsilon,t}} |\nabla \rho_\varepsilon| \cdot \int_{\Sigma_{\varepsilon,t}} |u_\varepsilon^2 \nabla S_\varepsilon| + \sup_{x \in \Sigma_{\varepsilon,t}} |\nabla \cdot J_{\rho_\varepsilon}| \cdot \int_{\Sigma_{\varepsilon,t}} dx \right]$$

$$\le 3\eta^{1/8} \left[const.\eta^{-1/4} \cdot \eta^{1/2} + const.\eta^{-1/4} \cdot |\Sigma_{\varepsilon,t}| \right]$$

$$\le const.\eta^{1/8} \left[\eta^{-1/4} \cdot \eta^{1/2} + \eta^{-1/4} \cdot \eta^{\frac{N+1}{8}} \right] \le const. \, \eta^{1/8}.$$

Then, by Lemma 72,

$$|K_\varepsilon(t)| \to 0 \qquad (3.157)$$

uniformly in t.

Now, let us estimate $|H_{1,\varepsilon}(t)|$; by (3.155) and (3.154) we have

$$\begin{aligned}
|H_{1,\varepsilon}(t)| &\leq \int_{\Sigma_{\varepsilon,t}} |u_\varepsilon^2 \nabla S_\varepsilon (\nabla \cdot J_{\rho_\varepsilon})| \, dx \\
&\leq \sup_{x \in \Sigma_{\varepsilon,t}} |\nabla \cdot J_{\rho_\varepsilon}| \cdot \int_{\Sigma_{\varepsilon,t}} |u_\varepsilon^2 \nabla S_\varepsilon| \\
&\leq const. \frac{1}{\eta^{1/4}} \cdot \eta^{\frac{1}{2}} = const. \eta^{1/4}.
\end{aligned}$$

By the above estimate,

$$|H_{1,\varepsilon}(t)| \to 0. \qquad (3.158)$$

We recall that

$$\int |\Psi_\varepsilon|^2 = 1 - o(1)$$

when $\varepsilon \to 0$, and that $\mathrm{supp}\Psi_\varepsilon \subset B(\hat{q}_\varepsilon, \, \eta^{1/8} + 2\eta^{1/4})$. We have

$$\nabla V(q_\varepsilon(t)) = (1 + o(1)) \int \nabla V(q_\varepsilon(t)) |\Psi_\varepsilon|^2$$

and so

$$\begin{aligned}
|H_2(t)| &= \left| \int \nabla V(x) |\Psi_\varepsilon|^2 \, dx - \nabla V(q_\varepsilon(t)) \right| \\
&= \int |\nabla V(x) - \nabla V(q_\varepsilon)| \, |\Psi_\varepsilon|^2 \, dx + o(1) \int \nabla V(q_\varepsilon(t)) |\Psi_\varepsilon|^2 \\
&\leq \|V''\|_{C^0(\mathbb{R}^N)} \int |x - q_\varepsilon| \, |\Psi_\varepsilon|^2 \, dx + o(1) \int \nabla V(q_\varepsilon(t)) |\Psi_\varepsilon|^2 \\
&\leq o(1) \left(\|V''\|_{C^0(\mathbb{R}^N)} + \|\nabla V\|_{C^0(\mathbb{R}^N)} \right) = o(1)
\end{aligned}$$

for all t. By the above Eqs. (3.157) and (3.158), (3.147) follows.

Let $\mathbf{P} = \mathbf{P}(\psi_\varepsilon)$ be defined by (3.127). By the definitions of p_ε, and (3.153), for every $t \in \mathbb{R}$, we have that

$$|p_\varepsilon(t) - \mathbf{P}(t)| = \left| \int \nabla S \left(|\Psi_\varepsilon|^2 - u_\varepsilon^2 \right) dx \right|$$

$$\leq \int_{\mathbb{R}^N \setminus B(\hat{q}_\varepsilon, \sqrt{\varepsilon})} |\nabla S|\, u_\varepsilon^2\, dx = o(1). \qquad (3.159)$$

By (3.126)

$$\dot{\mathbf{P}} = \int \left(-u_\varepsilon^2 \nabla V + \nabla \cdot T \right) dx$$

and since $T \in L^1(\mathbb{R}^N)$, $\dot{\mathbf{P}} = -\int u_\varepsilon^2 \nabla V\, dx$. So, by (3.134) and (3.148)

$$\dot{p}_\varepsilon - \dot{\mathbf{P}} = \int \nabla V \left(u_\varepsilon^2 - |\Psi_\varepsilon|^2 \right) dx - \int_{\Sigma_{\varepsilon,t}} \left[T \cdot \nabla \rho_\varepsilon + u_\varepsilon^2 \nabla S_\varepsilon \left(\nabla \cdot J_{\rho_\varepsilon} \right) \right] dx$$

$$= \int \nabla V \left(u_\varepsilon^2 - |\Psi_\varepsilon|^2 \right) dx + F_\varepsilon(q_\varepsilon) - \int_{\Sigma_{\varepsilon,t}} u_\varepsilon^2 \nabla S_\varepsilon \left(\nabla \cdot J_{\rho_\varepsilon} \right) dx.$$

Then, by (3.158) and Lemma 72,

$$\left| F_\varepsilon(q_\varepsilon) - \left(\dot{p}_\varepsilon - \dot{\mathbf{P}} \right) \right| = \left| \int_{\Sigma_{\varepsilon,t}} u_\varepsilon^2 \nabla S_\varepsilon \left(\nabla \cdot J_{\rho_\varepsilon} \right) dx - \int \nabla V \left(u_\varepsilon^2 - |\Psi_\varepsilon|^2 \right) dx \right|$$

$$\leq o(1) + \|\nabla V\|_{L^\infty} \int \left| u_\varepsilon^2 - |\Psi_\varepsilon|^2 \right|$$

$$\leq o(1) + \|\nabla V\|_{L^\infty} \int_{\mathbb{R}^N \setminus B(\hat{q}_\varepsilon, \sqrt{\varepsilon})} u_\varepsilon^2\, dx = c_1(\varepsilon)$$

where $c_1(\varepsilon) \to 0$ as $\varepsilon \to 0$. Finally by (3.159), $\forall \tau_0, \tau_1$,

$$\left| \int_{\tau_0}^{\tau_1} F_\varepsilon(q_\varepsilon)\, dt \right| = \left| \int_{\tau_0}^{\tau_1} \left(\dot{p}_\varepsilon - \dot{\mathbf{P}} \right) dt \right| + c_1(\varepsilon)(\tau_1 - \tau_0)$$

$$\leq |p_\varepsilon(\tau_1) - \mathbf{P}(\tau_1)| + |p_\varepsilon(\tau_0) - \mathbf{P}(\tau_0)| + c_1(\varepsilon)(\tau_1 - \tau_0)$$

$$\leq 2c_0(\varepsilon) + c_1(\varepsilon)(\tau_1 - \tau_0) \leq c(\varepsilon)(1 + |\tau_1 - \tau_0|)$$

with a suitable choice of $c(\varepsilon)$. $\qquad \square$

Collecting the previous results, we get our main theorem and Corollary 62:

Proof of Theorem 61. By the definitions (3.129) and (3.130), Lemma 72 and Theorems 74 and 61 holds with

$$R_\varepsilon = \eta^{\frac{1}{8}} \left(1 + 2\eta^{\frac{1}{8}} \right).$$

$\qquad \square$

Proof of Corollary 62. We rewrite (3.118), (3.120) and (3.121) in integral form and we get

$$\begin{cases} q_\varepsilon(t) = q_\varepsilon(0) + \int_0^t \frac{p_\varepsilon(s)}{m_\varepsilon(s)}\, ds + \int_0^t K_\varepsilon(s)ds \\ p_\varepsilon(t) = p_\varepsilon(0) - \int_0^t \nabla V(q_\varepsilon(s))ds + \int_0^t [F_\varepsilon(q_\varepsilon) + H_\varepsilon(s)]\, ds \end{cases} \tag{3.160}$$

$$\begin{cases} \mathfrak{q}(t) = q_\varepsilon(0) + \int_0^t \frac{\mathfrak{p}(s)}{m}\, ds \\ \mathfrak{p}(t) = p_\varepsilon(0) - \int_0^t \nabla V(\mathfrak{q}(s))ds \end{cases} \tag{3.161}$$

and hence, for any $|t| \leq T$

$$|q_\varepsilon(t) - \mathfrak{q}(t)| \leq \int_0^t \left| \frac{p_\varepsilon(s)}{m_\varepsilon(s)} - \frac{\mathfrak{p}(s)}{m} \right|\, ds + \int_0^t |K_\varepsilon(s)|\, ds$$

$$\leq L_1 \int_0^t |p_\varepsilon(s) - \mathfrak{p}(s)|\, ds + \alpha_1(\varepsilon)$$

where, by (3.147), $\alpha_1(\varepsilon) \to 0$ as $\varepsilon \to 0$ and

$$|p_\varepsilon(t) - \mathfrak{p}(t)| \leq \int_0^t |\nabla V(q_\varepsilon(s)) - \nabla V(\mathfrak{q}(s))|\, ds + \left| \int_0^t F_\varepsilon(q_\varepsilon)ds \right| + \int_0^t |H_\varepsilon(s)|\, ds$$

$$\leq L_2 \int_0^t |q_\varepsilon(s) - \mathfrak{q}(s)|\, ds + \alpha_2(\varepsilon)$$

where, by (3.147), $\alpha_2(\varepsilon) \to 0$ as $\varepsilon \to 0$. Then, setting $z_\varepsilon(t) = |q_\varepsilon(t) - \mathfrak{q}(t)| + |p_\varepsilon(t) - \mathfrak{p}(t)|$, we have

$$z_\varepsilon(t) \leq L \int_0^t z_\varepsilon(s)\, ds + \alpha(\varepsilon)$$

with a suitable choice of L and $\alpha(\varepsilon)$. Now, by the Gronwall inequality, we have

$$z_\varepsilon(t) \leq \alpha(\varepsilon)e^{Lt}$$

and from here, we get the conclusion. $\qquad\qquad\square$

Chapter 4
The Nonlinear Klein-Gordon Equation

This chapter deals with the Nonlinear Klein-Gordon Equation (NKG). After having analyzed the general features of NKG, we apply the abstract theory of Chap. 2 and we prove the existence of hylomorphic solitons. In the last part of this chapter, we show that some relativistic effects such as the space contraction, the time dilation, the Einstein equation, are consequences of the Poincarè invariance of NKG.

4.1 General Features of NKG

The D'Alambert equation,

$$\Box \psi = 0 \tag{4.1}$$

is the simplest equation invariant for the Poincaré group, moreover it is invariant for the "gauge" transformation

$$\psi \mapsto \psi + c, \ c \text{ constant.}$$

Also, if ψ is complex valued, it is invariant for the action (1.19). So clearly (4.1) satisfies assumptions **A-1–A-3** in the introduction. However (4.1) is linear and it does not produce solitary waves if the space dimension $N \geq 2$. In fact it is well known that, when $N \geq 2$, there exist only dispersive waves (see e.g. Theorem 1.15 of [93]). Let us add to (1.13) a nonlinear term:

$$\mathcal{L} = \frac{1}{2} |\partial_t \psi|^2 - \frac{1}{2} |\nabla \psi|^2 - W(\psi) \tag{4.2}$$

where

$$W : \mathbb{C} \to \mathbb{R}$$

© Springer International Publishing Switzerland 2014
V. Benci, D. Fortunato, *Variational Methods in Nonlinear Field Equations*,
Springer Monographs in Mathematics, DOI 10.1007/978-3-319-06914-2_4

satisfies the following assumption

$$W\left(e^{i\theta}\psi\right) = W\left(\psi\right) \text{ for all } \theta \in \mathbb{R}$$

namely $W\left(\psi\right) = F(|\psi|)$ for some function $F = \mathbb{R}^+ \to \mathbb{R}$.

The Lagrangian (4.2) is the "simplest" non-linear Lagrangian invariant for the Poincaré group and the trivial gauge action (1.19).

The equation of motion relative to the Lagrangian (4.2) is the following:

$$\Box\psi + W'\left(\psi\right) = 0 \qquad\qquad\qquad\text{(NKG)}$$

where

$$W'(\psi) = F'(|\psi|)\frac{\psi}{|\psi|}.$$

In the following sections we will see that Eq. (NKG), with suitable (but very general) assumptions on W, produces a very rich model in which there are solitary waves, solitons and vortices. Moreover we will see that these solitons behave as relativistic particles.

If $W'(\psi)$ is linear, namely $W'(\psi) = m^2\psi$, then Eq. (NKG) reduces to the Klein-Gordon equation

$$\Box\psi + m^2\psi = 0. \qquad\qquad\qquad\text{(4.3)}$$

Among the solutions of the Klein-Gordon equations there are the *wave packets* which behave as solitary waves but disperse in space as time goes on (see e.g. [93]). On the contrary, if W has a nonlinear suitable component, the wave packets do not disperse and give hylomorphic solitons.

Sometimes, it will be useful to write ψ in polar form, namely

$$\psi(t, x) = u(t, x)e^{iS(t,x)} \qquad\qquad\qquad\text{(4.4)}$$

where u, S are real functions.

In this case the Lagrangian is

$$\mathcal{L} = (\partial_t u)^2 - |\nabla u|^2 + \left[(\partial_t S)^2 - |\nabla S|^2\right]u^2 - W(u)$$

and the action takes the fom

$$S(u, S) = \frac{1}{2}\int (\partial_t u)^2 - |\nabla u|^2 + \left[(\partial_t S)^2 - |\nabla S|^2\right]u^2 dxdt - \int W(u)dxdt$$

and Eq. (NKG) becomes:

$$\Box u - \left[(\partial_t S)^2 - |\nabla S|^2 \right] u + W'(u) = 0 \tag{4.5}$$

$$\partial_t \left(u^2 \partial_t S \right) - \nabla \cdot \left(u^2 \nabla S \right) = 0. \tag{4.6}$$

4.1.1 Constants of Motion of NKG

For simplicity in this subsection we shall assume that the space dimension $N = 3$.

Since (4.2) is invariant under the Poincarè group, Eq. (NKG) has ten constants of motion

- **Energy.** We recall (see (1.38)) that the energy has the expression

$$E = \text{Re} \int \left(\frac{\partial \mathcal{L}}{\partial \psi_t} \overline{\partial_t \psi} - \mathcal{L} \right) dx.$$

Where $\bar{z}$ denotes the complex conjugate of z.

So, if we take the Lagrangian (4.2), we get

$$E = \int \left[\frac{1}{2} |\partial_t \psi|^2 + \frac{1}{2} |\nabla \psi|^2 + W(\psi) \right] dx. \tag{4.7}$$

Using (4.4) we get:

$$E = \int \left[\frac{1}{2} (\partial_t u)^2 + \frac{1}{2} |\nabla u|^2 + \frac{1}{2} \left[(\partial_t S)^2 + |\nabla S|^2 \right] u^2 + W(u) \right] dx. \tag{4.8}$$

- **Momentum.** We recall (see (1.40)) that the momentum along the x_i direction has the following expression

$$P_i = -\text{Re} \int \frac{\partial \mathcal{L}}{\partial \psi_t} \overline{\partial_i \psi} \, dx.$$

In particular, if we take the Lagrangian (4.2), we get

$$P_i = -\text{Re} \int \partial_t \psi \overline{\partial \psi_i} \, dx$$

and, setting $\mathbf{P} = (P_1, P_2, P_3)$, we can write

$$\mathbf{P} = -\text{Re} \int \partial_t \psi \overline{\nabla \psi} \, dx. \tag{4.9}$$

Using (4.4) we get:

$$\mathbf{P} = - \int \left(\partial_t u \, \nabla u + \partial_t S \, \nabla S \, u^2 \right) \, dx. \tag{4.10}$$

- **Angular momentum.** We recall (see (1.41)) that the angular momentum has the following expression

$$\mathbf{M} = \mathrm{Re} \int \frac{\partial \mathcal{L}}{\partial \psi_t} \overline{(\mathbf{x} \times \nabla \psi)} \, dx.$$

In particular, if we take the Lagrangian (4.2), we get

$$\mathbf{M} = \mathrm{Re} \int \mathbf{x} \times \nabla \psi \overline{\partial_t \psi} \, dx. \tag{4.11}$$

Using (4.4) we get:

$$\mathbf{M} = \int \left(\mathbf{x} \times \nabla S \partial_t S u^2 + \mathbf{x} \times \nabla u \, \partial_t u \right) \, dx. \tag{4.12}$$

- **Ergocenter velocity.** If we take the Lagrangian (4.2), the constant of motion related to the invariance with respect to the Lorentz boosts (1.10), is the following

$$\mathbf{K} = \int \mathbf{x} \left[\frac{1}{2} |\partial_t \psi|^2 + \frac{1}{2} |\nabla \psi|^2 + W(\psi) \right] dx - t\mathbf{P}. \tag{4.13}$$

First let us interpret it in a more meaningful way. If we derive the terms of the above equation with respect to t and since $\frac{d\mathbf{K}}{dt} = 0$, we get

$$\mathbf{P} = \frac{d}{dt} \int \mathbf{x} \left[\frac{1}{2} |\partial_t \psi|^2 + \frac{1}{2} |\nabla \psi|^2 + W(\psi) \right] dx. \tag{4.14}$$

Now, we define the *ergocenter* as follows

$$\mathbf{Q} := \frac{\int \mathbf{x} \left[\frac{1}{2} |\partial_t \psi|^2 + \frac{1}{2} |\nabla \psi|^2 + W(\psi) \right] dx}{\int \left[\frac{1}{2} |\partial_t \psi|^2 + \frac{1}{2} |\nabla \psi|^2 + W(\psi) \right] dx} = \frac{\int \mathbf{x} \left[\frac{1}{2} |\partial_t \psi|^2 + \frac{1}{2} |\nabla \psi|^2 + W(\psi) \right] dx}{E}; \tag{4.15}$$

then, by the conservation of E and Eq. (4.14), we get

$$\dot{\mathbf{Q}} = \frac{\mathbf{P}}{E}. \tag{4.16}$$

So, since also $\mathbf{P}$ is constant, we get that $\dot{\mathbf{Q}}$ is constant.

Concluding, the Poincaré group provides ten independent constants of motion which are E, $\mathbf{P}$, $\mathbf{M}$, $\mathbf{K}$; they can be replaced by the constants of motion E, $\mathbf{P}$, $\mathbf{M}$, $\dot{\mathbf{Q}}$ since also these quantities are independent.

Now we prove (4.13).

Proof. Let us compute K_i using Theorem 5; in this case the parameter λ is the velocity v_i which appears in (1.10); we have

$$\rho_{K_i} = \operatorname{Re}\left(\frac{\partial \mathcal{L}}{\partial \psi_t}\frac{\overline{\partial \psi}}{\partial v_i}\right) - \mathcal{L}\frac{\partial t}{\partial v_i}$$

$$= \operatorname{Re}\left(\partial_t \psi \overline{\left[\frac{\partial \psi}{\partial t}\frac{\partial t}{\partial v_i} - \sum_{k=1}^{3}\frac{\partial \psi}{\partial x_k}\frac{\partial x_k}{\partial v_i}\right]}\right) - \left(\frac{1}{2}|\partial_t \psi|^2 - \frac{1}{2}|\nabla \psi|^2 - W(\psi)\right)\frac{\partial t}{\partial v_i}$$

where the derivative with respect to v_i needs to be computed for $v_i = 0$.

Since for $k \neq i$, $\frac{\partial x_k}{\partial v_i} = 0$, we have that

$$\rho_{K_i} = |\partial_t \psi|^2\frac{\partial t}{\partial v_i} - \operatorname{Re}(\partial_t \psi \partial_{x_i}\psi)\cdot\frac{\partial x_i}{\partial v_i} - \left(\frac{1}{2}|\partial_t \psi|^2 - \frac{1}{2}|\nabla \psi|^2 - W(\psi)\right)\frac{\partial t}{\partial v_i}$$

$$= \left(\frac{1}{2}|\partial_t \psi|^2 + \frac{1}{2}|\nabla \psi|^2 + W(\psi)\right)\frac{\partial t}{\partial v_i} - \operatorname{Re}(\partial_t \psi \partial_{x_i}\psi)\cdot\frac{\partial x_i}{\partial v_i}$$

$$= \rho_E\frac{\partial t}{\partial v_i} - \rho_{P_i}\cdot\frac{\partial x_i}{\partial v_i}.$$

Also we have

$$\left(\frac{\partial t}{\partial v_i}\right)_{v_i=0} = \left(\frac{\partial}{\partial v_i}\frac{t - v_i x}{\sqrt{1 - v_i^2}}\right)_{v_i=0} = x$$

$$\left(\frac{\partial x}{\partial v_i}\right)_{v_i=0} = \left(\frac{\partial}{\partial v_i}\frac{x - v_i t}{\sqrt{1 - v_i^2}}\right)_{v_i=0} = t$$

so that

$$\rho_{K_i} = \rho_E x_i - \rho_{P_i} t.$$

$\square$

Finally, we have another constant of the motion relative to the action (1.19).

- **Hylenic Charge.** The charge, by definition, is the quantity which is preserved by the gauge action (1.19). The charge has the following expression (see (1.42))

$$C = \text{Im} \int \frac{\partial \mathcal{L}}{\partial \psi_t} \overline{\psi} \, dx.$$

In particular, if we take the Lagrangian (4.2), we get

$$C = \text{Im} \int \partial_t \psi \overline{\psi} \, dx.$$

Using (4.4) we get:

$$C = \int \partial_t S \, u^2 dx. \tag{4.17}$$

Proof. It is not necessary to make any computation since the conservation of (4.17) is a direct consequence of the continuity equation (4.6) and Lemma 6.   □

4.1.2   Swarm Interpretation of NKG

Before giving the swarm interpretation to Eq. (NKG), we will write it with the usual physical constants c, m and $\hbar$:

$$\frac{\partial^2 \psi}{\partial t^2} - c^2 \Delta \psi + W'(\psi) = 0$$

with

$$W(u) = \frac{m^2 c^4}{2\hbar^2} u^2 + N(u).$$

Here c has the dimension of a velocity (and it represents the speed of light), m has the dimension of *mass* and $\hbar$ is the Planck constant.

The polar form of ψ is written as follows

$$\psi(t, x) = u(t, x)e^{iS(t,x)/\hbar} \tag{4.18}$$

and Eqs. (4.5) and (4.6) become

$$\hbar^2 \left(\partial_t^2 u - c^2 \Delta u + N'(u) \right) + \left(-(\partial_t S)^2 + c^2 |\nabla S|^2 + m^2 c^4 \right) u = 0 \tag{4.19}$$

$$\partial_t \left(u^2 \partial_t S \right) - c^2 \nabla \cdot \left(u^2 \nabla S \right) = 0. \tag{4.20}$$

The continuity equation (2.79) for (NKG) is given by (4.20). This equation allows us to interprete the matter field to be a fluid composed by particles whose density is given by

$$\rho_C = u^2 \partial_t S$$

and which move in the velocity field

$$\mathbf{v} = -\frac{c^2 \nabla S}{\partial_t S}. \tag{4.21}$$

If

$$\hbar^2 \left(\partial_t^2 u - c^2 \Delta u + N'(u) \right) \ll \left(-(\partial_t S)^2 + c^2 |\nabla S|^2 + m^2 c^4 \right) u, \tag{4.22}$$

namely, if $\hbar$ is very small with respect to the other quantities involved, Eq. (4.19) can be approximated by

$$(\partial_t S)^2 = c^2 |\nabla S|^2 + m^2 c^4 \tag{4.23}$$

or

$$\partial_t S + \sqrt{m^2 c^4 + c^2 |\nabla S|^2} = 0. \tag{4.24}$$

This is the Hamilton-Jacobi equation of a free relativistic particle of mass m (cf. Eq. (1.59)) whose trajectory $q(t)$, by (4.21) satisfies the equation

$$\dot{q} = -\frac{c^2 \nabla S}{\partial_t S}. \tag{4.25}$$

If we do not assume (4.22), Eq. (4.24) needs to be replaced by

$$\partial_t S + \sqrt{m^2 c^4 + c^2 |\nabla S|^2 + \frac{Q(u)}{u}} = 0 \tag{4.26}$$

with

$$Q(u) = \hbar^2 \left(\partial_t^2 u - c^2 \Delta u + N'(u) \right).$$

The term $\frac{Q(u)}{u}$ can be regarded as a field describing a sort of interaction between particles.

Given a wave of the form (4.18), the local frequency and the local wave number are defined as follows:

$$\omega(t, x) = -\frac{\partial_t S(t, x)}{\hbar}$$

$$\mathbf{k}(t, x) = \frac{\nabla S(t, x)}{\hbar};$$

the energy of each particle moving according to (4.24), is given by (see (1.49))

$$E = -\partial_t S$$

and its momentum is given by

$$\mathbf{p} = \nabla S;$$

thus we have that

$$E = \hbar \omega$$

$$\mathbf{p} = \hbar \mathbf{k};$$

these two equations are the De Broglie relations.

Thus, Eq. (4.21) becomes

$$\mathbf{v} = c^2 \frac{\mathbf{k}}{\omega} = c^2 \frac{\mathbf{p}}{E}. \tag{4.27}$$

4.2   Existence Results for NKG

4.2.1   Existence of Solitary Waves

The easiest way to produce solitary waves of (NKG) consists in solving the static equation

$$-\Delta \psi + W'(\psi) = 0, \ \psi = \psi(x) \tag{4.28}$$

and setting

$$\psi_v(t, x) = \psi_v(t, x_1, \ldots, x_N) = \psi \left(\frac{x_1 - vt}{\sqrt{1 - v^2}}, x_2, \ldots, x_N \right). \tag{4.29}$$

By the Lorentz invariance of (NKG), $\psi_v(t, x)$ is still a solution of Eq. (NKG) which represents a bump which travels in the x_1-direction with speed v.

Thus by Theorem 144, we obtain the following result:

Theorem 75. *Let $N \geq 3$ and assume that W satisfies (G-i)–(G-iii) of Theorem 144. Then Eq. (NKG) has real valued solitary waves of the form (4.29).*

(G-iii) implies that W is not positive. However, it would be interesting to assume

$$W \geq 0. \qquad (4.30)$$

In fact the energy of a solution of Eq. (NKG)

$$E(\psi) = \int \left[\frac{1}{2} \left| \frac{\partial \psi}{\partial t} \right|^2 + \frac{1}{2} |\nabla \psi|^2 + W(\psi) \right] dx \qquad (4.31)$$

is positive if W is positive. The positivity of the energy is not only an important request for the physical models related to this equation, but it provides good a priori estimates for the solutions of the relative Cauchy problem. These estimates allow to prove the existence and the well-posedness results under very general assumptions on W.

Unfortunately Derrick [72], in a very well known paper, has proved that request (4.30) implies that Eq. (4.28), for $N \geq 3$, has only the trivial solution. His proof is based on an identity, which in a different form was also found by Pohozaev. The Derrick-Pohozaev identity (see Theorem 143 and in particular Eq. (A.3)) states that for any finite energy solution ψ of Eq. (4.28) it holds

$$\left(\frac{1}{2} - \frac{1}{N} \right) \int |\nabla \psi|^2 \, dx + \int W(\psi) dx = 0. \qquad (4.32)$$

Clearly the above equality and (4.30) imply that $\psi \equiv 0$.

However, we can try to prove the existence of solitary waves and solitons of (NKG) (with assumption (4.30)) exploiting the possible existence of *standing waves* (as defined by (3.30)), since this fact is not prevented by (4.32). Substituting

$$\psi(t, x) = u(x)e^{-i\omega_0 t}, u \geq 0$$

in (NKG), we get

$$-\Delta u + W'(u) = \omega_0^2 u. \qquad (4.33)$$

Since the Lagrangian (4.2) is invariant for the Lorentz group (see (1.10)), we can obtain other solutions $\psi_v(t, x)$ just making a Lorentz transformation on it. Namely, if $N = 3$, and if we take the velocity $\mathbf{v} = (v, 0, 0)$, $|v| < 1$, and set

$$t' = \gamma (t - vx_1), \ x_1' = \gamma (x_1 - vt), \ x_2' = x_2, \ x_3' = x_3 \quad \text{with} \quad \gamma = \frac{1}{\sqrt{1 - v^2}}$$

it turns out that also

$$\psi_\mathbf{v}(t, x) = \psi(t', x')$$

is a solution of (NKG).

So, given a standing wave $\psi(t, x) = u(x)e^{-i\omega_0 t}$, the function $\psi_\mathbf{v}(t, x) := \psi(t', x')$ is a solitary wave which travels with velocity $\mathbf{v}$. Thus, if $u(x) = u(x_1, x_2, x_3)$ is any solution of Eq. (4.33), then

$$\psi_\mathbf{v}(t, x_1, x_2, x_3) = u\left(\gamma\left(x_1 - vt\right), x_2, x_3\right) e^{i(\mathbf{k}\cdot\mathbf{x} - \omega t)}, \tag{4.34}$$

is a solution of Eq. (NKG) provided that

$$\omega = \gamma\omega_0 \text{ and } \mathbf{k} = \gamma\omega_0 \mathbf{v}. \tag{4.35}$$

Notice that (4.29) is a particular case of (4.34) when $\omega_0 = 0$.

As for the nonlinear Schrödinger equation (see Sect. 3.2.1), we confine ourselves to prove, under quite general assumptions on W, the existence of solitary waves only in the case $N \geq 3$. In the next subsection, where the existence of solitons will be studied, the case $N = 1, 2$ will be also considered (see Theorem 78).

Now, we make the following assumptions on W:

(WC-0)   **(Positivity)**

$$W(s) \geq 0. \tag{4.36}$$

(WC-i)   **(Nondegeneracy)** W is a C^2 function s.t. $W(0) = W'(0) = 0$ and

$$W''(0) = m^2 > 0. \tag{4.37}$$

(WC-ii)   **(Hylomorphy)** If we set

$$W(s) = \frac{1}{2}m^2 s^2 + N(s) \tag{4.38}$$

then

$$\exists s_0 \in \mathbb{R}^+ \text{ such that } N(s_0) < 0. \tag{4.39}$$

(WC-iii)   There exist $c > 0, 2 < p < 2^* = \frac{2N}{N-2}$ $(N \geq 3)$ s.t.

$$N'(s) \geq -cs^{p-1} \text{ for } s \geq 1. \tag{4.40}$$

Here are some remarks on these assumptions:

(WC-0)   This assumption implies that the energy is positive (see (4.31)); if (WC-0) does not hold, it is possible to have solitary waves, but not hylomorphic solitons (cf. Theorem 75).

(WC-i)   In order to have solitary waves it is necessary to have $W''(0) \geq 0$. There are some results also when $W''(0) = 0$ (null-mass case, see e.g. [46] and [11]), however the most interesting situations occur when $W''(0) > 0$.

(WC-ii)   Is the crucial assumption which characterizes the nonlinearity which might produce hylomorphic solitons and corresponds to (WA-ii) of page 69 and to (WB-ii) of page 71.

The hylomorphy condition (WC-ii) can also be written as follows:

$$\alpha_0 := \inf_{s \in \mathbb{R}^+} \frac{W(s)}{\frac{1}{2}|s|^2} < m^2. \tag{4.41}$$

(WC-iii)   Since $W \geq 0$, (WC-iii) is a mild technical assumption.

Theorem 76. *Let $N \geq 3$ and assume that (WC-0)–(WC-iii) hold. Then Eq. (NKG) has solitary waves of the form $\psi(t, x) = u(x)e^{-i\omega t}$ for every frequency $\omega \in (m_0, m)$ where*

$$m_0 = \inf \left\{ a \in \mathbb{R} : \exists u \in \mathbb{R}^+, \frac{1}{2}a^2 u^2 > W(u) \right\}.$$

Notice that by (WC-ii) $m_0 < m$.

Proof. By the previous discussion, it is sufficient to show that Eq. (4.33) has a solution u with finite energy. The solutions of finite energy of (4.33) are the critical points in the Sobolev space $H^1(\mathbb{R}^N)$ of the *reduced action* functional:

$$J(u) = \frac{1}{2}\int |\nabla u|^2 \, dx + \int G(u) dx, \tag{4.42}$$

where

$$G(s) = W(s) - \frac{1}{2}\omega_0^2 s^2 = \frac{1}{2}\left(m^2 - \omega_0^2\right)s^2 + N(s). \tag{4.43}$$

Now we want to apply Theorem 144. It is easy to check that for every frequency $\omega_0 \in (m_0, m)$, G defined in (4.43) satisfies assumptions (G-i) and (G-iii). Moreover it can be shown that G satisfies also assumption (G-ii) by following the same arguments used in proving Theorem 50. $\qquad\square$

4.2.2 Existence of Solitons

We set

$$X = H^1(\mathbb{R}^N, \mathbb{C}) \times L^2(\mathbb{R}^N, \mathbb{C})$$

and we will denote the generic element of X by $\mathbf{u} = (\psi(x), \hat{\psi}(x))$. We shall assume that the initial value problem for (NKG) is globally well posed. Then, for every $\mathbf{u} \in X$, there is a unique, global solution $\psi(t, x)$ of (NKG) such that

$$\psi(0, x) = \psi(x) \tag{4.44}$$

$$\partial_t \psi(0, x) = \hat{\psi}(x).$$

Using this notation, we can write Eq. (NKG) in Hamiltonian form:

$$\partial_t \psi = \hat{\psi} \tag{4.45}$$

$$\partial_t \hat{\psi} = \Delta \psi - W'(\psi). \tag{4.46}$$

The time evolution map $\gamma : \mathbb{R} \times X \to X$ is defined by

$$\gamma_t \mathbf{u}_0(x) = \mathbf{u}(t, x) \tag{4.47}$$

where $\mathbf{u}_0(x) = (\psi(x), \hat{\psi}(x)) \in X$ and $\mathbf{u}(t, x) = (\psi(t, x), \hat{\psi}(t, x))$ is the unique solution of (4.45) and (4.46) satisfying the initial conditions (4.44). The energy and the charge, as functionals defined in X, become

$$E(\mathbf{u}) = \int \left[\frac{1}{2} \left| \hat{\psi} \right|^2 + \frac{1}{2} |\nabla \psi|^2 + W(\psi) \right] dx \tag{4.48}$$

$$C(\mathbf{u}) = -\operatorname{Re} \int i \hat{\psi} \bar{\psi} \, dx. \tag{4.49}$$

Let

$$\Lambda = \frac{E(\mathbf{u})}{|C(\mathbf{u})|}$$

denote the hylenic ratio. We take the following norm in X. If $\mathbf{u} = (\psi, \hat{\psi})$, we set

$$\|\mathbf{u}\|^2 = \int \left[\left| \hat{\psi} \right|^2 + |\nabla \psi|^2 + m^2 |\psi|^2 \right] dx. \tag{4.50}$$

In order to prove the existence of solitons we need the following growth assumption on the function N defined by (4.38)

- (WC-iii′) There exist $q, r \in (2, 2^*)$, $2^* = \frac{2N}{N-2}$, $N \geq 3$ such that

$$|N'(s)| \leq c_1 s^{r-1} + c_2 s^{q-1}. \tag{4.51}$$

The existence of stable solutions for (NKG) has been largely investigate [10, 29, 47, 84, 85, 98, 135, 136]. In particular in [14] the existence of hylomorphic solitons for (NKG) has been proved. Here we prove a variant of this result.

The following theorem holds:

Theorem 77. *Assume $N \geq 3$ and that W satisfies (WC-0), (WC-i), (WC-ii) of page 120 and (WC-iii'). Then there exists $\delta_\infty > 0$ such that the dynamical system described by Eq. (NKG) has a family $\mathbf{u}_\delta$ ($\delta \in (0, \delta_\infty)$) of hylomorphic solitons (Definition 20) and $\delta_1 < \delta_2$ implies that*

(a) $E(\mathbf{u}_{\delta_1}) > E(\mathbf{u}_{\delta_2})$.
(b) $\Lambda(\mathbf{u}_{\delta_1}) < \Lambda(\mathbf{u}_{\delta_2})$.
(c) $|C(\mathbf{u}_{\delta_1})| > |C(\mathbf{u}_{\delta_2})|$.

In the case $N = 1, 2$ we have slightly different assumptions:

Theorem 78. *Assume $N = 1, 2$ and that W satisfies (WC-0)–(WC-ii) of page 120 and the following:*

$$\exists a > 0 : \ W(s) \geq a|s|^2 \tag{4.52}$$

$$\exists q, r \in (2, +\infty), \ |N'(s)| \leq c_1 s^{r-1} + c_2 s^{q-1}.$$

Then the same conclusions of Theorem 77 hold.

We recall that, following Coleman [60], the solitons in Theorems 77 and 78 are called Q-balls.

4.2.3   Coercivity

The proof is based on the abstract Theorem 34. First of all observe that E and C are invariant under translations i.e. under the representation T_z of the group $G = \mathbb{R}^N$

$$T_z \mathbf{u}(x) = \mathbf{u}(x + z), \quad z \in \mathbb{R}^N.$$

We need to show that assumptions (WC-0)–(WC-iii') imply that E and C satisfy assumptions (EC-0)–(EC-2) and (EC-3), the hylomorphy condition (2.16) and assumption (2.25) of Theorem 34.

Lemma 79. *Let the assumptions of Theorem 77 be satisfied, then E defined by (4.48) satisfies the coercitivity assumption (EC-3).*

Proof. Assumption (EC-3)(i) is clearly satisfied. Now we show that also (EC-3)(ii) and (EC-3)(iii) are satisfied. To this end we consider a sequence $\mathbf{u}_n = \left(\psi_n, \hat{\psi}_n \right)$ in X such that

$$E(\mathbf{u}_n) \to 0 \text{ (respectively } E(\mathbf{u}_n) \text{ bounded)} \tag{4.53}$$

and we shall prove that $\|\mathbf{u}_n\| \to 0$ (respectively $\|\mathbf{u}_n\|$ bounded), where $\|\cdot\|$ is defined in (4.50).

Since $W \geq 0$ and $E(\mathbf{u}_n) \to 0$ (respectively $E(\mathbf{u}_n)$ bounded) and comparing (4.48) with (4.50), in order to show that $\|\mathbf{u}_n\| \to 0$ (respectively $\|\mathbf{u}_n\|$ bounded) we have only to prove that

$$\|\psi_n\|_{L^2} \to 0 \text{ (respectively } \|\psi_n\|_{L^2} \text{ bounded)}. \tag{4.54}$$

Observe that, by using again $E(\mathbf{u}_n) \to 0$ (respectively $E(\mathbf{u}_n)$ bounded), we have that

$$\int W(\psi_n) \text{ and } \int |\nabla \psi_n|^2 \text{ converge to 0 (respectively bounded)}. \tag{4.55}$$

By (4.55) we have that

$$\int |\psi_n|^{2^*} \to 0 \text{ (respectively bounded)}. \tag{4.56}$$

Let $\varepsilon > 0$ and set

$$\Omega_n = \{x \in \mathbb{R}^N : |\psi_n(x)| > \varepsilon\} \text{ and } \Omega_n^c = \mathbb{R}^3 \backslash \Omega_n.$$

By (4.55) and since $W \geq 0$, we have

$$\int_{\Omega_n^c} W(u_n) \to 0 \text{ (respectively bounded)}. \tag{4.57}$$

By (4.38) we have

$$W(s) = \frac{m}{2} s^2 + o(s^2).$$

Then, if ε is small enough, there is a constant $c > 0$ such that

$$\int_{\Omega_n^c} W(u_n) \geq c \int_{\Omega_n^c} |\psi_n|^2. \tag{4.58}$$

From (4.57) and (4.58) we get

$$\int_{\Omega_n^c} |\psi_n|^2 \to 0 \text{ (respectively bounded)}. \tag{4.59}$$

By (4.56) we get that

$$\int_{\Omega_n} |\psi_n|^{2^*} \to 0 \text{ (respectively bounded)}. \tag{4.60}$$

On the other hand

$$\int_{\Omega_n} |\psi_n|^2 \leq \left(\int_{\Omega_n} |\psi_n|^{2^*} \right)^{\frac{N-2}{N}} \cdot m(\Omega_n)^{\frac{2}{N}} \tag{4.61}$$

where $m(\Omega_n)$ denotes the Lebesgue measure of Ω_n.

By (4.56) we have that

$$m(\Omega_n) \text{ is bounded.} \tag{4.62}$$

By (4.60)–(4.62) we get that

$$\int_{\Omega_n} |\psi_n|^2 \to 0 \text{ (respectively bounded).} \tag{4.63}$$

So (4.54) follows from (4.63) and (4.59). □

For dimensions $N = 1, 2$, we have the following:

Lemma 80. *Let the assumptions of Theorem 78 be satisfied, then E defined by (4.48) satisfies the coercitivity assumption (EC-3).*

Proof. The proof is the same than the proof of Lemma 79 except the proof of (4.54). In this case by assumption (4.52) we have that

$$E(\mathbf{u}_n) \geq \int W(\psi_n) \geq a \int |\psi_n|^2. \tag{4.64}$$

Clearly (4.54) follows from (4.53) and (4.64). □

4.2.4 Analysis of the Hylenic Ratio for NKG

Now we prove that also the hylomorphy condition (2.16) is satisfied. First of all, for $\mathbf{u} = (\psi, \hat{\psi}) \in X$, we set:

$$\|\mathbf{u}\|_\sharp = \max \left(\|\psi\|_{L^r}, \|\psi\|_{L^q} \right)$$

where r, q are introduced in (4.51). With some abuse of notation we shall write $\max \left(\|\psi\|_{L^r}, \|\psi\|_{L^q} \right) = \|\psi\|_\sharp$.

Lemma 81. *The seminorm $\|\mathbf{u}\|_\sharp$ satisfies the property (2.19), namely if $\mathbf{u}_n$ is a vanishing sequence (see Definition 28) then $\|\psi_n\|_\sharp \to 0$.*

Proof. Let $\{\psi_n\} \subset H^1(\mathbb{R}^N, \mathbb{C})$ be a vanishing sequence. May be taking a subsequence, we have that at least one of the following holds:

(i) $\|\psi_n\|_\sharp = \|\psi_n\|_{L^r}$.
(ii) $\|\psi_n\|_\sharp = \|\psi_n\|_{L^q}$.

Suppose that (i) holds. Then, we argue as in Lemma 55 and show that $\|\psi_n\|_{L^r} \to 0$. If (ii) holds, we argue in the same way replacing r with q. $\square$

Now we set

$$\Lambda_0 := \inf\{\liminf \Lambda(\mathbf{u}_n) \mid \mathbf{u}_n \text{ is a vanishing sequence}\}$$

$$\Lambda_\sharp = \lim_{\|\mathbf{u}\|_\sharp \to 0} \inf \Lambda(\mathbf{u}) = \lim_{\varepsilon \to 0} \inf\{\Lambda(\psi, \hat{\psi}) \mid \hat{\psi} \in L^2; \psi \in H^1; \|\psi\|_\sharp < \varepsilon\}.$$

By Lemma 81 and by the definitions of Λ_0 and $\Lambda_\sharp$, we have that

$$\Lambda_0 \geq \Lambda_\sharp. \tag{4.65}$$

Now we evaluate $\Lambda_\sharp$.

Lemma 82. *If N, defined by (4.38), satisfies assumption (4.51), then the following inequality holds*

$$\Lambda_\sharp \geq m.$$

Proof. Clearly we have

$$\Lambda_\sharp = \lim_{\varepsilon \to 0} \inf\{\Lambda(\varepsilon\psi, \hat{\psi}) \mid \hat{\psi} \in L^2, \psi \in H^1, \|\psi\|_\sharp = 1\}. \tag{4.66}$$

So we need to analyse $\Lambda(\varepsilon\psi, \hat{\psi})$ for $\|\psi\|_\sharp = 1$ and $\varepsilon > 0$.
First of all observe that by (4.51) we have

$$\left| \int N(|\psi|) dx \right| \leq k_1 \int |\psi|^r + k_2 \int |\psi|^q$$

$$\leq k_1 \|\psi\|_\sharp^r + k_2 \|\psi\|_\sharp^q.$$

So, if we take $\|\psi\|_\sharp = 1$ and $\varepsilon > 0$, we get

$$\left| \int N(|\varepsilon\psi|) dx \right| \leq k_1 \varepsilon^r + k_2 \varepsilon^q. \tag{4.67}$$

By the Sobolev embeddings, there is $k_3 > 0$ such that

$$\int \left(|\nabla\psi|^2 + m^2 |\psi|^2 \right) dx \geq k_3 \|\psi\|_\sharp^2. \tag{4.68}$$

Now, choose

$$2 < s < \min(r, q).$$

Since $r, q > s$, we have, by (4.68), (4.67) and taking $\varepsilon > 0$ small enough, that

$$\varepsilon^s \int \left(|\nabla \psi|^2 + m^2 |\psi|^2 \right) dx - \left| \int N(|\varepsilon \psi|) dx \right|$$

$$\geq \varepsilon^s k_3 \|\psi\|_\sharp^2 - k_1 \varepsilon^r - k_2 \varepsilon^q = k_3 \varepsilon^s - k_1 \varepsilon^r - k_2 \varepsilon^q \geq 0.$$

So

$$\left| \int N(\varepsilon |\psi|) dx \right| \leq \varepsilon^s \int \left(|\nabla \psi|^2 + m^2 |\psi|^2 \right) dx. \tag{4.69}$$

Using (4.69), for any $\psi \in H^1$ with $\|\psi\|_\sharp = 1$, $\hat{\psi} \in L^2$ and $\varepsilon > 0$ small enough, we have:

$$\Lambda(\varepsilon \psi, \hat{\psi}) = \frac{\frac{1}{2} \int \left(\left| \hat{\psi} \right|^2 + |\nabla \varepsilon \psi|^2 + m^2 |\varepsilon \psi|^2 \right) dx + \int N(|\varepsilon \psi|) dx}{\left| \mathrm{Re} \int i \hat{\psi} \varepsilon \overline{\psi} \, dx \right|}$$

$$\geq \frac{\frac{1}{2} \int \left| \hat{\psi} \right|^2 + \left(\frac{\varepsilon^2}{2} - \varepsilon^s \right) \int \left(|\nabla \psi|^2 + m^2 |\psi|^2 \right)}{\varepsilon \left| \mathrm{Re} \int i \hat{\psi} \overline{\psi} \, dx \right|}$$

$$\geq \frac{\frac{1}{2} \int \left| \hat{\psi} \right|^2 + \frac{\varepsilon^2}{2} \left(1 - 2\varepsilon^{s-2} \right) m^2 \int |\psi|^2}{\varepsilon \left(\int \left| \hat{\psi} \right|^2 dx \right)^{1/2} \left(\int |\psi|^2 dx \right)^{1/2}}$$

$$\geq \frac{\left(\int \left| \hat{\psi} \right|^2 dx \right)^{1/2} \cdot \varepsilon m \sqrt{1 - 2\varepsilon^{s-2}} \left(\int |\psi|^2 dx \right)^{1/2}}{\varepsilon \left(\int \left| \hat{\psi} \right|^2 dx \right)^{1/2} \left(\int |\psi|^2 dx \right)^{1/2}} = m \sqrt{1 - 2\varepsilon^{s-2}}.$$

Then

$$\Lambda_\sharp = \lim_{\varepsilon \to 0} \inf \left\{ \Lambda(\varepsilon \psi, \hat{\psi}) \mid \hat{\psi} \in L^2, \psi \in H^1, \|\psi\|_\sharp = 1 \right\} \geq m. \tag{4.70}$$

$\square$

Next we will show that the hylomorphy assumption (2.16) is satisfied.

Lemma 83. *Assume that W satisfies the assumptions of Theorem 77, then*

$$\inf_{\mathbf{u} \in X} \Lambda(\mathbf{u}) < \Lambda_0.$$

Proof. Let $R > 0$; set

$$u_R = \begin{cases} s_0 & if \ |x| < R \\ 0 & if \ |x| > R + 1 \\ \frac{|x|}{R} s_0 - (|x| - R)\frac{R+1}{R} s_0 \ if \ R < |x| < R + 1 \end{cases} . \tag{4.71}$$

By the hylomorphy assumption (WC-ii) there exists $0 < \beta < m$ such that

$$W(s_0) \le \frac{\beta^2 s_0^2}{2}. \tag{4.72}$$

We set $\psi = u_R$, and $\hat{\psi} = \beta u_R$.

Clearly we have:

$$\inf_{u \in X} \Lambda(\mathbf{u}) = \inf_{\psi, \hat{\psi}} \frac{\int \left(\frac{1}{2} \left| \hat{\psi} \right|^2 + \frac{1}{2} |\nabla \psi|^2 + W(\psi) \right) dx}{\left| \mathrm{Re} \int i \hat{\psi} \overline{\psi} \, dx \right|} \le \Lambda(u_R, \beta u_R)$$

$$\le \frac{\int \left(\frac{1}{2} \beta^2 |u_R|^2 + \frac{1}{2} |\nabla u_R|^2 + W(u_R) \right) dx}{\beta \int |u_R|^2 \, dx}$$

$$\le \frac{\int_{|x|<R} \left(\frac{1}{2} \beta^2 |u_R|^2 + W(u_R) \right) dx}{\beta \int_{|x|<R} |u_R|^2 \, dx}$$

$$+ \frac{\int_{R<|x|<R+1} \left(\frac{1}{2} \beta^2 |u_R|^2 + \frac{1}{2} |\nabla u_R|^2 + W(u_R) \right) dx}{\beta \int_{|x|<R} |u_R|^2 \, dx}$$

$$= \frac{1}{2} \beta + \frac{\int_{|x|<R} W(s_0) dx}{\beta \int_{|x|<R} |s_0|^2 \, dx} + O\left(\frac{1}{R} \right).$$

Then, by (4.72), we have

$$\inf_{u \in X} \Lambda(\mathbf{u}) \le \frac{1}{2} \beta + \frac{\int_{|x|<R} \frac{1}{2} s_0^2 \beta^2}{\beta \int_{|x|<R} |s_0|^2 \, dx} + O\left(\frac{1}{R} \right) = \beta + O\left(\frac{1}{R} \right).$$

So

$$\inf_{u \in X} \Lambda(\mathbf{u}) \le \beta. \tag{4.73}$$

Then, since $\beta < m$, by (4.65), Lemma 82 and (4.73), we have that

$$\Lambda_0 \ge \Lambda_\sharp \ge m > \beta \ge \inf_{u \in X} \Lambda(\mathbf{u})$$

and so the conclusion easily follows.   □

Proof of Theorem 77. Assumptions (EC-0) and (EC-1) of Theorem 34 are clearly satisfied. The proof that E and C satisfy the splitting property (EC-2) follows, with few small changes, the same arguments as those used in proving Lemma 53. By Lemmas 79 and 83 also the coercitivity assumption (EC-3) and hylomorphy condition (2.16) are satisfied.

Finally it remains to show that also (2.25) is satisfied. To show this, observe that

$$E'(\mathbf{u}) = 0 \text{ implies that } \hat{\psi} = 0 \text{ and } - \Delta\psi + W'(\psi) = 0.$$

Then, since $W \geq 0$, by the Derrick-Pohozaev identity (Theorem 143), we deduce that $\psi = 0$. We conclude that $\mathbf{u} = \left(\psi, \hat{\psi}\right) = 0$.

So all the assumptions of the Theorem 34 are satisfied and the conclusion follows.

$\square$

4.2.5 Symmetry, Travelling Solitary Waves and Solitons in NKG

We begin this section with the following theorem which gives more information on the structure of solitons:

Theorem 84. *Let* $\mathbf{u}_0(x) = (\psi_0(x), \hat{\psi}_0(x)) \in X = H^1(\mathbb{R}^N, \mathbb{C}) \times L^2(\mathbb{R}^N, \mathbb{C})$ *be a hylomorphic soliton relative to Eq. (NKG). Then there exist a real function* $u_0(x) \in H^1$ *and constants* $\omega, \theta_0 \in \mathbb{R}$ *such that*

$$\psi_0(x) = u_0(x)e^{-i\theta_0} \tag{4.74}$$

$$\hat{\psi}_0(x) = -i\omega u_0(x)e^{-i\theta_0}. \tag{4.75}$$

Moreover

$$\psi(t, x) := u_0(x)e^{-i\theta_0 - i\omega t}, \tag{4.76}$$

is a solution of (NKG).

Proof. Since $\mathbf{u}_0 = (\psi_0(x), \hat{\psi}_0(x)) \in X$ is a hylomorphic soliton it is a critical point of E constrained on the manifold

$$\mathfrak{M}_c = \left\{\mathbf{u} = (\psi(x), \hat{\psi}(x)) \in X : C(\mathbf{u}) = c\right\}. \tag{4.77}$$

Clearly

$$E'(\mathbf{u}_0) = -\omega C'(\mathbf{u}_0) \tag{4.78}$$

where $-\omega$ is a Lagrange multiplier. We now compute the derivatives $E'(\mathbf{u}_0)$, $C'(\mathbf{u}_0)$.
For all $(v_0, v_1) \in X$, we have

$$E'(\mathbf{u}_0) \begin{bmatrix} v_0 \\ v_1 \end{bmatrix} = \mathrm{Re} \int \left[\hat{\psi}_0 \overline{v_1} + \nabla \psi_0 \overline{\nabla v_0} + W'(\psi_0) \overline{v_0} \right] dx$$

$$C'(\mathbf{u}_0) \begin{bmatrix} v_0 \\ v_1 \end{bmatrix} = -\mathrm{Re} \int \left(i \hat{\psi}_0 \overline{v_0} + i v_1 \overline{\psi_0} \right) dx$$

$$= -\mathrm{Re} \int \left(i \hat{\psi}_0 \overline{v_0} + \overline{i v_1 \overline{\psi_0}} \right) dx$$

$$= -\mathrm{Re} \int \left(i \hat{\psi}_0 \overline{v_0} - i \psi_0 \overline{v_1} \right) dx.$$

Then (4.78) can be written as follows:

$$\mathrm{Re} \int \left[\nabla \psi_0 \overline{\nabla v_0} + W'(\psi_0) \overline{v_0} \right] dx = \omega \, \mathrm{Re} \int i \hat{\psi}_0 \overline{v_0} \, dx$$

$$\mathrm{Re} \int \hat{\psi}_0 \overline{v_1} \, dx = -\omega \, \mathrm{Re} \int i \psi_0 \overline{v_1} \, dx.$$

Then

$$-\Delta \psi_0 + W'(\psi_0) = i \omega \hat{\psi}_0 \qquad\qquad (4.79)$$

$$\hat{\psi}_0 = -i \omega \psi_0. \qquad\qquad (4.80)$$

By these identities we easily deduce that

$$\psi(t, x) := \psi_0(x) e^{-i\omega t}$$

is a solution of (NKG).

So, in order to prove the theorem, it remains to show that there exist a real
function $u_0(x)$ and a real constant θ_0 such that

$$\psi_0(x) = u_0(x) e^{-i\theta_0}. \qquad\qquad (4.81)$$

Set

$$X_\omega = \left\{ (\psi, \hat{\psi}) \in X : \hat{\psi} = -i \omega \psi \right\}.$$

By (4.48) and (4.49), for any $\mathbf{u} = (\psi, \hat{\psi}) \in X_\omega$ we have that

$$E(\mathbf{u}) = \int \left[\frac{1}{2} \left| \hat{\psi} \right|^2 + \frac{1}{2} |\nabla \psi|^2 + W(\psi) \right] dx$$

$$= \int \left[\frac{1}{2} |\nabla \psi|^2 + W(\psi) + \frac{1}{2} \omega^2 |\psi|^2 \right] dx$$

$$C(\mathbf{u}) = -\operatorname{Re} \int i \hat{\psi} \overline{\psi} \, dx$$

$$= \int \omega |\psi|^2 \, dx.$$

For $\psi \in H^1(\mathbb{R}^N, \mathbb{C})$, we set

$$\mathcal{E}(\psi, \omega) = \int \left[\frac{1}{2} |\nabla \psi|^2 + W(\psi) + \frac{1}{2} \omega^2 |\psi|^2 \right] dx$$

and

$$C(\psi, \omega) = \int \omega |\psi|^2 \, dx.$$

Moreover we set

$$\mathcal{M}_c = \left\{ \psi \in H^1(\mathbb{R}^N, \mathbb{C}), \omega \in \mathbb{R} : C(\psi, \omega) = c \right\}.$$

Clearly $\psi_0(x)$ minimizes $\mathcal{E}(\psi, \omega)$ on $\mathcal{M}_c$. Now we set

$$\psi(x) = u(x) e^{-i\theta(x)}, \ u(x) \in \mathbb{R}$$

$$\psi_0(x) = u_0(x) e^{-i\theta_0(x)}, u_0(x) \in \mathbb{R}. \tag{4.82}$$

So we have

$$\mathcal{E}(\psi, \omega) = \int \left[\frac{1}{2} |\nabla u|^2 + \frac{1}{2} u^2 |\nabla \theta|^2 + W(u) + \frac{1}{2} \omega^2 |u|^2 \right] dx,$$

$$C(\psi, \omega) = \omega \int u^2 \, dx$$

and

$$\mathcal{E}(u_0, \omega) = \int \left[\frac{1}{2} |\nabla u_0|^2 + W(u_0) + \frac{1}{2}\omega^2 u_0^2 \right] dx$$

$$\leq \int \left[\frac{1}{2} |\nabla u_0|^2 + \frac{1}{2} u_0^2 |\nabla \theta_0|^2 + W(u_0) + \frac{1}{2}\omega^2 u_0^2 \right] dx = \mathcal{E}(\psi_0, \omega).$$

So, since ψ_0 is a minimizer of $\mathcal{E}(\psi, \omega)$ on $\mathcal{M}_c$ and $u_0 \in \mathcal{M}_c$, we easily get that θ_0 is constant.

So by (4.82)

$$\psi_0(x) = u_0(x)e^{-i\theta_0}$$

where $u_0(x)$ is real and θ_0 constant. Then (4.81) holds.   □

From the proof of the previous theorem we get the following:

Theorem 85. *Let* $\mathbf{u}_0(x) = (\psi_0(x), \hat{\psi}_0(x))$, ω *be as in Theorem 84. Then* $\psi_0(x)$ *satisfies the equation*

$$- \Delta\psi + W'(\psi) = \omega^2\psi \tag{4.83}$$

and it is a ground state solution of the equation

$$- \Delta\psi + W'(\psi) + \omega^2\psi = 2\lambda\omega\psi \tag{4.84}$$

(where $\lambda = \omega$*) with respect to the functional*

$$\mathcal{E}(\psi, \omega) = \int \left[\frac{1}{2} |\nabla\psi|^2 + W(\psi) + \frac{1}{2}\omega^2 |\psi|^2 \right] dx$$

on the manifold

$$\mathcal{M}_c = \left\{ \psi \in H^1(\mathbb{R}^N, \mathbb{C}), \omega \in \mathbb{R} : \omega \int |\psi|^2 = c \right\}.$$

Remark 86. If ψ is a solution of Eq. (4.83) then $(\psi(x)e^{-i\omega t}, -i\omega\psi(x)e^{-i\omega t})$ is a solution of (NKG) but not a soliton since in general it is not stable. For example in order to get solutions of Eq. (4.83), we can minimize the functional

$$\int \left[\frac{1}{2} |\nabla\psi|^2 + W(\psi) \right] dx \tag{4.85}$$

over tha manifold

$$\left\{ \psi \in H^1(\mathbb{R}^N, \mathbb{C}) : \int |\psi|^2 = k \right\}.$$

For a suitable choice of k, this minimizer ψ_G solves (4.83) and it is a ground state solution with respect to the energy (4.85). In general, we have that $\psi_G \neq \psi_0$ and in this case we do not have any guarantee that $(\psi_G(x)e^{-i\omega t}, -i\omega\psi_G(x)e^{-i\omega t})$ is a soliton. So in order to get solitons it is not sufficient to get ground state solutions with respect to an arbitrary functional, but respect the energy functional.

Remark 87. By the proof of the above theorem we have that $u_0(x)$ is a ground state solution of Eq. (4.84) and hence, arguing as in Remark 60, $u_0(x)$ is radially symmetric and it does not change sign.

Now we will exploit the symmetries of the Poincaré group to produce other soliton solutions of Eq. (NKG) as we did for NSE in Sect. 3.2.5 exploiting the Galileo group. We have already discussed this issue for the solitary wave (see Sect. 4.2.1); so this section is a partial repetition of some arguments in that section.

Assume that

$$\psi(t, x) = u(x)e^{-i\omega t}, u \in \mathbb{R}$$

is a solution of (NKG). First of all, since (NKG) is invariant for translations, for any $\bar{x} \in \mathbb{R}^3$, the function

$$\psi_{\bar{x}}(t, x) = u(x - \bar{x})e^{-i\omega t} \tag{4.86}$$

is a standing wave concentrated around the point $\bar{x}$. The space rotations do not produce other solutions since $u(x)$ is rotationally invariant by Remark 87. Since Eq. (NKG) is invariant for a Lorentz boost, we get other solutions; namely we can produce travelling waves just applying the transformation (1.10) with $\mathbf{v} = (v, 0, 0)$ and $\bar{x} = (\bar{x}_1, \bar{x}_2, \bar{x}_3)$ to (4.86)

$$\psi_{\bar{x}, \mathbf{v}}(t, x) = u(x_1' - \bar{x}_1, x_2 - \bar{x}_2, x_3 - \bar{x}_3)e^{-i\omega t'}$$
$$= u(\gamma(x_1 - vt) - \bar{x}_1, x_2 - \bar{x}_2, x_3 - \bar{x}_3)e^{-i\gamma\omega(t - vx_1)},$$
$$= u(\gamma(x_1 - vt) - \bar{x}_1, x_2 - \bar{x}_2, x_3 - \bar{x}_3)e^{i(\mathbf{p}\cdot x - Et)},$$

where

$$E = \gamma\omega$$

and

$$\mathbf{p} = \gamma\omega(v, 0, 0) = \gamma\omega\mathbf{v}.$$

The meaning of these relation has been analysed in Sect. 4.1.2.

$\psi_{\bar{x}, \mathbf{v}}(t, x)$ is a solitary wave concentrated in the point $\bar{x}\gamma^{-1} + \mathbf{v}t$, and hence it travels with velocity $\mathbf{v}$. Notice that it experiences the relativistic contraction of a factor γ in the direction of motion x_1.

Finally, other solutions can be produced by the invariance (1.19); for $\theta \in [0, 2\pi)$, we have the solutions

$$\psi_{\bar{x},\mathbf{v},\theta} = u(\gamma (x_1 - vt) - \bar{x}_1, \ x_2 - \bar{x}_2, \ x_3 - \bar{x}_3) e^{i(\mathbf{p}\cdot x - Et + \theta)}. \tag{4.87}$$

The invariance by time translations do not produce new solutions, since a time translation on $\psi_{\bar{x},\mathbf{v},\theta}$ produces a space and phase translation.

Concluding, we have a radially symmetric solution of the form (4.86); by the invariance of the equation, this solution produces a seven parameters family of solutions given by (4.87). The parameters are $\bar{x}, \mathbf{v}, \theta$.

4.3 Dynamical Properties of Solitary Waves in NKG

In this section we show that some relativistic effects as the space contraction, the time dilation, the Einstein equation, are consequences of assumptions **A-1** and **A-2** (see Introduction) and in particular of the Poincarè invariance of Eq. (NKG). Probably NKG is the simplest nonlinear equation which satisfies **A-1** and **A-2** and hence the solitary waves and the Q-balls must satisfy the relativistic effects (page 123).

4.3.1 Space Contraction and Time Dilation of Solitary Waves

Observe that, by Eqs. (4.34) or (4.87), the following theorem follows:

Theorem 88. *Any moving solitary wave experiences a contraction in the direction of its movement of a factor $1/\gamma$ with $\gamma = \frac{1}{\sqrt{1-v^2}}$.*

Now the standing waves of Eq. (NKG) can be considered as a clock. Let us denote by $\mathbf{q}(t)$ the position of our clock at the time t.

If we assume that at $t = 0$, $\mathbf{q}(0) = (0,0,0)$, the motion of the clock is given by

$$\mathbf{q}(t) = (vt, 0, 0);$$

then the behavior of the moving clock at the point $\mathbf{q}(t)$ is obtained replacing x by $\mathbf{q}(t)$ in Eq. (4.34):

$$\psi_v(t, \mathbf{q}(t)) = \psi_v(t, vt, 0, 0)$$
$$= u(0,0,0) e^{i(\mathbf{k}\cdot\mathbf{q}(t) - \omega t)};$$

taking into account Eq. (4.35), we get

$$\mathbf{k} \cdot \mathbf{q}(t) - \omega t = (k_1 v - \omega)t$$
$$= (\gamma \omega_0 v^2 - \gamma \omega_0)t$$

$$= \gamma \omega_0 \left(v^2 - 1 \right) t$$
$$= \frac{\omega_0}{\gamma} t.$$

Then

$$\psi_v(t, \mathbf{q}(t)) = u(0, 0, 0) e^{-i \frac{\omega_0}{\gamma} t}$$

this equation shows that our moving clock is vibrating with a frequency

$$\frac{\omega_0}{\gamma} = \omega_0 \sqrt{1 - v^2}.$$

Since the intervals of time measured by a clock are inversely proportional to the frequency of the vibrations, we get that

$$\Delta T = \frac{\Delta T_0}{\sqrt{1 - v^2}}.$$

Then we get the following

Theorem 89. *A moving clock moves slower than a resting clock by a factor* γ^{-1}.

4.3.2   The Mass of Solitary Waves

In classical mechanics the mass is a symmetric tensor m_{ij} which relates the momentum $\mathbf{P} = (P_1, P_2, P_3)$ to the velocity $\mathbf{v} = (v_1, v_2, v_3)$ by the following formula

$$P_i = m_{ij} v_j.$$

Since the momentum of a solitary wave is defined by (4.9), it is possible to define the mass of a solitary wave by the above formula and to compute it. As we will see it turns out that it is a scalar, namely $m_{ij} = m \delta_{ij}$. Thus the solitons behave as material particles with a well defined mass.

Theorem 90. *Let* ψ_v *be defined by (4.34) with* $u \in H^1(\mathbb{R}^3)$; *then its momentum is given by*

$$\mathbf{P}(\psi_v) = \mathbf{v} \gamma \int \left((\partial_n u)^2 + \omega_0^2 u^2 \right) dx_1 dx_2 dx_3$$

were ∂_n *denotes the directional derivative in the direction* $\mathbf{n} = \mathbf{v}/|\mathbf{v}|$ *(4.9).*

Proof. It is not restrictive to choose $\mathbf{v} = (v, 0, 0)$; then, by (4.10) and (4.34), we have that the component P_1 of $\mathbf{P}(\psi_{\mathbf{v}})$ along the x_1 axis is given by

$$P_1 = -\int \left(\frac{\partial}{\partial t} u' \frac{\partial}{\partial x} u' + \omega k_1 u'^2 \right) dx$$

where we have set

$$u'(t, x_1, x_2, x_3) = u(\gamma(x_1 - vt), x_2, x_3); \quad \mathbf{k} = (k_1, 0, 0).$$

Then, performing the derivatives, we get

$$P_1 = -\int \left[\gamma^2 v \left(\frac{\partial u}{\partial x} \right)^2 + \omega k_1 u^2 \right]_{x_1 = \gamma(x_1 - vt)} dx.$$

By (4.35) we have that

$$P_1 = v \int \gamma^2 \left[\left(\frac{\partial u}{\partial x} \right)^2 + \omega_0^2 u^2 \right]_{x_1 = \gamma(x_1 - vt)} dx.$$

Finally, making the change of variables $y = (\gamma(x_1 - v_1 t), v_2, v_1)$, we get

$$P_1 = v\gamma \int \left((\partial_1 u(y))^2 + \omega_0^2 u(y)^2 \right) dy.$$

It is immediate to check that

$$P_2 = P_3 = 0.$$

Thus the theorem is proved.   □

The vector $\mathbf{P}(\psi_{\mathbf{v}})$ is parallel to $\mathbf{v}$. Next we will prove that its norm does not depend on the direction of $\mathbf{v}$.

Since u solves Eq. (4.33), u is a solution of (A.2) (see Theorem 143) where $H(u) = W(u) - \frac{1}{2}\omega_0^2 u^2$. Since $u \in H^1(\mathbb{R}^3)$, by Theorem 143 and Eq. (A.4), we have that

$$\mathbf{P}(\psi_{\mathbf{v}}) = v\gamma \int \left(\frac{1}{3} |\nabla u|^2 + \omega_0^2 u^2 \right) dx_1 dx_2 dx_3. \qquad (4.88)$$

Then the mass of a solitary wave $\psi_{\mathbf{v}}$ is well defined and we have

$$m(\psi_{\mathbf{v}}) = \gamma \int \left(\frac{1}{3} |\nabla u|^2 + \omega_0^2 u^2 \right) dx. \qquad (4.89)$$

From here we get the second remarkable fact of theory of relativity:

Theorem 91. *The mass of a solitary wave increases with velocity by the factor*

$$\gamma = \frac{1}{\sqrt{1-v^2}}.$$

Remark 92. By Eq. (A.3), with $H(u) = W(u) - \frac{1}{2}\omega_0^2 u^2$, we have that

$$\frac{1}{3}\int |\nabla u|^2 \, dx = \omega_0^2 \int u^2 dx - 2\int W(u)dx$$

then, the rest mass of a solitary waves can be expressed also in the following ways:

$$m\,(\psi_0) = 2\int \left(\omega_0^2 u^2 - W(u)\right) dx \tag{4.90}$$

$$= \int \left(\frac{2}{3}|\nabla u|^2 + 2W\right) dx. \tag{4.91}$$

4.3.3 The Einstein Equation

First we will give an explicit formula for the energy of a solitary wave:

Theorem 93. *Let ψ_v in (4.34) be solution of (NKG); then, if $u \in H^1(\mathbb{R}^3)$, its energy $E\,(\psi_v)$ is given by*

$$E\,(\psi_v) = \gamma \int \left(\frac{1}{3}|\nabla u|^2 + \omega_0^2 u^2\right) dx.$$

Proof. By (4.8) we have that

$$E(\psi_v) = \int \left[\frac{1}{2}\left(\partial_t u'\right)^2 + \frac{1}{2}|\nabla u'|^2 + \frac{1}{2}\left(\omega^2 + k^2\right)u'^2 + W(u')\right]dx$$

where we have set

$$u'\,(t, x_1, x_2, x_3) = u\,(\gamma\,(x_1 - vt)\,, x_2, x_3)\,, \quad k = |\mathbf{k}|\,.$$

Then, performing the derivatives, we get

$$E(\psi_v) = \int \left[\frac{1}{2} \left((\gamma^2 v^2 + \gamma^2) |\partial_1 u|^2 \right. \right.$$

$$\left. \left. + \sum_{i \neq 1} |\partial_i u|^2 + \frac{1}{2} \left(\omega^2 + k^2 \right) u^2 \right) + W(u) \right]_{x_1 = \gamma(x_1 - vt)} dx.$$

Making the change of variables $y = (\gamma (x_1 - v_1 t), x_2, x_3)$, we get

$$E(\psi_v) = \frac{1}{2\gamma} \int \left[(\gamma^2 v^2 + \gamma^2) |\partial_1 u|^2 + \sum_{i \neq 1} |\partial_i u|^2 + (k^2 + \omega^2) u^2 \right] dy$$

$$+ \frac{1}{\gamma} \int W(u(y)) dy. \tag{4.92}$$

Since u solves Eq. (A.2) with $H(u) = W(u) - \frac{1}{2}\omega_0^2 u^2$, then, by (A.3), we have

$$\int W(u) = \frac{1}{2} \int \omega_0^2 u^2 - \frac{1}{6} \int |\nabla u|^2. \tag{4.93}$$

Moreover, by (A.4)

$$\int |\partial_i u|^2 \, dx = \frac{1}{3} \int |\nabla u|^2 \, dx, \ i = 1, 2, 3. \tag{4.94}$$

Substituting (4.94) and (4.93) in (4.92), we have

$$E(\psi_v) = \frac{1}{2\gamma} \int \frac{\gamma^2 v^2 + \gamma^2 + 2}{3} |\nabla u|^2 + (k^2 + \omega^2) u^2 dx + \frac{1}{\gamma} \int W(u)$$

$$= \frac{1}{2\gamma} \int \frac{\gamma^2 v^2 + \gamma^2 + 1}{3} |\nabla u|^2 + (k^2 + \omega^2 + \omega_0^2) u^2 dx.$$

By the definition of γ, we have that

$$\gamma^2 v^2 + \gamma^2 + 1 = \frac{v^2 + 1 + 1 - v^2}{1 - v^2} = 2\gamma^2,$$

moreover, by (4.35),

$$k^2 + \omega^2 + \omega_0^2 = \gamma^2 \omega_0^2 v^2 + \gamma^2 \omega_0^2 + \omega_0^2$$

$$= \omega_0^2 \left(\gamma^2 v^2 + \gamma^2 + 1 \right) = 2\omega_0^2 \gamma^2.$$

Then

$$E(\psi_v) = \frac{1}{2\gamma} \int \frac{2\gamma^2}{3} |\nabla u|^2 + 2\omega_0^2 \gamma^2 u^2 dx$$

$$= \gamma \int \frac{1}{3} |\nabla u|^2 + \omega_0^2 u^2 dx.$$

$\square$

So, by Theorem 93 and Eq. (4.89), we get the Einstein equation:

Theorem 94. *The energy of a solitary wave equals its mass:*

$$E(\psi_v) = m(\psi_v).$$

By the above theorem, it turns out that the ergocenter $\mathbf{Q}$, defined by Eq. (4.15), actually is the center of mass. Then, the conservation of $\dot{\mathbf{Q}}$ implies that the center of mass moves along a straight line.

Remark 95. Theorem 94 could have been proved by Eq. (4.16) just proving that $\dot{\mathbf{Q}} = \mathbf{v}$. On the contrary, we have proved it using Theorems 90 and 93 since we wanted to have an explicit formula for the momentum and the energy of a solitary wave.

4.3.4 The Energy-Momentum 4-Vector

The expression

$$\mathfrak{v} = \begin{bmatrix} \gamma \\ \gamma \mathbf{v} \end{bmatrix}$$

is a 4-vector in Minkowsky space, with Minkowsky norm

$$\|\mathfrak{v}\|_M^2 = -\gamma^2 + \|\gamma \mathbf{v}\|^2 = -\frac{1}{1-v^2} + \frac{v^2}{1-v^2} = -1.$$

Actually, if $\mathbf{v}$ varies among all the Euclidean vectors in space, $\mathfrak{v}$ describes all the possible time-like vectors in the positive *time-cone*. The vector $\mathfrak{v}$ is called 4-velocity. If $\mathbf{v}$ represent the space velocity of a material point, $\mathfrak{v}$ represents this velocity in the space time. If $\mathbf{v} = \mathbf{0}$, then

$$\mathfrak{v} = \begin{bmatrix} 1 \\ 0 \end{bmatrix}$$

namely, the material point moves only in time.

By (4.88), (4.89) and Theorem 93, it follows that also the 4-uple $(E\,(\psi_v)\,,\mathbf{P}\,(\psi_v))$, is a 4-vector. In fact

$$\begin{bmatrix} E\,(\psi_v) \\ \mathbf{P}\,(\psi_v) \end{bmatrix} = m\,(\psi_0) \cdot \begin{bmatrix} \gamma \\ \gamma\mathbf{v} \end{bmatrix}. \tag{4.95}$$

So, the 4-uple $(E\,(\psi_v)\,,\mathbf{P}\,(\psi_v))$, is a time like 4-vector of Minkowsky norm $-m\,(\psi_0)$. Also $(\omega,\mathbf{k})$ is a 4-vector since, by (4.35), we have

$$\begin{bmatrix} \omega \\ \mathbf{k} \end{bmatrix} = \omega_0 \begin{bmatrix} \gamma \\ \gamma\mathbf{v} \end{bmatrix}.$$

Then, comparing the above formula with (4.95), we get

$$\begin{bmatrix} E\,(\psi_v) \\ \mathbf{P}\,(\psi_v) \end{bmatrix} = \frac{m\,(\psi_0)}{\omega_0} \begin{bmatrix} \omega \\ \mathbf{k} \end{bmatrix}. \tag{4.96}$$

In the preceding subsections we have seen that solutions of type (4.34) behave like relativistic particles; nevertheless they have the structure of traveling waves: it reminds us of the big discussion in the beginning of the last century about the dualism between waves and particles, one of the starting points of Quantum Mechanics.

If we look at a solution ψ_v of (NKG) of the form (4.34) with an apparatus at rest with dimension of the order of a wavelength, it is not adequate to consider only what happens at the "particle"; we need to consider also the space-time vibrations.

By (4.96), we have that

$$E\,(\psi_v) = \frac{m\,(\psi_0)}{\omega_0}\,\omega \tag{4.97}$$

and

$$\mathbf{P}\,(\psi_v) = \frac{m\,(\psi_0)}{\omega_0}\,\mathbf{k}. \tag{4.98}$$

The quantity $m\,(\psi_0)\,/\omega_0 = E\,(\psi_0)\,/\omega_0$ has the dimension of an action and is reminiscent of the Planck constant. More or less this is the argument which De Broglie used to deduce his celebrated relations

$$E = \hbar\omega \tag{4.99}$$

$$\mathbf{P} = \hbar\mathbf{k}.$$

Here, it is important to emphasize that in this model $E\,(\psi_0)\,/\omega_0$ is *not* a universal constant since it depends on the form of the rest soliton ψ_0. Different solutions of (NKG) of type (4.34) might give different values of $E\,(\psi_0)\,/\omega_0$.

4.3.5   Remarks on Fields and Particles

Newtonian Physics describes the universe as being composed of particles interacting with each other via an action at distance; the Dynamics is described by a Lagrangian as (1.1) in which the potential energy $V(t, x_1, \ldots, x_k)$ depends instantaneously on the positions $x_1, \ldots, x_k$ of the particles.

In last century, thanks to the studies in electromagnetism, people introduced the notion of field and gave up the notion of instantaneous action: in this new paradigm the particles do not act directly on each other, but act on a field, and the modification of this field influences the motion of the other particles.

Moreover, if we assume that the equations are invariant for the action of the Poincaré group, the modification of any field cannot propagate faster than the constant c appearing in the Poincaré group.

The first consequence of this fact is that rigid bodies cannot exist (see e.g. [101] ch. III); therefore, if we want to admit the existence of entities which are different from a field (such as particles), they must be pointwise entities: material points.

The notion of material point not only presents a lot of philosophical difficulties, but it is also inconsistent with assumptions **A-1** and **A-2** in the introduction. We will try to explain this inconsistency, making a very simple example: a non-relativistic charged material point which moves in an electric field $\mathbf{E} = -\nabla\varphi$. For simplicity we ignore the magnetic field. The action of the particle is given by

$$S_p = \int \frac{1}{2} m_0 \, |\dot{x}|^2 \, dt - e \int \varphi(x) \, dt$$

where $x(t)$ is the trajectory of the material point, m_0 is its mass and e is its charge. The variation of $x(t)$ of the above functional gives the correct equation:

$$m_0 \ddot{x} = -e\nabla\varphi.$$

The action of the field is given by

$$S_f = \frac{1}{8\pi} \int \int |\nabla\varphi|^2 \, dx \, dt.$$

The variation of the field gives us the equation

$$\frac{1}{4\pi} \Delta\varphi = 0$$

which is the correct one.

However, if we want the field to interact with matter we have the action

$$S = S_p + S_f = \int \frac{1}{2} m_0 \, |\dot{x}|^2 \, dt - e \int \varphi(x(t)) \, dt - \frac{1}{8\pi} \int \int |\nabla\varphi|^2 \, dx \, dt.$$

Using the Dirac δ-measure, the action can be written as follows:

$$\mathcal{S} = \int \frac{1}{2} m_0 |\dot{x}|^2 \, dt - e \int \int \varphi(x) \, \delta(x(t)) \, dx \, dt - \frac{1}{8\pi} \int \int |\nabla \varphi|^2 \, dx \, dt.$$

(4.100)

The variation of the field in the above functional gives the equation

$$\Delta \varphi = 4\pi e \delta(x(t))$$

(4.101)

which has the solution

$$\varphi(x) = 4\pi e \frac{1}{|x - x(t)|}.$$

But this solution is not acceptable since the constant $\frac{1}{8\pi} \int \int |\nabla \varphi|^2 \, dx \, dt$ diverges, and the constant $\int \int \varphi(x) \, \delta(x(t)) \, dx \, dt$ does not even makes sense. If we add the magnetic field and if we use a relativistic Lagrangian instead of $\frac{1}{2} m_0 |\dot{x}|^2$ things do not get better. The strange thing about this story is that the variational principles give all the fundamental equations of physics in a natural way and are used in most text books (see e.g. [101]); nevertheless the Mathematics used does not make sense if we let fields and pointwise particles interact. In other words up to now we do not know how to couple the Maxwell equation with matter within a mathematically consistent theory (see chapter 17 in [89]).

If we abandon the idea of using a variational principle, Eq. (4.101) makes sense, but we are left with the problem of pointwise particles having infinite energy. Given the relativistic equivalence between mass and energy, pointwise particles would have infinite mass and of course this is not a secondary problem in building any consistent theory (see e.g. [101] ch. V for a brief discussion on this point). Even if Quantum Mechanics in the last century made people forget this problem, it seems interesting to know if a theoretically consistent theory called Classical Physics exists.

Concluding this discussion, if we want a mathematically simple theory consistent with Assumptions **A-1**and **A-2** in the introduction, it is better to give up the obscure notion of material point and to assume the following point of view:

material particles are bumps of the field   (4.102)

Following this point of view, material particles, and more in general material bodies are nothing else but regions of space-time where the field energy concentrates and forms a *stable* structure.

Thus, elementary particles can be modelled by solitons; material bodies can be considered as solitary waves.

Chapter 5
The Nonlinear Klein-Gordon-Maxwell Equations

In this chapter, the Maxwell equations are deduced from the general principles of Chap. 1. If these equations are coupled with the nonlinear Klein-Gordon equation, we get the simplest gauge theory with "matter". After having analyzed the general features of these equations, we apply the abstract theory of Chap. 2 and we prove the existence of hylomorphic solitons.

5.1 General Feature of the Klein-Gordon-Maxwell Equations

5.1.1 The Maxwell Equations in Empty Space

In Chap. 2 we introduced the Lagrangian (1.13) which gives rise to the simplest Poincarè invariant equation for a scalar field ψ, namely Eq. (1.11). In order to generalize this equation we will use the language of the forms.

If we regard ψ as a zero form, then (1.12) can be written

$$S_0 = \int \mathcal{L}_0 dx dt = -\frac{1}{2} \int \langle d\psi, d\psi \rangle_M \tag{5.1}$$

where $\langle \xi, \eta \rangle_M$ is the Minkowskian scalar product on the space $\Lambda^k (\mathbb{R}^4)$ of the k-forms. We recall the definition of $\langle \cdot, \cdot \rangle_M$ for the 1-form and the 2-form: if

$$\xi = \sum_{i=0}^{3} \xi_i dx^i \quad and \quad \eta = \sum_{i=0}^{3} \eta_i dx^i$$

then

$$\langle \xi, \eta \rangle_M = \sum_{i,j=0}^{3} g^{ij} \xi_i \eta_j,$$

© Springer International Publishing Switzerland 2014
V. Benci, D. Fortunato, *Variational Methods in Nonlinear Field Equations*,
Springer Monographs in Mathematics, DOI 10.1007/978-3-319-06914-2_5

where

$$[g^{ij}] = \begin{bmatrix} -1 & 0 & 0 & 0 \\ 0 & 1 & 0 & 0 \\ 0 & 0 & 1 & 0 \\ 0 & 0 & 0 & 1 \end{bmatrix}.$$

If

$$\xi = \sum_{i=0}^{3} \xi_{ik} dx^i \wedge dx^k \quad and \quad \eta = \sum_{i=0}^{3} \eta_{ik} dx^i \wedge dx^k$$

then

$$\langle \xi, \eta \rangle_M = \sum_{i,j,k,l=0}^{3} g^{ij} g^{kl} \xi_{ik} \eta_{jl}.$$

Using the form language, the D'Alembert equation (1.11) becomes:

$$\delta d\psi = 0 \tag{5.2}$$

where $\delta : \Lambda^k \to \Lambda^{k-1}$ is the functional adjoint operator of $d : \Lambda^{k-1}(\mathbb{R}^4) \to \Lambda^k(\mathbb{R}^4)$, namely it is the operator defined by the following equation:

$$\int \langle \xi, d\eta \rangle_M \, dxdt = \int \langle \delta\xi, \eta \rangle_M \, dxdt$$

where we have assumed $\xi \in \Lambda^k(\mathbb{R}^4)$, $\eta \in \Lambda^{k-1}(\mathbb{R}^4)$, $k = 1, \ldots, 4$, and η with compact support.

The action $S[\xi]$ is invariant for the "trivial gauge group" $\xi \to \xi + c$ where $c \in \mathbb{C}$ is a constant, namely $S[\xi] = S[\xi + c]$. Then if ξ is a solution of (5.2), also $\xi + c$ solves (5.2).

One of the most natural generalization of (5.1) is given by

$$S_1[\xi] = \int \mathcal{L}_1 dxdt = -\frac{1}{2} \int \langle dA, dA \rangle_M \, dxdt \tag{5.3}$$

where A is a 1-form:

$$A = \sum_{j=0}^{3} A_j dx^j. \tag{5.4}$$

The variation of the action (5.3) gives the following Euler-Lagrange equation:

$$\delta dA = 0. \tag{5.5}$$

This simple generalization gives a much richer structure; in fact the action (5.3) is invariant for the gauge transformation $A \to A + d\chi$ where $\chi \in \mathfrak{G} = \mathcal{C}^1 \left(\mathbb{R}^4, \mathbb{R} \right)$; namely the gauge group is an infinite dimensional group. However, in most physical interpretations of this theory it is assumed that A and $A + d\chi$ give the same experimental results, namely χ has no physical meaning. For this reason, we can introduce the quantity

$$F = dA \tag{5.6}$$

which does not depend on χ (since $dd\chi = 0$) and which is considered the physically measurable quantity.

By Eq. (5.5), and the fact $ddA = 0$, we have that F satisfies the following equations:

$$dF = 0 \tag{5.7}$$

$$\delta F = 0. \tag{5.8}$$

These are the Maxwell equations in the empty space.

Now let us write Eqs. (5.7) and (5.8) using the vector notation. We denote by

$$j : \mathbb{R}^{3+1} \to \Lambda^1 \left(\mathbb{R}^4 \right)$$

the duality map which associate to a 4-vector $(v_0, \mathbf{v})$ the 1-form $j(v_0, \mathbf{v})$ defined by

$$j(v_0, \mathbf{v}) [(w_0, \mathbf{w})] = -v_0 w_0 + \mathbf{v} \cdot \mathbf{w}.$$

Then, if A is the 1-form (5.4), we set $(\varphi, \mathbf{A}) = j^{-1}(A)$, namely

$$\varphi := A^0 = -A_0, \quad \mathbf{A} := (A^1, A^2, A^3) = (A_1, A_2, A_3). \tag{5.9}$$

Then the Lagrangian (5.3) becomes

$$\langle dA, dA \rangle_M = \frac{1}{2} \left[\sum_{i,j=1}^{3} (\partial_i A_j - \partial_j A_i)^2 - \sum_{j=1}^{3} (\partial_0 A_j - \partial_j A_0)^2 - \sum_{i=1}^{3} (\partial_i A_0 - \partial_0 A_i)^2 \right]$$

$$= \frac{1}{2} \left[\sum_{i,j=1}^{3} (\partial_i A^j - \partial_j A^i)^2 - \sum_{j=1}^{3} (\partial_t A^j + \partial_j \varphi)^2 - \sum_{i=1}^{3} (\partial_i \varphi + \partial_t A^i)^2 \right]$$

$$= |\nabla \times \mathbf{A}|^2 - |\partial_t \mathbf{A} + \nabla \varphi|^2.$$

So (5.3) takes the following form:

$$\mathcal{S}_1 [(\varphi, \mathbf{A})] = \frac{1}{2} \int \left(|\partial_t \mathbf{A} + \nabla \varphi|^2 - |\nabla \times \mathbf{A}|^2 \right) dx dt.$$

Making the variation of S with respect to φ and $\mathbf{A}$ we get

$$\nabla \cdot (\partial_t \mathbf{A} + \nabla \varphi) = 0 \tag{5.10}$$

$$\nabla \times (\nabla \times \mathbf{A}) + \frac{\partial}{\partial t} (\partial_t \mathbf{A} + \nabla \varphi) = 0 . \tag{5.11}$$

If we make the following change of variables:

$$\mathbf{E} = - (\partial_t \mathbf{A} + \nabla \varphi) \tag{5.12}$$

$$\mathbf{H} = \nabla \times \mathbf{A} \tag{5.13}$$

we have that

$$\nabla \times \mathbf{E} + \partial_t \mathbf{H} = 0 \tag{5.14}$$

$$\nabla \cdot \mathbf{H} = 0 \tag{5.15}$$

and (5.10), (5.11) become

$$\nabla \cdot \mathbf{E} = 0 \tag{5.16}$$

$$\nabla \times \mathbf{H} - \partial_t \mathbf{E} = 0. \tag{5.17}$$

Thus we have obtained the Maxwell equations in the empty space (5.14)–(5.17) in the usual 3-vector notation. In this case, the action (5.3) can be written as follows

$$S_1 = \frac{1}{2} \int \left(\mathbf{E}^2 - \mathbf{H}^2 \right) dx dt = \frac{1}{2} \int \left(|\partial_t \mathbf{A} + \nabla \varphi|^2 - |\nabla \times \mathbf{A}|^2 \right) dx dt. \tag{5.18}$$

Moreover we can give a physical meaning to the two form in (5.6)

$$F = \sum_{i<j} F_{ij} dx^i dx^j .$$

By duality, we can associate to it a antisymmetric 4-tensor $\{F^{ij}\}$. Direct computations show that

$$\{F^{ij}\} = \begin{bmatrix} 0 & -E^1 & -E^2 & -E^3 \\ E^1 & 0 & H^3 & -H^2 \\ E^2 & -H^3 & 0 & H^1 \\ E^3 & H^2 & -H^1 & 0 \end{bmatrix} . \tag{5.19}$$

Thus the electromagnetic field $\{F^{ij}\}$ is a 4-tensor composed by the 3-vector "electric field"

$$\mathbf{E} = \begin{bmatrix} E_1 \\ E_2 \\ E_3 \end{bmatrix}$$

and an antisymmetric 3-tensor "magnetic field"

$$\mathcal{H} = \begin{bmatrix} 0 & -H_3 & H_2 \\ H_3 & 0 & -H_1 \\ -H_2 & H_1 & 0 \end{bmatrix}.$$

It is well known that it is possible to associate a pseudovector $\mathbf{H}$ to any antisymmetric 3-tensor $\mathcal{H}$ by the equation

$$\mathcal{H}\mathbf{v} = \mathbf{H} \times \mathbf{v}, \ \mathbf{v} \in \mathbb{R}^3.$$

In this case, $\mathbf{H}$ takes the form

$$\mathbf{H} = \begin{bmatrix} H_1 \\ H_2 \\ H_3 \end{bmatrix}.$$

Remark 96. The fact that the unknown A is a differential form implies that the gauge group $\mathfrak{G}$ is non-trivial. So the Maxwell equations are the simplest equations, with a nontrivial gauge group, which satisfy **A-1–A-3** in the introduction.

5.1.2 Gauge Theories

A gauge theory provides a very elegant way to combine the action functionals (5.1) and (5.3) and to obtain an action functional invariant for the gauge group $\mathfrak{G}$ (1.18).

Let G be a subgroup of $U(N)$, the unitary group in $\mathbb{C}^N$ and denote by $\Lambda^k(\mathbb{R}^4, \mathfrak{g})$ the set of k-forms defined in $\mathbb{R}^4$ with values in the Lie algebra $\mathfrak{g}$ of the group G. A 1-form

$$\Gamma = \sum_{j=0}^{3} \Gamma_j dx_j \in \Lambda^1(\mathbb{R}^4, \mathfrak{g})$$

is called *connection form*. The operator

$$d_\Gamma = \Lambda^k(\mathbb{R}^4, \mathfrak{g}) \to \Lambda^{k+1}(\mathbb{R}^4, \mathfrak{g})$$

defined by

$$dr = d + \Gamma = \sum_{j=0}^{3} \left(\partial_j + \Gamma_j \right) dx_j$$

is called *covariant differential* and the operators

$$D_j = \frac{\partial}{\partial x_j} + \Gamma_j : C^1 \left(\mathbb{R}^4, \mathbb{C}^N \right) \to C^0 \left(\mathbb{R}^4, \mathbb{C}^N \right), \quad j = 0, \dots, 3$$

are called *covariant derivatives*. Let $[\cdot, \cdot]$ denote the commutator of $\mathfrak{g}$. The 2-form

$$\mathcal{F} = d_\Gamma \Gamma = \sum_{i,j=0}^{3} \left(\partial_i \Gamma_j + \left[\Gamma_i, \Gamma_j \right] \right) dx^i \wedge dx^j$$

is called *curvature*.

We set

$$\mathcal{L}_0 = \frac{1}{2} \langle d_\Gamma \psi, d_\Gamma \psi \rangle_M ,$$

where $\psi \in \mathbb{C}^N$. Moreover we set

$$\mathcal{L}_G = \frac{1}{2q^2} \langle d_\Gamma \Gamma, d_\Gamma \Gamma \rangle_M ,$$

where q is a parameter which controls the coupling of $\mathcal{L}_G$ with $\mathcal{L}_0$.

A *gauge field* by definition (see e.g. [124, 147]), is a critical point of the action functional

$$S = \int \mathcal{L} \, dxdt, \tag{5.20}$$

where the Lagrangian is given by

$$\mathcal{L} = \mathcal{L}_0 + \mathcal{L}_G - W(\psi),$$

and $W : \mathbb{C}^N \to \mathbb{R}$ is a function which is assumed to be G-invariant, namely

$$W(g\psi) = W(\psi), \quad g \in G. \tag{5.21}$$

5.1.3 Maxwell Equations and Matter

We are interested in the Abelian gauge theory, namely in the case in which

$$G = U(1) = S^1 = \{ z \in \mathbb{C} : |z| = 1 \}.$$

It this case the $\Gamma_j(x,t)$ are imaginary numbers; if we set

$$A_j = -\frac{1}{iq}\Gamma_j, \quad j = 0,\ldots,3 \tag{5.22}$$

it turns out that $A = \sum_{j=0}^{3} A_j dx^j$ is a real valued 4-form. Moreover, since in this case $[\Gamma_i, \Gamma_j] = 0$, we have that

$$d_\Gamma\Gamma = \sum_{i,j=0}^{3} \partial_i\Gamma_j dx^i \wedge dx^j = d\Gamma = -iqdA.$$

In this case, it turns out that

$$\mathcal{L}_G = \frac{1}{2q^2}\langle d_\Gamma\Gamma, d_\Gamma\Gamma\rangle_M = -\frac{1}{2}\langle dA, dA\rangle_M = \mathcal{L}_1$$

as defined in Eq. (5.3). By (5.9) and (5.22), the covariant derivatives take the form

$$D_\varphi = \frac{\partial}{\partial t} + iq\varphi, \quad D_{A_j} = \frac{\partial}{\partial x^j} - iqA_j, \quad j = 1,2,3 \tag{5.23}$$

and, for $q = 0$, they reduce to the usual derivatives. Using the above notation, the Lagrangian density $\mathcal{L}_0$, can be written as follows

$$\begin{aligned}\mathcal{L}_0 &= \frac{1}{2}\left|D_\varphi\psi\right|^2 - \frac{1}{2}\left|\mathbf{D_A}\psi\right|^2 \\ &= \frac{1}{2}\left[\left|\left(\frac{\partial}{\partial t} + iq\varphi\right)\psi\right|^2 - \left|(\nabla - iq\mathbf{A})\,\psi\right|^2\right]\end{aligned}$$

where $\mathbf{D_A}\psi = (D_{A_1}\psi, D_{A_2}\psi, D_{A_3}\psi)$ and, using (5.18), the action (5.20) takes the form

$$\mathcal{S} = \frac{1}{2}\int\left[\left|D_\varphi\psi\right|^2 - \left|\mathbf{D_A}\psi\right|^2 + \left|\frac{\partial\mathbf{A}}{\partial t} + \nabla\varphi\right|^2 - \left|\nabla\times\mathbf{A}\right|^2\right]dxdt - \int W(\psi)dxdt. \tag{5.24}$$

Making the variation of $\mathcal{S}$ with respect to ψ, φ and $\mathbf{A}$ we get the following system of equations

$$D_\varphi^2\psi - \mathbf{D_A^2}\psi + W'(\psi) = 0 \tag{5.25}$$

$$\nabla\cdot\left(\frac{\partial\mathbf{A}}{\partial t} + \nabla\varphi\right) = q\left(\text{Im}\,\frac{\partial_t\psi}{\psi} + q\varphi\right)|\psi|^2 \tag{5.26}$$

$$\nabla \times (\nabla \times \mathbf{A}) + \frac{\partial}{\partial t}\left(\frac{\partial \mathbf{A}}{\partial t} + \nabla\varphi\right) = q\left(\mathrm{Im}\,\frac{\nabla\psi}{\psi} - q\mathbf{A}\right)|\psi|^2 . \tag{5.27}$$

The Abelian gauge theory, namely Eqs. (5.25)–(5.27), provides a very elegant way to couple the Maxwell equation with matter if we interpret ψ as a matter field.

In order to give a more meaningful form to these equations, we will write ψ in polar form

$$\psi(x,t) = u(x,t)\,e^{iS(x,t)}, \quad u \in \mathbb{R}, \quad S \in \mathbb{R}/2\pi\mathbb{Z}.$$

So (5.20) takes the following form

$$\mathcal{S}(u, S, \varphi, \mathbf{A}) = \int\int\left[\frac{1}{2}\left(\frac{\partial u}{\partial t}\right)^2 - \frac{1}{2}|\nabla u|^2 - W(u)\right]dxdt +$$

$$+ \frac{1}{2}\int\int\left[\left(\frac{\partial S}{\partial t} + q\varphi\right)^2 - |\nabla S - q\mathbf{A}|^2\right]u^2 dxdt$$

$$+ \frac{1}{2}\int\int\left(\left|\frac{\partial \mathbf{A}}{\partial t} + \nabla\varphi\right|^2 - |\nabla \times \mathbf{A}|^2\right)dxdt$$

and the Eqs. (5.25)–(5.27) take the form:

$$\Box u + W'(u) + \left[|\nabla S - q\mathbf{A}|^2 - \left(\frac{\partial S}{\partial t} + q\varphi\right)^2\right]u = 0 \tag{5.28}$$

$$\frac{\partial}{\partial t}\left[\left(\frac{\partial S}{\partial t} + q\varphi\right)u^2\right] - \nabla \cdot \left[(\nabla S - q\mathbf{A})\,u^2\right] = 0 \tag{5.29}$$

$$\nabla \cdot \left(\frac{\partial \mathbf{A}}{\partial t} + \nabla\varphi\right) = q\left(\frac{\partial S}{\partial t} + q\varphi\right)u^2 \tag{5.30}$$

$$\nabla \times (\nabla \times \mathbf{A}) + \frac{\partial}{\partial t}\left(\frac{\partial \mathbf{A}}{\partial t} + \nabla\varphi\right) = q\,(\nabla S - q\mathbf{A})\,u^2 . \tag{5.31}$$

We will refer to these equations as the Nonlinear Klein-Gordon-Maxwell equations (NKGM).

In order to show the relation of the above equations with the Maxwell equations written in the usual way, we make the following change of variables:

$$\mathbf{E} = -\left(\frac{\partial \mathbf{A}}{\partial t} + \nabla\varphi\right) \tag{5.32}$$

$$\mathbf{H} = \nabla \times \mathbf{A} \tag{5.33}$$

$$\rho = -\left(\frac{\partial S}{\partial t} + q\varphi\right) qu^2 \tag{5.34}$$

$$\mathbf{j} = (\nabla S - q\mathbf{A}) qu^2. \tag{5.35}$$

So (5.30) and (5.31) are the second couple of the Maxwell equations with respect to a matter distribution whose charge and current density are respectively ρ and $\mathbf{j}$:

$$\nabla \cdot \mathbf{E} = \rho \qquad\qquad \text{(GAUSS)}$$

$$\nabla \times \mathbf{H} - \frac{\partial \mathbf{E}}{\partial t} = \mathbf{j} \qquad\qquad \text{(AMPERE)}$$

(5.32) and (5.33) give rise to the first couple of the Maxwell equations

$$\nabla \times \mathbf{E} + \frac{\partial \mathbf{H}}{\partial t} = 0 \qquad\qquad \text{(FARADAY)}$$

$$\nabla \cdot \mathbf{H} = 0. \qquad\qquad \text{(NOMONOPOLE)}$$

Using the variables $\mathbf{j}$ and ρ, Eq. (5.28) can be written as follows

$$\Box u + W'(u) + \frac{\mathbf{j}^2 - \rho^2}{q^2 u^3} = 0 \qquad\qquad \text{(MATTER)}$$

and finally Eq. (5.29) is the charge continuity equation

$$\frac{\partial}{\partial t}\rho + \nabla \cdot \mathbf{j} = 0. \tag{5.36}$$

Notice that Eq. (5.36) is a consequence of (GAUSS) and (AMPERE). In conclusion, an Abelian gauge theory, via Eqs. (GAUSS)–(MATTER), provides a model of interaction of the matter field ψ with the electromagnetic field $(\mathbf{E}, \mathbf{H})$.

The gauge group is given by $\mathfrak{G} \cong C^\infty(\mathbb{R}^4)$ with the additive structure and it acts on the variables $\psi, \varphi, \mathbf{A}$ as follows

$$T_\chi \psi = \psi e^{i\chi}; \tag{5.37}$$

$$T_\chi \varphi = \varphi - \frac{\partial \chi}{\partial t}; \tag{5.38}$$

$$T_\chi \mathbf{A} = \mathbf{A} + \nabla \chi; \tag{5.39}$$

with $\chi \in C^\infty(\mathbb{R}^4)$. Our equations are gauge invariant by the way they have been constructed. However, if we use the variable $u, \rho, \mathbf{j}, \mathbf{E}, \mathbf{H}$ this fact can be checked directly since these variable are gauge invariant.

5.1.4   Constants of Motion of NKGM

Let us examine the constants of the motion for NKGM which are relevant for us. In this subsection we compute these constants using the gauge invariant variables $u, \rho, \mathbf{j}, \mathbf{E}, \mathbf{H}$.

- **Energy.** Energy, by definition, is the quantity which is preserved by the time invariance of the Lagrangian; using the gauge invariant variables, the energy E, calculated along the solutions of Eq. (GAUSS), takes the following form

$$E = E_m + E_f \tag{5.40}$$

where

$$E_m = \int \left[\frac{1}{2}\left(\left(\frac{\partial u}{\partial t} \right)^2 + |\nabla u|^2 \right) + W(u) + \frac{\rho^2 + \mathbf{j}^2}{2q^2 u^2} \right] dx$$

and

$$E_f = \frac{1}{2} \int \left(\mathbf{E}^2 + \mathbf{H}^2 \right) dx.$$

In Sect. 5.2.3 we shall give a proof of (5.40) (see Proposition 102) which exploits the Hamiltonian formulation of (NKGM). Here we give a direct proof of (5.40) by applying Noether's theorem to the Lagrangian related to the system (5.28)–(5.31).

Proof. By Theorem 5 and by analogous arguments as those used to deduce (1.38), we have that the density of energy ρ_E for the system (5.28)–(5.31) is given by:

$$\rho_E = \frac{\partial \mathcal{L}}{\partial \left(\frac{\partial u}{\partial t} \right)} \cdot \frac{\partial u}{\partial t} + \frac{\partial \mathcal{L}}{\partial \left(\frac{\partial S}{\partial t} \right)} \cdot \frac{\partial S}{\partial t} + \frac{\partial \mathcal{L}}{\partial \left(\frac{\partial \varphi}{\partial t} \right)} \cdot \frac{\partial \varphi}{\partial t} + \frac{\partial \mathcal{L}}{\partial \left(\frac{\partial \mathbf{A}}{\partial t} \right)} \cdot \frac{\partial \mathbf{A}}{\partial t} - \mathcal{L} \tag{5.41}$$

where $\mathcal{L}$ is the Lagrangian related to the system (5.28)–(5.31)

$$\mathcal{L} = \frac{1}{2}\left(\frac{\partial u}{\partial t} \right)^2 - \frac{1}{2}|\nabla u|^2 - W(u) +$$

$$+ \frac{1}{2}\left(\frac{\partial S}{\partial t} + q\varphi \right)^2 - \frac{1}{2}|\nabla S - q\mathbf{A}|^2 \, u^2$$

$$+ \frac{1}{2}\left| \frac{\partial \mathbf{A}}{\partial t} + \nabla \varphi \right|^2 - \frac{1}{2}|\nabla \times \mathbf{A}|^2 . \tag{5.42}$$

Now we will compute each term. We have:

$$\frac{\partial \mathcal{L}}{\partial \left(\frac{\partial u}{\partial t} \right)} \cdot \frac{\partial u}{\partial t} = \left(\frac{\partial u}{\partial t} \right)^2 \tag{5.43}$$

$$\frac{\partial \mathcal{L}}{\partial \left(\frac{\partial S}{\partial t} \right)} \cdot \frac{\partial S}{\partial t} = \left(\frac{\partial S}{\partial t} + q\varphi \right) \frac{\partial S}{\partial t} u^2$$

$$= \left(\frac{\partial S}{\partial t} + q\varphi \right) \frac{\partial S}{\partial t} u^2 + \left(\frac{\partial S}{\partial t} + q\varphi \right) q\varphi u^2 - \left(\frac{\partial S}{\partial t} + \varphi \right) q\varphi u^2$$

$$= \left(\frac{\partial S}{\partial t} + q\varphi \right)^2 u^2 - \left(\frac{\partial S}{\partial t} + q\varphi \right) q\varphi u^2$$

$$= \frac{\rho^2}{q^2 u^2} + \rho\varphi. \tag{5.44}$$

By the Gauss equation (GAUSS), multiplying by φ and integrating, we get

$$-\int \mathbf{E} \cdot \nabla\varphi = \int \rho\varphi. \tag{5.45}$$

Thus by (5.44) and (5.45) we get

$$\int \frac{\partial \mathcal{L}}{\partial \left(\frac{\partial S}{\partial t} \right)} \cdot \frac{\partial S}{\partial t} = \int \frac{\rho^2}{q^2 u^2} - \mathbf{E} \cdot \nabla\varphi. \tag{5.46}$$

Also we have

$$\frac{\partial \mathcal{L}}{\partial \left(\frac{\partial \varphi}{\partial t} \right)} \cdot \frac{\partial \varphi}{\partial t} = 0 \tag{5.47}$$

and

$$\frac{\partial \mathcal{L}}{\partial \left(\frac{\partial \mathbf{A}}{\partial t} \right)} \cdot \frac{\partial \mathbf{A}}{\partial t} = \left(\frac{\partial \mathbf{A}}{\partial t} + \nabla\varphi \right) \cdot \frac{\partial \mathbf{A}}{\partial t} = -\mathbf{E} \cdot \frac{\partial \mathbf{A}}{\partial t}. \tag{5.48}$$

Moreover, using the notation (5.12), (5.13), (5.34), and (5.35), the Lagrangian (5.42) becomes

$$\mathcal{L} = \frac{1}{2} \left(\frac{\partial u}{\partial t} \right)^2 - \frac{1}{2} |\nabla u|^2 - W(u) + \frac{\rho^2 - \mathbf{j}^2}{2q^2 u^2} + \frac{\mathbf{E}^2 - \mathbf{H}^2}{2}. \tag{5.49}$$

Then, by (5.43)–(5.49) and (5.41), we get

$$E = \int \rho_E dx = \int \frac{\partial \mathcal{L}}{\partial \left(\frac{\partial u}{\partial t} \right)} \cdot \frac{\partial u}{\partial t} + \frac{\partial \mathcal{L}}{\partial \left(\frac{\partial S}{\partial t} \right)} \cdot \frac{\partial S}{\partial t} + \frac{\partial \mathcal{L}}{\partial \left(\frac{\partial \mathbf{A}}{\partial t} \right)} \cdot \frac{\partial \mathbf{A}}{\partial t} - \mathcal{L}$$

$$= \int \left(\frac{\partial u}{\partial t} \right)^2 + \frac{\rho^2}{q^2 u^2} - \mathbf{E} \cdot \nabla\varphi - \mathbf{E} \cdot \frac{\partial \mathbf{A}}{\partial t} - \mathcal{L}$$

$$= \int \left(\frac{\partial u}{\partial t}\right)^2 + \frac{\rho^2}{q^2 u^2} + \mathbf{E}^2 \tag{5.50}$$

$$- \int \left[\frac{1}{2}\left(\frac{\partial u}{\partial t}\right)^2 - \frac{1}{2}|\nabla u|^2 - W(u) + \frac{\rho^2 - \mathbf{j}^2}{2q^2 u^2} + \frac{\mathbf{E}^2 - \mathbf{H}^2}{2}\right]$$

$$= \int \left[\frac{1}{2}\left(\frac{\partial u}{\partial t}\right)^2 + \frac{1}{2}|\nabla u|^2 + W(u) + \frac{\rho^2 + \mathbf{j}^2}{2q^2 u^2} + \frac{\mathbf{E}^2 + \mathbf{H}^2}{2}\right].$$

□

- **Momentum.** The momentum **P**, by definition, is the quantity which is preserved by the space invariance of the Lagrangian. By Theorem 5 and by analogous arguments as those used to deduce (1.40), we have that the momentum **P** related to Lagrangian (5.42) takes the following form

$$\mathbf{P} = \mathbf{P}_m + \mathbf{P}_f \tag{5.51}$$

where

$$\mathbf{P}_m = \int \left[-(\partial_t u \, \nabla u) + \frac{\rho \mathbf{j}}{q^2 u^2}\right] dx$$

and

$$\mathbf{P}_f = \int \mathbf{E} \times \mathbf{H} \, dx.$$

- **Angular momentum.** The angular momentum **M**, by definition, is the quantity which is preserved by virtue of the invariance under space rotations of the Lagrangian (5.42) with respect to the origin. It can be shown that:

$$\mathbf{M} = \mathbf{M}_m + \mathbf{M}_f \tag{5.52}$$

where

$$\mathbf{M}_m = \int \left[-\mathbf{x} \times (\nabla u \, \partial_t u) + \mathbf{x} \times \frac{\rho \mathbf{j}}{q^2 u^2}\right] dx \tag{5.53}$$

and

$$\mathbf{M}_f = \int \mathbf{x} \times (\mathbf{E} \times \mathbf{H}) \, dx.$$

Notice that each of the constants of motion $E, \mathbf{P}, \mathbf{M}$ can be splitted in two parts (see (5.40), (5.51), and (5.52)). The first one refers to the "matter field" and the second to the "electromagnetic field".

- **Electric charge.** The electric charge is the quantity Q which is preserved by the gauge action (5.37)–(5.39). Using (5.36) and (5.34), we see that it has the following expression

$$Q = \int \rho dx = -q \int (\partial_t S + q\varphi) u^2 dx .$$ (5.54)

5.1.5 Swarm Interpretation of NKGM

As usual, we start from the continuity equation (5.29) to define the hylenic density ρ_C and the hylenic flow $\mathbf{j}_C$ as follows:

$$\rho_C = -\left(\frac{\partial S}{\partial t} + q\varphi \right) u^2$$

$$\mathbf{j}_C = (\nabla S - q\mathbf{A}) u^2.$$

The hylenic charge is the constant of the motion

$$C = \int \rho_C dx.$$

If we compare the above equations with (5.34) and (5.35) we have that

$$\rho = q\rho_C,$$ (5.55)

$$\mathbf{j} = q\mathbf{j}_C;$$

so q can be interpreted as the absolute value of the electric charge carried by each particle; namely all the particles have the same electric charge up to a sign; the "antiparticles" carry a negative charge since the hylenic charge ρ_C is negative.

The velocity field is given by

$$\mathbf{v} = \frac{\nabla S - q\mathbf{A}}{-\left(\frac{\partial S}{\partial t} + q\varphi \right)}.$$

If we set

$$\Omega = -(\partial_t S + q\varphi)$$ (5.56)

$$\mathbf{K} = \nabla S - q\mathbf{A}$$ (5.57)

$$W(u) = \frac{m^2 u^2}{2} + N(u)$$ (5.58)

Eqs. (5.28) and (5.29) take the form:

$$\Box u + N'(u) + \left(m^2 + \mathbf{K}^2 - \Omega^2\right) u = 0 \tag{5.59}$$

$$\partial_t \left(\Omega u^2\right) + \nabla \cdot \left(\mathbf{K} u^2\right) = 0. \tag{5.60}$$

If

$$\Box u + N'(u) \ll \left(m^2 + \mathbf{K}^2 - \Omega^2\right) u \tag{5.61}$$

Eq. (5.59) can be approximated by

$$\Omega^2 = m^2 + \mathbf{K}^2$$

or

$$\partial_t S + q\varphi + \sqrt{m^2 + |\nabla S - q\mathbf{A}|^2} = 0. \tag{5.62}$$

This is the Hamilton-Jacobi equation of a free relativistic particle of mass m and charge q. If we do not assume (5.61), Eq. (5.62) needs to be replaced by

$$\partial_t S + q\varphi + \sqrt{m^2 + |\nabla S - q\mathbf{A}|^2 + \frac{Q(u)}{u}} = 0$$

with

$$Q(u) = \Box u + N'(u).$$

The term $\frac{Q(u)}{u}$ can be regarded as a field describing a sort of interaction between particles.

5.2   NKGM as a Dynamical System

5.2.1   The Modified Lagrangian

Following [37], in this section and in the next one, we shall show that equations NKGM present a Hamiltonian structure and hence they can be written as equations of first order in t (see Eq. 5.88). This fact is not trivial since the Lagrangian does not depend on $\partial_t \varphi$ (see Sect. 5.2.1) and a suitable study is necessary.

This study gives an insight to the dynamics of the electromagnetic field interacting with matter via NKGM. A particular role is played by the gauge invariance and the phase space can be described in the Lorentz gauge by a manifold of states which satisfy the Gauss equation (GAUSS) (see (5.79)).

The Lagrangian in (5.24) has the following form:

$$\mathcal{L}(\psi, \partial_t \psi, \mathbf{A}, \partial_t \mathbf{A}, \varphi) = \frac{1}{2} \left(|D_\varphi \psi|^2 - |D_\mathbf{A} \psi|^2 \right) - W(\psi) \tag{5.63}$$
$$+ \frac{1}{2} \left(|\partial_t \mathbf{A} + \nabla \varphi|^2 - |\nabla \times \mathbf{A}|^2 \right).$$

Since $\mathcal{L}$ does not depend on $\partial_t \varphi$ it is not possible to make the Legendre transformation with respect to $\partial_t \varphi$ and hence to get an Hamiltonian formulation of the dynamics.

To overcome this difficulty, we set

$$\mathbf{Q} = (\psi, \mathbf{A}, \varphi), \quad \dot{\mathbf{Q}} = (\partial_t \psi, \partial_t \mathbf{A}, \partial_t \varphi)$$

and

$$M_L = \left\{ (\mathbf{Q}, \dot{\mathbf{Q}}) : \partial_t \varphi + \nabla \cdot \mathbf{A} = 0, \nabla \cdot (\partial_t \mathbf{A} + \nabla \varphi) = q \operatorname{Re} \left(i D_\varphi \psi \overline{\psi} \right) \right\} \tag{5.64}$$

where

$$\nabla \cdot (\partial_t \mathbf{A} + \nabla \varphi) = q \operatorname{Re} \left(i D_\varphi \psi \overline{\psi} \right)$$

is the Gauss equation and

$$\partial_t \varphi + \nabla \cdot \mathbf{A} = 0,$$

is called the Lorentz condition and it determines the choice of a particular gauge called *Lorentz gauge*.

We consider the *modified Lagrangian* defined as follows:

$$\hat{\mathcal{L}}(\mathbf{Q}, \dot{\mathbf{Q}}) = \frac{1}{2} \left[|D_\varphi \psi|^2 - |D_\mathbf{A} \psi|^2 \right] - W(\psi) \tag{5.65}$$
$$+ \frac{1}{2} \left[|\partial_t \mathbf{A}|^2 - |\nabla \mathbf{A}|^2 - (\partial_t \varphi)^2 + |\nabla \varphi|^2 \right].$$

The dynamics induced by $\hat{\mathcal{L}}$ is given by the following equations:

$$D_\varphi^2 \psi - D_\mathbf{A}^2 \psi + W'(\psi) = 0 \tag{5.66}$$
$$\Box \mathbf{A} - q \operatorname{Re} \left(i D_\mathbf{A} \psi \overline{\psi} \right) = 0 \tag{5.67}$$
$$\Box \varphi + q \operatorname{Re} \left(i D_\varphi \psi \overline{\psi} \right) = 0. \tag{5.68}$$

The next theorem shows the relation of the modified Lagrangian with the usual Lagrangian.

Theorem 97. *The set M_L is invariant for the dynamics induced by Eqs. (5.66)–(5.68). Moreover, if the initial data are in M_L and $(\psi, \mathbf{A}, \varphi)$ is a smooth solution of Eqs. (5.66)–(5.68) then it is also a solution of (5.25)–(5.27).*

Proof. The charge and current density (5.34) and (5.35), expressed in terms of the variable ψ, take the form

$$\rho = -q \operatorname{Re}\left(i D_\varphi \psi \overline{\psi}\right)$$

$$\mathbf{j} = q \operatorname{Re}\left(i D_{\mathbf{A}} \psi \overline{\psi}\right).$$

Let us first show that M_L is invariant for the dynamics induced by (5.66)–(5.68).
Let

$$\mathbf{Q}(x, t) = (\psi(x, t), \mathbf{A}(x, t), \varphi(x, t))$$

be the solution of (5.66)–(5.68) with initial data

$$(\mathbf{Q}(x, t), \dot{\mathbf{Q}}(x, t))_{t=0} \in M_L.$$

So

$$\partial_t \varphi(x, 0) + \nabla \cdot \mathbf{A}(x, 0) = 0, \ \nabla \cdot (\partial_t \mathbf{A}(x, 0) + \nabla \varphi(x, 0)) = -\rho_0 \qquad (5.69)$$

where

$$\rho_0 = \rho(x, 0) = -q \operatorname{Re}\left(i D_\varphi \psi \overline{\psi}\right)_{t=0}.$$

We want to show that for all $t \geq 0$

$$(\mathbf{Q}(x, t), \dot{\mathbf{Q}}(x, t)) \in M_L$$

namely that for all $t \geq 0$

$$\partial_t \varphi(x, t) + \nabla \cdot \mathbf{A}(x, t) = 0, \qquad\qquad (5.70)$$

$$\nabla \cdot (\partial_t \mathbf{A}(x, t) + \nabla \varphi(x, t)) = q \operatorname{Re}\left(i D_\varphi \psi \overline{\psi}\right). \qquad (5.71)$$

We set for all x, t

$$F = F(x, t) = \partial_t \varphi(x, t) + \nabla \cdot \mathbf{A}(x, t).$$

Then

$$\Box F = \partial_t (\Box \varphi) + \nabla \cdot (\Box \mathbf{A}) = \text{(by 5.67) and (5.68)}$$

$$= \partial_t \rho + \nabla \cdot \mathbf{j} \ \text{(by 5.36)}$$

$$= 0$$

and so

$$\Box F = 0. \tag{5.72}$$

By (5.69)

$$F(x,0) = \partial_t \varphi(x,0) + \nabla \cdot \mathbf{A}(x,0) = 0. \tag{5.73}$$

Moreover

$$\partial_t F = \partial_t^2 \varphi + \nabla \cdot (\partial_t \mathbf{A}) = \text{(by 5.68)} = \nabla \cdot (\nabla \varphi) + \rho + \nabla \cdot (\partial_t \mathbf{A}(x,t)). \tag{5.74}$$

Then

$$\partial_t F(x,0) = \nabla \cdot (\nabla \varphi(x,0) + \partial_t \mathbf{A}(x,0)) + \rho_0 = \text{(by (5.69))} = 0. \tag{5.75}$$

By (5.72), (5.73), and (5.75) we get

$$\Box F = 0$$
$$F = 0 \text{ for } t = 0$$
$$\partial_t F = 0 \text{ for } t = 0.$$

So,

$$F = 0.$$

Then (5.70) is proved. Now in order to prove (5.71), we observe that by (5.70) we get

$$\partial_t F = \partial_t^2 \varphi(x,t) + \nabla \cdot (\partial_t \mathbf{A}(x,t)) = 0 \text{ for all } t.$$

From which, using (5.74), we have

$$\nabla \cdot (\nabla \varphi) + \rho + \nabla \cdot (\partial_t \mathbf{A}(x,t)) = 0 \text{ for all } t.$$

Then (5.71) clearly holds.

Let us now prove the second part of the theorem. Let $(\psi, \mathbf{A}, \varphi)$ be a smooth solution of Eqs. (5.66)–(5.68) with initial data in M_L. We show that it is a solution of (5.25)–(5.27).

Since (5.66) coincides with (5.25), we are reduced to show that $(\psi, \mathbf{A}, \varphi)$ satisfies (5.26) and (5.27).

Clearly we have

$$\mathbf{j} = \text{(by 5.67)}$$
$$= \partial_t^2 \mathbf{A} - \Delta \mathbf{A}$$
$$= \partial_t^2 \mathbf{A} - \nabla (\nabla \cdot \mathbf{A}) + \nabla \times (\nabla \times \mathbf{A}) \text{ (by (5.70))}$$
$$= \partial_t (\partial_t \mathbf{A}) + \nabla (\partial_t \varphi) + \nabla \times (\nabla \times \mathbf{A})$$
$$= \partial_t (\partial_t \mathbf{A} + \nabla \varphi) + \nabla \times (\nabla \times \mathbf{A}).$$

So (5.27) is satisfied. On the other hand

$$\rho = \text{(by (5.68))}$$
$$= \partial_t (\partial_t \varphi) - \nabla \cdot \nabla \varphi \text{ (by (5.70))}$$
$$= -\partial_t \nabla \cdot \mathbf{A} - \nabla \cdot \nabla \varphi$$
$$= -\nabla \cdot (\partial_t \mathbf{A} + \nabla \varphi).$$

Then also (5.26) is satisfied.   □

Remark 98. By the proof of the above theorem, we can easily deduce that any smooth solution $(\psi, \mathbf{A}, \varphi)$ of Eqs. (5.25)–(5.27) belonging to M_L, is also a solution of (5.66)–(5.68).

5.2.2   Hamiltonian Formulation

The Lagrangian (5.65) depends on the configuration variable $\mathbf{Q} = (\psi, \mathbf{A}, \varphi)$ and its time derivative $\dot{\mathbf{Q}} = (\partial_t \psi, \partial_t \mathbf{A}, \partial_t \varphi)$. Now it is possible to define the conjugate variable of $\mathbf{Q}$ via the Legendre transform. The conjugate variable will be denoted by $\mathbf{P} = \left(\hat{\psi}, \hat{\mathbf{A}}, \hat{\varphi}\right)$. We have:

$$\hat{\psi} = \frac{\partial \hat{\mathcal{L}}}{\partial (\partial_t \psi)} = \partial_t \psi + i q \varphi \psi = D_\varphi \psi \tag{5.76}$$

$$\hat{\mathbf{A}} = \frac{\partial \hat{\mathcal{L}}}{\partial (\partial_t \mathbf{A})} = \partial_t \mathbf{A} \tag{5.77}$$

$$\hat{\varphi} = \frac{\partial \hat{\mathcal{L}}}{\partial (\partial_t \varphi)} = -\partial_t \varphi. \tag{5.78}$$

We denote by **u** the state of our dynamical system described by the canonical variables

$$\mathbf{u} = (\mathbf{Q}, \mathbf{P}) = (\psi, \mathbf{A}, \varphi, \hat{\psi}, \hat{\mathbf{A}}, \hat{\varphi}).$$

The invariant set M_L (see (5.64)), expressed in the canonical variables $(\mathbf{Q}, \mathbf{P})$ is given by

$$M_H = \left\{ (\mathbf{Q}, \mathbf{P}) : -\hat{\varphi} + \nabla \cdot \mathbf{A} = 0, \nabla \cdot \left(\hat{\mathbf{A}} + \nabla \varphi \right) = q \operatorname{Re} \left(i \hat{\psi} \overline{\psi} \right) \right\}. \qquad (5.79)$$

We notice that the equation

$$-\hat{\varphi} + \nabla \cdot \mathbf{A} = 0 \qquad (5.80)$$

defines the Lorentz gauge in the canonical variables, while

$$\nabla \cdot \left(\hat{\mathbf{A}} + \nabla \varphi \right) = q \operatorname{Re} \left(i \hat{\psi} \overline{\psi} \right) \qquad (5.81)$$

is nothing else but the Eq. (GAUSS) in these variables.

The duality between **P** and $\dot{\mathbf{Q}}$ is given by

$$\left\langle \mathbf{P}, \dot{\mathbf{Q}} \right\rangle = \operatorname{Re} \left(\hat{\psi} \overline{\partial_t \psi} \right) + \hat{\mathbf{A}} \cdot \partial_t \mathbf{A} + \hat{\varphi} \partial_t \varphi, \qquad (5.82)$$

then, the Hamiltonian density takes the form

$$H(\mathbf{Q}, \mathbf{P}) = \left[\left\langle \mathbf{P}, \dot{\mathbf{Q}} \right\rangle - \hat{\mathcal{L}} \left(\mathbf{Q}, \dot{\mathbf{Q}} \right) \right]_{\dot{\mathbf{Q}} = \dot{\mathbf{Q}}(\mathbf{u})}. \qquad (5.83)$$

By (5.23) and (5.76), we have

$$\partial_t \psi = D_\varphi \psi - i q \varphi \psi = \hat{\psi} - i q \varphi \psi \qquad (5.84)$$

and hence, inserting (5.84), (5.77) and (5.78) in (5.82), we have

$$\left\langle \mathbf{P}, \dot{\mathbf{Q}} \right\rangle_{\dot{\mathbf{Q}} = \dot{\mathbf{Q}}(\mathbf{u})} = \left| \hat{\psi} \right|^2 + q \varphi \operatorname{Re} \left(i \hat{\psi} \overline{\psi} \right) + \left| \hat{\mathbf{A}} \right|^2 - \hat{\varphi}^2. \qquad (5.85)$$

Moreover the Lagrangian $\hat{\mathcal{L}}$ (see (5.65)), expressed in the canonical variables $\mathbf{u} = (\mathbf{Q}, \mathbf{P})$, takes the form

$$\hat{\mathcal{L}}(\mathbf{Q}, \mathbf{P}) = \hat{\mathcal{L}} \left(\mathbf{Q}, \dot{\mathbf{Q}}(\mathbf{u}) \right) \qquad (5.86)$$

$$= \frac{1}{2} \left(\left| \hat{\psi} \right|^2 - |D_\mathbf{A} \psi|^2 \right) + \frac{1}{2} \left(\left| \hat{\mathbf{A}} \right|^2 - |\nabla \mathbf{A}|^2 - \hat{\varphi}^2 + |\nabla \varphi|^2 \right) - W(\psi).$$

Then by (5.85) and (5.86) the Hamiltonian density (5.83) becomes

$$H(\mathbf{Q}, \mathbf{P}) = \frac{1}{2} \left(\left| \hat{\psi} \right|^2 + \left| \hat{\mathbf{A}} \right|^2 - \hat{\varphi}^2 + |D_{\mathbf{A}} \psi|^2 + |\nabla \mathbf{A}|^2 - |\nabla \varphi|^2 \right)$$

$$+ q\varphi \operatorname{Re} \left(i \hat{\psi} \overline{\psi} \right) + W(\psi). \tag{5.87}$$

The action (in a bounded domain Ω) is given by:

$$\mathcal{S}(\mathbf{Q}, \mathbf{P}) = \int_{\Omega} \left(\left\langle \mathbf{P}, \dot{\mathbf{Q}} \right\rangle - H(\mathbf{Q}, \mathbf{P}) \right) dx dt$$

$$= \int_{\Omega} \left(\operatorname{Re} \left(\hat{\psi} \overline{\partial_t \psi} \right) + \hat{\mathbf{A}} \cdot \partial_t \mathbf{A} + \hat{\varphi} \partial_t \varphi \right) dx dt$$

$$- \int_{\Omega} \left[\frac{1}{2} \left(\left| \hat{\psi} \right|^2 + \left| \hat{\mathbf{A}} \right|^2 - \hat{\varphi}^2 + |D_{\mathbf{A}} \psi|^2 + |\nabla \mathbf{A}|^2 - |\nabla \varphi|^2 \right) \right] dx dt$$

$$- \int_{\Omega} \left[q\varphi \operatorname{Re} \left(i \hat{\psi} \overline{\psi} \right) + W(\psi) \right] dx dt.$$

Making the variations of $\mathcal{S}(\mathbf{Q}, \mathbf{P})$ with respect $\hat{\psi}, \psi, \hat{\mathbf{A}}, \mathbf{A}, \hat{\varphi}, \varphi$ respectively, we get the canonical equations of motion:

$$\partial_t \psi = \hat{\psi} - i q \varphi \psi$$

$$\partial_t \hat{\psi} = -i q \varphi \hat{\psi} + D_{\mathbf{A}}^2 \psi - W'(\psi)$$

$$\partial_t \mathbf{A} = \hat{\mathbf{A}} \tag{5.88}$$

$$\partial_t \hat{\mathbf{A}} = \Delta \mathbf{A} + q \operatorname{Re} \left(i D_{\mathbf{A}} \psi \overline{\psi} \right)$$

$$\partial_t \varphi = -\hat{\varphi}$$

$$\partial_t \hat{\varphi} = -\Delta \varphi - q \operatorname{Re} \left(i \hat{\psi} \overline{\psi} \right).$$

Remark 99. If we use the covariant derivative D_{φ} (5.23), the Hamilton equations take a simpler form:

$$D_{\varphi} \psi = \hat{\psi} \tag{5.89}$$

$$D_{\varphi} \hat{\psi} = D_{\mathbf{A}}^2 \psi - W'(\psi) \tag{5.90}$$

$$\partial_t \mathbf{A} = \hat{\mathbf{A}} \tag{5.91}$$

$$\partial_t \hat{\mathbf{A}} = \Delta \mathbf{A} + q \operatorname{Re} \left(i D_{\mathbf{A}} \psi \overline{\psi} \right) \tag{5.92}$$

$$\partial_t \varphi = -\hat{\varphi} \tag{5.93}$$

$$\partial_t \hat{\varphi} = -\Delta \varphi - q \operatorname{Re} \left(i D_{\varphi} \psi \overline{\psi} \right).$$

Remark 100. In the following we shall assume that the Cauchy problem for (5.88) is globally well posed in a suitable function space (namely $\overline{M}_H$ with the distance (5.108), see page 167). Actually, in the literature there are few results relative to this problem (we know only [75, 97, 108, 119]) and we do not know which are the assumptions that W should satisfy.

5.2.3 The Phase Space of NKGM

Until now, we have assumed that M_H defined in (5.79) (or equivalently M_L) consist of C^∞ functions; in order to apply the theory of Sect. 2.2, we would like to define a *natural* metric and to take the completion of M_H with respect to this metric.

The choice of the "right" metric is a delicate problem which depends on mathematical and physical considerations. In many problems, we have that

$$\{\text{energy}\} = \{\text{positive quadratic form}\} + \{\text{higher order terms}\}. \tag{5.94}$$

In these problems, usually, the "norm of the energy" is a good choice:

$$\|\cdot\| := \sqrt{\{\text{positive quadratic form}\}}. \tag{5.95}$$

In NKGM the energy is just the Hamiltonian $\mathcal{H}(\mathbf{Q}, \mathbf{P})$ and by (5.87), if the energy is finite, we have

$$\mathcal{H}(\mathbf{Q}, \mathbf{P}) = \int H(\mathbf{Q}, \mathbf{P}) dx$$

$$= \frac{1}{2} \int \left(\left|\hat{\psi}\right|^2 + \left|\hat{\mathbf{A}}\right|^2 - \hat{\varphi}^2 + |D_{\mathbf{A}}\psi|^2 + |\nabla\mathbf{A}|^2 - |\nabla\varphi|^2 \right) dx +$$

$$+ \int \left[q\varphi \operatorname{Re}\left(i\hat{\psi}\overline{\psi} \right) + W(\psi) \right] dx.$$

We observe that the Hamiltonian $\mathcal{H}(\mathbf{Q}, \mathbf{P})$ in general is not positive. However, if $(\mathbf{Q}, \mathbf{P}) \in M_H$ and if $W \geq 0$, $\mathcal{H}(\mathbf{Q}, \mathbf{P})$ is positive as the following proposition shows:

Proposition 101. *If $(\mathbf{Q}, \mathbf{P}) \in M_H$, and the energy is finite, then*

$$\mathcal{H}(\mathbf{Q}, \mathbf{P}) = \frac{1}{2} \int \left(\left|\hat{\psi}\right|^2 + |D_{\mathbf{A}}\psi|^2 + \left|\nabla\varphi + \hat{\mathbf{A}}\right|^2 + |\nabla \times \mathbf{A}|^2 \right) + \int W(\psi). \tag{5.96}$$

Proof. We recall that, if $(\mathbf{Q}, \mathbf{P}) \in M_H$, then by Eq. (5.80), we have that

$$|\hat{\varphi}|^2 = |\nabla \cdot \mathbf{A}|^2$$

and, recalling that

$$|\nabla \mathbf{A}|^2 = |\nabla \cdot \mathbf{A}|^2 + |\nabla \times \mathbf{A}|^2,$$

we get

$$\mathcal{H}(\mathbf{Q}, \mathbf{P}) = \frac{1}{2} \int \left(\left| \hat{\psi} \right|^2 + \left| \hat{\mathbf{A}} \right|^2 + |D_{\mathbf{A}} \psi|^2 + |\nabla \times \mathbf{A}|^2 - |\nabla \varphi|^2 \right)$$
$$+ \int q\varphi \operatorname{Re} \left(i \hat{\psi} \overline{\psi} \right) + W(\psi). \tag{5.97}$$

Moreover, by (5.81), multiplying by φ and integrating by parts, we get

$$\int |\nabla \varphi|^2 + \nabla \varphi \cdot \hat{\mathbf{A}} = q \int \varphi \operatorname{Re} \left(i \hat{\psi} \overline{\psi} \right).$$

Thus, replacing $q \int \varphi \operatorname{Re} \left(i \hat{\psi} \overline{\psi} \right)$ in Eq. (5.97), we have

$$\mathcal{H}(\mathbf{Q}, \mathbf{P}) = \frac{1}{2} \int \left(\left| \hat{\psi} \right|^2 + \left| \hat{\mathbf{A}} \right|^2 + |D_{\mathbf{A}} \psi|^2 + |\nabla \times \mathbf{A}|^2 + |\nabla \varphi|^2 + 2\nabla \varphi \cdot \hat{\mathbf{A}} \right) + \int W(\psi)$$
$$= \frac{1}{2} \int \left(\left| \hat{\psi} \right|^2 + |D_{\mathbf{A}} \psi|^2 + \left| \nabla \varphi + \hat{\mathbf{A}} \right|^2 + |\nabla \times \mathbf{A}|^2 \right) + \int W(\psi).$$

$\square$

Now we want to express the energy in the gauge invariant variables (GIV) $u, v, \rho, \mathbf{j}, \mathbf{E}, \mathbf{H}$ defined by

$$u = |\psi|$$
$$v = \frac{\operatorname{Re}(\hat{\psi} \overline{\psi})}{|\psi|}$$
$$\rho = -q \operatorname{Re} \left(i \hat{\psi} \overline{\psi} \right) \tag{5.98}$$
$$\mathbf{j} = q \operatorname{Re} \left(i D_{\mathbf{A}} \psi \overline{\psi} \right)$$
$$\mathbf{E} = - \left(\nabla \varphi + \hat{\mathbf{A}} \right)$$
$$\mathbf{H} = \nabla \times \mathbf{A}.$$

We notice that $\mathbf{E}$ and $\mathbf{H}$ coincide with the electromagnetic field defined by Eqs. (5.32) and (5.33), while ρ and $\mathbf{j}$ are the electric and the current density. We will denote by

$$\mathcal{G} : \mathbf{u} = (\psi, \hat{\psi}, \mathbf{A}, \hat{\mathbf{A}}, \varphi, \hat{\varphi}) \mapsto \mathcal{G}(\mathbf{u}) = (u, v, \rho, \mathbf{j}, \mathbf{E}, \mathbf{H})$$

the transformation (5.98), so that $\mathbf{U} := \mathcal{G}(\mathbf{u})$ represents a state in the gauge invariant variables.

We now set

$$M_\mathcal{G} = \left\{ \mathbf{U} = (u, v, \rho, \mathbf{j}, \mathbf{E}, \mathbf{H}) \in C^\infty(\mathbb{R}^3, \mathbb{R}^{12}) \mid \nabla \cdot \mathbf{E} = \rho, \ \nabla \cdot \mathbf{H} = 0, \ E(\mathbf{U}) < +\infty \right\}$$

where $E(\mathbf{U})$ is the energy as function of $\mathbf{U}$. The expression of the energy is given by the following proposition:

Proposition 102. *If* $\mathbf{U} \in M_\mathcal{G}$*, then the energy takes the following expression:*

$$E(\mathbf{U}) = \frac{1}{2} \int \left(v^2 + |\nabla u|^2 + \frac{\rho^2 + \mathbf{j}^2}{q^2 u^2} + \mathbf{E}^2 + \mathbf{H}^2 \right) dx + \int W(u) dx.$$

Proof. By using the polar form $\psi = u e^{iS}$ and by the first equation of (5.88), we have

$$v = \frac{\mathrm{Re}(\hat{\psi}\overline{\psi})}{|\psi|} = \frac{\mathrm{Re}\left[(\partial_t \psi + iq\varphi\psi) \overline{\psi} \right]}{u}$$

$$= \frac{\mathrm{Re}\left[\partial_t \psi \overline{\psi} \right]}{u} = \frac{\mathrm{Re}\left[(\partial_t u e^{iS} + i\partial_t S u e^{iS}) u e^{-iS} \right]}{u}$$

$$= \partial_t u.$$

Then

$$v = \partial_t u. \tag{5.99}$$

Using again the first equation of (5.88) and the polar form of ψ, we have

$$\hat{\psi} = \partial_t \psi + iq\varphi\psi = [\partial_t u + i\,(\partial_t S + q\varphi)\,u]\,e^{iS}. \tag{5.100}$$

Moreover, by (5.34),

$$\rho = -q\,(\partial_t S + q\varphi)\,u^2. \tag{5.101}$$

Thus, by (5.100), (5.101) and (5.99), we have

$$\left|\hat{\psi}\right|^2 = |\partial_t u|^2 + |\partial_t S + q\varphi|^2\,u^2 = |v|^2 + \frac{\rho^2}{q^2 u^2}. \tag{5.102}$$

Similarly, writing ψ in polar form, we have

$$D_\mathbf{A}\psi = \nabla\psi - iq\mathbf{A}\psi = [\nabla u - i\,(\nabla S - q\mathbf{A})\,u]\,e^{iS}$$

and using (5.35)

$$\mathbf{j} = q\,(\nabla S - q\mathbf{A})\,u^2,$$

we get

$$|D_{\mathbf{A}}\psi|^2 = |\nabla u|^2 + |\nabla S - q\mathbf{A}|^2\,u^2 = |\nabla u|^2 + \frac{\mathbf{j}^2}{q^2 u^2}. \tag{5.103}$$

Thus, by (5.96), (5.102), (5.103) and by the last two equations of GIV

$$E\,(\mathbf{u}) = \frac{1}{2}\int \left(\left|\hat{\psi}\right|^2 + |D_{\mathbf{A}}\psi|^2 + \left|\nabla\varphi + \hat{\mathbf{A}}\right|^2 + |\nabla \times \mathbf{A}|^2\right) + \int W(\psi)$$

$$E\,(\mathbf{U}) = \frac{1}{2}\int \left[v^2 + \frac{\rho^2}{q^2 u^2} + |\nabla u|^2 + \frac{\mathbf{j}^2}{q^2 u^2} + \mathbf{E}^2 + \mathbf{H}^2\right] dx + \int W(u)dx.$$

$$\square$$

As usual, we write W as follows

$$W(s) = \frac{m^2}{2}s^2 + N(s),\;\; N(s) = o(s^2)$$

and we will assume $m > 0$. Then we have

$$E(\mathbf{U}) = \frac{1}{2}\int \left[v^2 + |\nabla u|^2 + m^2 u^2 + \frac{\rho^2 + \mathbf{j}^2}{q^2 u^2} + \mathbf{E}^2 + \mathbf{H}^2\right] dx + \int N(u)dx.$$

The term $\left(\rho^2 + \mathbf{j}^2\right)/u^2$ is singular and the energy does not have the form (5.94). In order to avoid this problem, it is convenient to introduce new gauge invariant variables which eliminate this singularity:

$$\theta = \frac{-\rho}{qu} = \frac{1}{q}\left(\frac{\partial S}{\partial t} + q\varphi\right)u;\;\; \Theta = \frac{\mathbf{j}}{qu} = \frac{1}{q}\left(\nabla S - q\mathbf{A}\right)u. \tag{5.104}$$

We denote this new change of variables by

$$\mathcal{Q} : (\psi,\hat{\psi},\mathbf{A},\hat{\mathbf{A}},\varphi,\hat{\varphi}) \mapsto (u,v,\theta,\Theta,\mathbf{j},\mathbf{E},\mathbf{H})$$

and we set $\mathbf{V} = \mathcal{Q}(\mathbf{u}) = (u,v,\theta,\Theta,\mathbf{j},\mathbf{E},\mathbf{H})$.

Using these new variables the energy takes the form:

$$E\,(\mathbf{V}) = \frac{1}{2}\int \left[v^2 + |\nabla u|^2 + m^2 u^2 + \theta^2 + \Theta^2 + \mathbf{E}^2 + \mathbf{H}^2\right] + \int N(u). \tag{5.105}$$

Moreover, by (5.104) and (5.55) the charge C takes the form

$$C = C(u, \theta) = -\int \theta u \, dx. \tag{5.106}$$

Thus, we can take a norm having the form (5.95), namely

$$\|\mathbf{V}\| = \left(\int \left[v^2 + |\nabla u|^2 + m^2 u^2 + \theta^2 + \Theta^2 + \mathbf{E}^2 + \mathbf{H}^2 \right] dx \right)^{\frac{1}{2}}. \tag{5.107}$$

This norm will induce a distance on the set M_H defined by (5.79):

$$d_{M_H}(\mathbf{u}_1, \mathbf{u}_2) = \|\mathcal{Q}(\mathbf{u}_1) - \mathcal{Q}(\mathbf{u}_2)\|. \tag{5.108}$$

We will denote by $\overline{M}_H$ the completion of M_H with respect to the above distance.

As pointed out in a preceding remark, in the following we shall assume that the Cauchy problem for Eq. (5.88) is globally well posed in $\overline{M}_H$ with respect to the distance (5.108). So NKGM can be considered as a dynamical system $(\overline{M}_H, \gamma_t)$ whose dynamics is given by (5.88).

So the definition of hylomorphic soliton in this situation becomes the following one:

Definition 103. A state $\mathbf{u} = (\psi, \hat{\psi}, \mathbf{A}, \hat{\mathbf{A}}, \varphi, \hat{\varphi}) \in \overline{M}_H$ is called hylomorphic soliton with respect to the dynamics given by (5.88) if it satisfies the properties of Definitions 16 and 20 where the linear space X is replaced by $\overline{M}_H$ equipped with the metric (5.108).

5.3 Existence of Charged Solitons for NKGM

5.3.1 Statement of the Existence Result

As usual, the first tentative to find solitary waves of NKGM is to look for standing waves. Thus let us look for solutions of Eqs. (5.28)–(5.31) of the type

$$u(t, x) = u(x)$$

$$S(t, x) = -\omega t$$

$$\mathbf{A} = 0$$

$$\varphi(t, x) = \varphi(x).$$

Replacing the above variables in Eqs. (5.28) and (5.30) we get

$$-\Delta u + W'(u) - (q\varphi - \omega)^2 u = 0 \tag{5.109}$$

$$\Delta\varphi - (q\varphi - \omega)\, qu^2 = 0 \qquad\qquad (5.110)$$

while Eqs. (5.29) and (5.31) are identically satisfied.

The above equations have a variational structure, namely they are the Euler-Lagrange equations of the functional

$$\frac{1}{2}\int \left(|\nabla u|^2 - |\nabla\varphi|^2 - (q\varphi - \omega)^2\, u^2\right) dx + \int W(u) dx.$$

Existence results for Eqs. (5.109) and (5.110), under various assumptions on W, are stated in many papers [8, 9, 28, 31, 52, 64, 65, 69, 112]. However there are not many results for the stability.

The problem with the stability of electrically charged matter fields is that the electric charge tends to brake them, since charges of the same sign repel each other. In this respect Coleman in his celebrated paper [60] says "*I have been unable to construct Q-balls when the continuous symmetry is gauged. I think what is happening physically is that the long-range force caused by the gauge field forces the charge inside the Q-ball to migrate to the surface, and this destabilizes the system, but I am not sure of this*".

We have the following result:

Theorem 104. *Assume that W satisfies the same assumptions of Theorem 77, namely assumptions (WC-0),(WC-i),(WC-ii) (page 120) and (WC-iii') (page 122). Then there exists $\bar{q} > 0$ such that for every $q \in [0, \bar{q}]$ the dynamical system described by (5.88) has a family $\mathbf{u}_\delta$ ($\delta \in (0, \delta_\infty)$, $\delta_\infty > 0$) of hylomorphic solitons in the sense of Definition 103. The evolution of $\mathbf{u}_\delta$ is given by*

$$\mathbf{u}_\delta(t) = (\psi_\delta(t),\ \hat{\psi}_\delta(t),\ \mathbf{A}_\delta(t),\ \hat{\mathbf{A}}_\delta(t),\ \varphi_\delta(t),\ \hat{\varphi}_\delta(t))$$
$$= (u_\delta(x)e^{-i\omega_\delta t},\ D_\varphi(u_\delta(x)e^{-i\omega_\delta t}),\ \mathbf{0}, \mathbf{0},\ \varphi_\delta(x), 0).$$

More exactly, $u_\delta(x), \varphi_\delta(x), \omega_\delta$ satisfy Eqs. (5.109) (5.110) and hence, $\mathbf{u}_\delta(t)$ satisfies (5.88). Moreover if $\delta_1 < \delta_2$ we have that

(a) $\Lambda(\mathbf{u}_{\delta_1}) < \Lambda(\mathbf{u}_{\delta_2})$.
(b) $|C(\mathbf{u}_{\delta_1})| > |C(\mathbf{u}_{\delta_2})|$.
(c) $E(\mathbf{u}_{\delta_1}) > E(\mathbf{u}_{\delta_2})$.

A variant of Theorem 104 was first proved in [36]. This theorem establishes that stable charged Q-balls namely charged solitons do exist, provided that the interaction constant q between matter and gauge field is sufficiently small. Thus, this result gives a partial answer to the problem risen by Coleman. In order to give a complete answer, we should know what happens if the interaction is strong. We conjecture that, in this case, charged stable Q-balls do not exist, but a complete proof does not exist.

5.3.2 Working with the Gauge Invariant Variables.

In this and in the next subsections, we shall prove the existence of charged hylomorphic solitons by using the abstract Theorems 34 and 38. In order to do this we will use the variables $(u, v, \theta, \Theta, \mathbf{E}, \mathbf{H})$ defined by (5.98) and (5.104). At the end, in the proof of Theorem 104 we will go back to the variables $\hat{\psi}, \psi, \hat{\mathbf{A}}, \mathbf{A}, \hat{\varphi}, \varphi$.

Consider the set

$$M = \{\mathbf{V} = (u, v, \theta, \Theta, \mathbf{E}, \mathbf{H}) : \|\mathbf{V}\| < \infty, \ \nabla \cdot \mathbf{E} = -q\theta u, \ \nabla \cdot \mathbf{H} = 0\} \quad (5.111)$$

where $\|\mathbf{V}\|$ is defined in (5.107). The energy and the charge are defined by (5.105), (5.106).

Unfortunately, M is not a linear space and Theorem 34 cannot be applied. However, we can reduce our problem to the situation described in Sect. 2.2.2 by virtue of the following remarks:

- If $(u, v, \theta, \Theta, \mathbf{E}, \mathbf{H})$ minimizes the energy E in (5.105) on M, then we can take $\mathbf{H} = \Theta = \mathbf{0}$ and $v = 0$.
- Since $\mathbf{H} = \mathbf{0}$ we can take $\mathbf{E} = -\nabla \varphi$.

Then the constraint

$$\nabla \cdot \mathbf{E} = -q\theta u$$

permits to write $\mathbf{E}$ as a function of $(u, \theta) \in H^1(\mathbb{R}^3) \times L^2(\mathbb{R}^3)$, namely

$$\mathbf{E} = -\nabla \Phi(u, \theta) \quad (5.112)$$

where $\varphi = \Phi(u, \theta) \in \mathcal{D}^{1,2}$[1] is the unique solution (see proof of Proposition 105) of the equation

$$\Delta \varphi = qu\theta. \quad (5.113)$$

Clearly we can write also

$$\mathbf{E}(u, \theta) = q\nabla \left(\int \frac{\theta(y)u(y)}{|x - y|} \, dy \right).$$

We have taken $\mathbf{H} = \Theta = \mathbf{0}, v = 0$ and the electric field $\mathbf{E} = \mathbf{E}(u, \theta)$ is a function of u, θ. Then the energy E in (5.105), which we want to minimize on M, will depend only on the variables u, θ. Then the generic point in the phase space can be identified with

---

[1] $\mathcal{D}^{1,2}$ is the closure of C_0^∞ with respect to the norm $\|\nabla \varphi\|_{L^2}$.

$$\mathbf{u} = (u, \theta) \in X$$

where

$$X = H^1\left(\mathbb{R}^3\right) \times L^2\left(\mathbb{R}^3\right) \tag{5.114}$$

equipped with the following norm:

$$\|\mathbf{u}\|^2 = \|(u, \theta)\|^2 = \int \left[|\nabla u|^2 + m^2 u^2 + \theta^2\right] dx. \tag{5.115}$$

Now the energy is given by the expression:

$$E\left(\mathbf{u}\right) = \frac{1}{2} \|\mathbf{u}\|^2 + K\left(\mathbf{u}\right) + \int N(u) dx \tag{5.116}$$

where

$$K\left(\mathbf{u}\right) = \frac{1}{2} \int \mathbf{E}^2(u, \theta) dx = \frac{1}{2} \int |\nabla \Phi(u, \theta)|^2. \tag{5.117}$$

Proposition 105. *The function* $K : X \to \mathbb{R}$ *as defined above is differentiable.*

Proof. Let $u \in H^1\left(\mathbb{R}^3\right)$ and $\theta \in L^2\left(\mathbb{R}^3\right)$. Then, by Sobolev embedding, $u \in L^2\left(\mathbb{R}^3\right) \cap L^6\left(\mathbb{R}^3\right)$ and by interpolation $u \in L^3\left(\mathbb{R}^3\right)$. Hence, by the Hölder inequality, $u\theta \in L^{\frac{6}{5}}\left(\mathbb{R}^3\right)$. Since $\mathcal{D}^{1,2}$ is embedded into $L^6\left(\mathbb{R}^3\right)$, we have

$$u\theta \in L^{\frac{6}{5}}\left(\mathbb{R}^3\right) = \left(L^6\left(\mathbb{R}^3\right)\right)' \subset \left(\mathcal{D}^{1,2}\right)'$$

Δ is an isomorphism between $\mathcal{D}^{1,2}$ and its dual $\left(\mathcal{D}^{1,2}\right)'$; then there exists a unique solution $\varphi = \Phi(u, \theta) \in \mathcal{D}^{1,2}$ of the equation

$$\Delta \varphi = qu\theta. \tag{5.118}$$

Explicitly, we have

$$\varphi = \Phi(u, \theta) = q\Delta^{-1}(u\theta) = -q \int \frac{\theta(y)u(y)}{|x - y|} dy. \tag{5.119}$$

By using the above expression it can be directly verified that the map

$$(u, \theta) \in X \to \varphi = \Phi(u, \theta)$$

is differentiable and for any $h \in H^1(\mathbb{R}^3)$ and $\eta \in L^2(\mathbb{R}^3)$ we have

$$\frac{\partial \Phi(u, \theta)}{\partial u}(h) = q\Delta^{-1}(\theta h) \tag{5.120}$$

$$\frac{\partial \Phi(u, \theta)}{\partial \vartheta}(\eta) = q\Delta^{-1}(u\eta). \tag{5.121}$$

For any $h \in H^1(\mathbb{R}^3)$, by (5.120) and (5.118), we have that

$$\left\langle \frac{\partial K(u, \theta)}{\partial u}, h \right\rangle = \left\langle -\Delta\varphi, \frac{\partial \Phi(u, \theta)}{\partial u}(h) \right\rangle = \left\langle -\Delta\varphi, \, q\Delta^{-1}(\theta h) \right\rangle$$

$$= \left\langle -qu\theta, q\Delta^{-1}(\theta h) \right\rangle = \left\langle -\Delta^{-1}qu\theta, q\theta h \right\rangle = \left\langle -\varphi, q\theta h \right\rangle$$

$$= -\int q\theta\varphi h dx = -\left\langle \varphi q\theta, h \right\rangle.$$

So we get

$$\frac{\partial K(u, \theta)}{\partial u} = -q\theta\varphi. \tag{5.122}$$

Analogously we have

$$\frac{\partial K(u, \theta)}{\partial \theta} = -qu\varphi. \tag{5.123}$$

Then the conclusion easily follows. $\square$

5.3.3 Proof of the Existence Result

Now, we can apply the theory of Sect. 2.2.2 to the functions E and C defined on X by (5.116) and (5.106) respectively. As usual, Λ will denote the hylenic ratio

$$\Lambda = \frac{E(\mathbf{u})}{|C(\mathbf{u})|}.$$

First of all observe that the energy E and the hylenic charge C, defined in (5.116) and (5.106) are invariant under translations i. e. under the representation T_z of the group $G = \mathbb{R}^3$

$$T_z\mathbf{u}(x) = \mathbf{u}(x + z), \quad z \in \mathbb{R}^3.$$

In order to prove Theorem 104 we shall use Theorem 38.

First we show that the splitting property (EC-2) is satisfied.

Lemma 106. *The energy E (5.116) and the hylenic charge C (5.106) satisfy the splitting property (EC-2).*

Proof. The hylenic charge C satisfies the splitting property since it is quadratic. In order to prove that the energy

$$E\,(\mathbf{u}) = \frac{1}{2}\,\|\mathbf{u}\|^2 + K\,(\mathbf{u}) + \int N(u)dx$$

satisfies the splitting property it is sufficient to prove that each piece does it. The term $\frac{1}{2}\,\|\mathbf{u}\|^2$ satisfies the splitting property since it is quadratic; also $\int N(u)dx$ satisfies this property as it has been proved in Lemma 53. It remains to prove that the splitting property is satisfied by the functional

$$K\,(u,\theta) = \frac{1}{2}\int |\nabla\Phi(u,\theta)|^2\,.$$

So let $\{u_n\} \subset H^1(\mathbb{R}^3)$ and $\{\theta_n\} \subset L^2(\mathbb{R}^3)$ be sequences such that

$$u_n \rightharpoonup u \text{ weakly in } H^1(\mathbb{R}^3),\ \theta_n \rightharpoonup \theta \text{ weakly in } L^2(\mathbb{R}^3)$$

and set

$$\tilde{u}_n = u_n - u,\ \tilde{\theta}_n = \theta_n - \theta.$$

Moreover set $\tilde{\varphi}_n = \Phi(\tilde{u}_n, \tilde{\theta}_n)$, $\varphi_n = \Phi(u_n, \theta_n)$, $\varphi = \Phi(u,\theta)$, namely $\tilde{\varphi}_n$, φ_n, φ are the solutions in $\mathcal{D}^{1,2}(\mathbb{R}^3)$ of the following equations:

$$\Delta\tilde{\varphi}_n = \tilde{u}_n\tilde{\theta}_n$$

$$\Delta\varphi_n = u_n\theta_n$$

$$\Delta\varphi = u\theta.$$

Notice that here, in order to simplify the notation, we have put $q = 1$.

It is easy to see that $u_n\theta_n$ is bounded in $L^{\frac{6}{5}}(\mathbb{R}^3) \subset \left(\mathcal{D}^{1,2}\right)'$. Then φ_n is bounded in $\mathcal{D}^{1,2}(\mathbb{R}^3)$ and consequently also in $L^6(\mathbb{R}^3)$.

We need to prove that

$$K\,(u_n,\theta_n) = K(u,\theta) + K(\tilde{u}_n, \tilde{\theta}_n) + o(1).$$

Let us estimate $K\,(u_n,\theta_n)$:

$$2K\,(u_n,\theta_n) = \int |\nabla\Phi(u_n,\theta_n)|^2 = \int |\nabla\varphi_n|^2 = \int -\Delta\varphi_n \cdot \varphi_n = -\int u_n\theta_n\varphi_n$$

$$= -\int (\tilde{u}_n + u)\left(\tilde{\theta}_n + \theta\right)\varphi_n$$

$$= -\int u\theta\varphi_n - \int u\tilde{\theta}_n\varphi_n - \int \tilde{u}_n\theta\varphi_n - \int \tilde{u}_n\tilde{\theta}_n\varphi_n$$

$$= -\int u\theta\varphi - \int u\theta\,(\varphi_n - \varphi) - \int u\tilde{\theta}_n\varphi_n$$

$$\quad - \int \tilde{u}_n\theta\varphi_n - \int \tilde{u}_n\tilde{\theta}_n\tilde{\varphi}_n - \int \tilde{u}_n\tilde{\theta}_n\,(\varphi_n - \tilde{\varphi}_n)$$

$$= 2K\,(u, \theta) + 2K\left(\widetilde{u}_n, \tilde{\theta}_n\right) - A_n - B_n - C_n - D_n$$

where

$$A_n = \int u\theta\,(\varphi_n - \varphi)$$

$$B_n = \int u\tilde{\theta}_n\varphi_n$$

$$C_n = \int \tilde{u}_n\theta\varphi_n$$

$$D_n = \int \tilde{u}_n\tilde{\theta}_n\,(\varphi_n - \tilde{\varphi}_n)\,.$$

It remains to show that these four quantities vanish. We estimate each piece separately:

$$A_n = \int u\theta\,(\varphi_n - \varphi) = \int \Delta\varphi\,(\varphi_n - \varphi) = \int \varphi\Delta\,(\varphi_n - \varphi) \quad (5.124)$$

$$= \int \varphi\,(u_n\theta_n - u\theta) = \int \varphi u_n\,(\theta_n - \theta) + \int \varphi\,(u_n - u)\,\theta$$

$$= \int \varphi u_n\tilde{\theta}_n + \int \varphi\tilde{u}_n\theta$$

$$= \int_{B_R} \varphi u_n\tilde{\theta}_n + \int_{B_R^c} \varphi u_n\tilde{\theta}_n + \int \varphi\tilde{u}_n\theta. \quad (5.125)$$

We fix $\varepsilon > 0$. We can choose R so large that $\|\varphi\|_{L^6(B_R^c)} < \varepsilon$, and so

$$\left|\int_{B_R^c} \varphi u_n\tilde{\theta}_n\right| \le \varepsilon \left\|\tilde{\theta}_n\right\|_{L^2(B_R^c)} \|u_n\|_{L^3(B_R^c)}$$

$$\le c_1\varepsilon \left\|\tilde{\theta}_n\right\|_{L^2(B_R^c)} \|u_n\|_{H^1(B_R^c)} \le c_2\varepsilon \quad (5.126)$$

where c_1 and c_2 do not depend on n. Moreover $u_n \to u$ in $L^3(B_R)$ and hence $\varphi u_n \to \varphi u$ in $L^2(B_R)$. So, since $\tilde{\theta}_n \rightharpoonup 0$ in L^2, it follows that

$$\int_{B_R} \varphi u_n \tilde{\theta}_n \to 0. \tag{5.127}$$

Finally it is immediate to see that

$$\int \varphi \tilde{u}_n \theta \to 0. \tag{5.128}$$

So by (5.125)–(5.128) we get

$$|A_n| \leq c_2 \varepsilon + o(1). \tag{5.129}$$

Now let us consider B_n. We have that

$$B_n = \int u \tilde{\theta}_n \varphi_n = \int_{B_R} u \tilde{\theta}_n \varphi_n + \int_{B_R^c} u \tilde{\theta}_n \varphi_n.$$

We can choose R so large that $\|u\|_{L^3(B_R^c)} < \varepsilon$. Then, since φ_n is bounded in $L^6(\mathbb{R}^3)$, we have

$$\left| \int_{B_R^c} u \tilde{\theta}_n \varphi_n \right| \leq \varepsilon \left\| \tilde{\theta}_n \right\|_{L^2(B_R^c)} \|\varphi_n\|_{L^6(B_R^c)} \leq c_3 \varepsilon,$$

where c_3 does not depend on n. The sequence φ_n is bounded in $\mathcal{D}^{1,2}$, then it is bounded also in $H^1(B_R)$. Then, up to a subsequence, $\varphi_n \rightharpoonup \varphi$ in $H^1(B_R)$. So $u\varphi_n \to u\varphi$ in $L^2(B_R)$. Since $\tilde{\theta}_n \rightharpoonup 0$ in L^2, it follows that

$$\int_{B_R} \varphi_n u \tilde{\theta}_n \to 0.$$

Then

$$|B_n| \leq c_3 \varepsilon + o(1). \tag{5.130}$$

Now let us consider C_n; we have that

$$C_n = \int \tilde{u}_n \theta \varphi_n = \int_{B_R} \tilde{u}_n \theta \varphi_n + \int_{B_R^c} \tilde{u}_n \theta \varphi_n.$$

We can choose R so large that $\|\theta\|_{L^2(B_R^c)} < \varepsilon$, and so

$$\left| \int_{B_R^c} \tilde{u}_n \theta \varphi_n \right| \leq \varepsilon \|\tilde{u}_n\|_{L^3(B_R^c)} \|\varphi_n\|_{L^6(B_R^c)} \leq c_4 \varepsilon$$

where c_4 does not depend on n. Moreover $\tilde{u}_n \varphi_n \to 0$ in $L^2(B_R)$; then, it follows that $\int_{B_R} \tilde{u}_n \theta \varphi_n \to 0$. Then

$$|C_n| \leq c_4 \varepsilon + o(1). \tag{5.131}$$

Let us now consider D_n. We have that

$$
\begin{aligned}
D_n &= \int \tilde{u}_n \tilde{\theta}_n \left(\varphi_n - \tilde{\varphi}_n\right) = \int \Delta \tilde{\varphi}_n \left(\varphi_n - \tilde{\varphi}_n\right) = \int \tilde{\varphi}_n \Delta \left(\varphi_n - \tilde{\varphi}_n\right) \\
&= \int \tilde{\varphi}_n \left(\theta_n u_n - \tilde{\theta}_n \tilde{u}_n\right) = \int \tilde{\varphi}_n \left(\theta_n u_n - \tilde{\theta}_n u_n + \tilde{\theta}_n u_n - \tilde{\theta}_n \tilde{u}_n\right) \\
&= \int \tilde{\varphi}_n u_n \left(\theta_n - \tilde{\theta}_n\right) + \int \tilde{\varphi}_n \tilde{\theta}_n \left(u_n - \tilde{u}_n\right) \\
&= \int \tilde{\varphi}_n u_n \theta + \int \tilde{\varphi}_n \tilde{\theta}_n u = \int \tilde{\varphi}_n u \theta + \int \tilde{\varphi}_n \left(u_n - u\right) \theta + \int \tilde{\varphi}_n \tilde{\theta}_n u \\
&= \int \tilde{\varphi}_n u \theta + \int \tilde{\varphi}_n \tilde{u}_n \theta + \int \tilde{\varphi}_n \tilde{\theta}_n u \\
&= \int \tilde{\varphi}_n u \theta + \int_{B_R} \tilde{\varphi}_n \tilde{u}_n \theta + \int_{B_R^c} \tilde{\varphi}_n \tilde{u}_n \theta + \int_{B_R} \tilde{\varphi}_n \tilde{\theta}_n u + \int_{B_R^c} \tilde{\varphi}_n \tilde{\theta}_n u.
\end{aligned}
$$

We shall first show that

$$\int \tilde{\varphi}_n u \theta = o(1). \tag{5.132}$$

To this end it will be enough to show that

$$\tilde{\varphi}_n \rightharpoonup 0 \text{ in } L^6. \tag{5.133}$$

Observe that $\tilde{\varphi}_n = \Delta^{-1}(\tilde{u}_n \tilde{\theta}_n)$, where $\tilde{u}_n \tilde{\theta}_n$ is bounded in $L^{\frac{6}{5}}$. So, up to a subsequence, we have

$$\tilde{u}_n \tilde{\theta}_n \rightharpoonup \chi \text{ in } L^{\frac{6}{5}}. \tag{5.134}$$

Now $\chi = 0$. In fact, for any ball $B \subset \mathbb{R}^3$, we have

$$\tilde{u}_n \to 0 \text{ in } L^2(B).$$

Then

$$\tilde{u}_n \tilde{\theta}_n \to 0 \text{ in } L^1(B).$$

So, up to a subsequence,

$$\tilde{u}_n \tilde{\theta}_n \to 0 \text{ a.e. in } \mathbb{R}^3$$

and, by (5.134), we get

$$\tilde{u}_n \tilde{\theta}_n \rightharpoonup 0 \text{ in } L^{\frac{6}{5}}.$$

Then

$$\tilde{\varphi}_n = \Delta^{-1}(\tilde{u}_n \tilde{\theta}_n) \rightharpoonup 0 \text{ in } L^6$$

and (5.133) is proved.

Now we can choose R so large that $\|\theta\|_{L^3(B_R^c)} < \varepsilon$. Then

$$\left| \int_{B_R^c} \tilde{\varphi}_n \tilde{u}_n \theta \right| \le \varepsilon \|\tilde{u}_n\|_{L^3(B_R^c)} \|\tilde{\varphi}_n\|_{L^6(B_R^c)} \le c_5 \varepsilon \qquad (5.135)$$

where c_5 does not depend on n. Since $\|\tilde{u}_n\|_{L^3(B_R)} \to 0$ and $\left| \int_{B_R} \tilde{\varphi}_n \tilde{u}_n \theta \right| \le$ $\|\tilde{\varphi}_n\|_{L^6(B_R)} \|\tilde{u}_n\|_{L^3(B_R)} \|\theta\|_{L^2(B_R)}$, we have

$$\int_{B_R} \tilde{\varphi}_n \tilde{u}_n \theta = o(1). \qquad (5.136)$$

Finally we can choose R so large that $\|u\|_{L^3(B_R^c)} < \varepsilon$. Then

$$\left| \int_{B_R^c} \tilde{\varphi}_n \tilde{\theta}_n u \right| \le \varepsilon \left\| \tilde{\theta}_n \right\|_{L^2(B_R^c)} \|\tilde{\varphi}_n\|_{L^6(B_R^c)} \le c_6 \varepsilon \qquad (5.137)$$

where c_6 does not depend on n. Since $\tilde{\theta}_n \rightharpoonup 0$ in $L^2(B_R)$ and $\tilde{\varphi}_n u \to \tilde{\varphi} u$ in $L^2(B_R)$ we have

$$\int_{B_R} \tilde{\varphi}_n \tilde{\theta}_n u = o(1). \qquad (5.138)$$

By (5.132) and (5.135)–(5.138) we get

$$|D_n| \le c_6 \varepsilon + o(1). \qquad (5.139)$$

Concluding by (5.129)–(5.131) and (5.139) we have

$$|A_n| + |B_n| + |C_n| + |D_n| \le (c_2 + \ldots + c_6)\varepsilon + o(1)$$

and the conclusion follows by the arbitrariness of ε.   □

In the following we shall analyse the coercitivity and the hylenic ratio for NKGM.

The next lemma states that the coercivity assumption (EC-3) is satisfied.

Lemma 107. *Let the assumptions of Theorem 104 be satisfied, then E defined by (5.116) satisfies (EC-3), namely for any sequence $\mathbf{u}_n = (u_n, \theta_n)$ in X such that $E(\mathbf{u}_n) \to 0$ (respectively $E(\mathbf{u}_n)$ bounded), we have $\|\mathbf{u}_n\| \to 0$ (respectively $\|\mathbf{u}_n\|$ bounded), where $\|\cdot\|$ is defined in (5.115).*

Proof. The proof follows the same arguments used in proving Lemma 79. □

Next we shall show that the hylomorphy condition (2.16) holds. First of all we set:

$$\|\mathbf{u}\|_\sharp = \|(u, \theta)\|_\sharp = \max \left(\|u\|_{L^r} , \|u\|_{L^q} \right) \tag{5.140}$$

where r, q are introduced in (4.51). With some abuse of notation we shall write

$$\max \left(\|u\|_{L^r} , \|u\|_{L^q} \right) = \|u\|_\sharp .$$

Lemma 108. $\|\mathbf{u}\|_\sharp$ *defined in (5.140) satisfies the property (2.19), namely*

$$\{\mathbf{u}_n \text{ is a vanishing sequence}\} \Rightarrow \|\mathbf{u}_n\|_\sharp \to 0.$$

Proof. The proof follows the same arguments as in the proof of Lemma 81. □

Now, as usual, we set

$$\Lambda_0 := \inf \{\liminf \Lambda(\mathbf{u}_n) \mid \mathbf{u}_n \text{ is a vanishing sequence}\}$$

$$\Lambda_\sharp = \lim_{\|\mathbf{u}\|_\sharp \to 0} \inf \Lambda(\mathbf{u}) = \tag{5.141}$$

$$\lim_{\varepsilon \to 0} \inf \left\{ \Lambda(\varepsilon u, \theta) \mid u \in H^1, \ \theta \in L^2; \ \|u\|_\sharp = 1 \right\}.$$

By the definitions of Λ_0 and $\Lambda_\sharp$ and Lemma 108, we have that

$$\Lambda_0 \geq \Lambda_\sharp. \tag{5.142}$$

The following lemma holds:

Lemma 109. *Let W satisfy assumption (4.51), then the following inequality holds*

$$\Lambda_\sharp \geq m. \tag{5.143}$$

Proof. By the same arguments used in the proof of Lemma 82 (see (4.69)), we have for $\|u\|_\sharp = 1$ and $\varepsilon > 0$ small

$$\left| \int N(\varepsilon |u|) dx \right| \leq \varepsilon^s \int \left(|\nabla u|^2 + m^2 u^2 \right) dx \text{ where } s > 2. \tag{5.144}$$

Then, by using (5.144), for any $\mathbf{u} = (u, v, \theta, \Theta)$, with $u \in H^1$, $\|u\|_\sharp = 1$ and any $(v, \theta, \Theta) \in \left(L^2\right)^5$, for $\varepsilon > 0$ small we have

$$\Lambda(\varepsilon u, \theta) \geq \frac{\frac{1}{2} \int \left(\theta^2 + \varepsilon^2 |\nabla u|^2 + \varepsilon^2 m^2 |u|^2\right) dx + \int N(|\varepsilon u|) dx}{\varepsilon \left|\int \theta u\right|}$$

$$\geq \frac{\frac{1}{2} \int \theta^2 + \left(\frac{\varepsilon^2}{2} - \varepsilon^s\right) \int \left(|\nabla u|^2 + m^2 |u|^2\right)}{\varepsilon \|\theta\|_{L^2} \|u\|_{L^2}}$$

$$\geq \frac{\left(\int \theta^2 dx\right)^{1/2} \cdot \varepsilon m \sqrt{1 - 2\varepsilon^{s-2}} \left(\int u^2 dx\right)^{1/2}}{\varepsilon \|\theta\|_{L^2} \|u\|_{L^2}} = m\sqrt{1 - 2\varepsilon^{s-2}}.$$

Then, since $s > 2$, we have

$$\Lambda_\sharp = \lim_{\varepsilon \to 0} \inf \left\{\Lambda(\varepsilon u, \theta) \mid u \in H^1, \ \theta \in L^2; \ \|u\|_\sharp = 1\right\} \geq m. \tag{5.145}$$

$\square$

Next we will show that the hylomorphy condition (2.16) is satisfied.

Lemma 110. *Assume that W satisfies the same assumptions of Theorem 77. Then*

$$\inf_{\mathbf{u} \in X} \Lambda(\mathbf{u}) < \Lambda_0. \tag{5.146}$$

Proof. Clearly, by (5.142) and (5.143), in order to prove (5.146) it will be enough to show that

$$\Lambda_* = \inf_{\mathbf{u} \in X} \Lambda(\mathbf{u}) < m, \tag{5.147}$$

Let $R > 0$; set

$$u_R = \begin{cases} s_0 & if \ |x| < R \\ 0 & if \ |x| > R + 1 \\ \frac{|x|}{R} s_0 - (|x| - R)\frac{R+1}{R} s_0 & if \ R < |x| < R + 1 \end{cases} \tag{5.148}$$

where $R > 1$.

By the hylomorphy assumption (4.41) there exists $\alpha \in (0, m)$ such that

$$W(s_0) \leq \frac{1}{2}\alpha^2 s_0^2. \tag{5.149}$$

Now let $\varphi_R = \Phi(u_R, \alpha u_R) \in \mathcal{D}^{1,2}$ be the solution of the equation

$$\Delta \varphi = q\alpha u_R^2. \tag{5.150}$$

Clearly $(u_R, \alpha u_R) \in X$ and we have

$$\Lambda_* = \inf_{(u,\theta) \in X} \frac{E(u,\theta)}{|C(u,\theta)|} \leq \frac{E(u_R, \alpha u_R)}{|C(u_R, \alpha u_R)|} =$$

$$\frac{\frac{1}{2} \int \left[|\nabla u_R|^2 + \alpha^2 u_R^2 + |\nabla \varphi_R|^2 \right] dx + \int W(u_R) dx}{\alpha \int u_R^2 \, dx}$$

$$\leq \frac{\frac{1}{2} \int_{|x|<R} \left[|\nabla u_R|^2 + \alpha^2 u_R^2 \right] + \int_{|x|<R} W(u_R)}{\alpha \int_{|x|<R} u_R^2 \, dx}$$

$$+ \frac{\frac{1}{2} \int_{R<|x|<R+1} \left[|\nabla u_R|^2 + \alpha^2 u_R^2 \right] + \int_{R<|x|<R+1} W(u_R)}{\alpha \int_{|x|<R} u_R^2} + \frac{\frac{1}{2} \int |\nabla \varphi_R|^2}{\alpha \int_{|x|<R} u_R^2}$$

$$= \frac{\frac{1}{2} \int_{|x|<R} \alpha^2 s_0^2 + \int_{|x|<R} W(s_0)}{\alpha \int_{|x|<R} s_0^2} + \frac{c_1 R^2}{\alpha \int_{|x|<R} s_0^2} + \frac{\frac{1}{2} \int |\nabla \varphi_R|^2}{\alpha \int_{|x|<R} s_0^2}$$

$$\leq \alpha + \frac{c_2}{\alpha R} + \frac{\frac{1}{2} \int |\nabla \varphi_R|^2}{\frac{4}{3} \pi \alpha s_0^2 R^3} \tag{5.151}$$

where in the last inequality we have used (5.149).

In order to estimate the term containing φ_R in (5.151), we remember that φ_R is the solution of (5.150). Observe that u_R^2 has radial symmetry and that the electric field outside any spherically symmetric charge distribution is the same as if all of the charge were concentrated into a point. So $|\nabla \varphi_R(r)|$ corresponds to the strength of an electrostatic field at distance r, created by an electric charge given by

$$|C_{el}| = \int_{|x| \leq r} q \alpha u_R^2 dx = 4\pi \int_0^r q \alpha u_R^2 v^2 dv$$

and located at the origin. So we have

$$|\nabla \varphi_R(r)| = \frac{|C_{el}|}{r^2} \begin{cases} = \frac{4}{3}\pi q \alpha s_0^2 r & \text{if } r < R \\ \leq \frac{4}{3}\pi q \alpha s_0^2 \frac{(R+1)^3}{r^2} & \text{if } r \geq R \end{cases}.$$

Then

$$\int |\nabla \varphi_R|^2 \, dx \leq c_3 q^2 \alpha^2 s_0^4 \left(\int_{r<R} r^4 dr + \int_{r>R} \frac{(R+1)^6}{r^2} dr \right)$$

$$\leq c_4 q^2 \alpha^2 s_0^4 \left(R^5 + \frac{(R+1)^6}{R} \right) \leq c_5 q^2 \alpha^2 s_0^4 R^5.$$

Then

$$\frac{\frac{1}{2} \int |\nabla \varphi_R|^2}{\frac{4}{3} \pi \alpha s_0^2 R^3} \leq c_6 q^2 \alpha s_0^2 R^2. \tag{5.152}$$

By (5.151) and (5.152), we get

$$\Lambda_* \leq \alpha + \frac{c_1}{\alpha R} + c_6 q^2 \alpha s_0^2 R^2. \tag{5.153}$$

Now set

$$m - \alpha = 2\varepsilon$$

and take

$$R = \frac{c_1}{\alpha \varepsilon}, \ 0 < q < \sqrt{\frac{\varepsilon^3 \alpha}{s_0^2 c_1^2 c_6}}.$$

With these choices of R and q, a direct calculation shows that

$$\alpha + \frac{c_1}{\alpha R} + c_6 q^2 \alpha s_0^2 R^2 < m. \tag{5.154}$$

Then, by (5.153) and (5.154), we get that there exists a positive constant c such that, for $0 < q < \frac{c}{s} \sqrt{(m-\alpha)^3 \alpha}$, we have

$$\Lambda_* < m. \tag{5.155}$$

$\square$

Finally we are ready to prove Theorem 104.

Proof of Theorem 104. We will first apply Theorem 38. Assumption (EC-0) follows from Proposition 105 and the assumptions on W. Moreover (EC-1) is clearly satisfied. The proof of the splitting property (EC-2) follows from Lemma 106. By Lemmas 107 and 110 also the coercitivity assumption (EC-3) and hylomorphy condition (2.16) are satisfied. Then by Theorem 38 there exists a minimizer

$$\mathbf{u}_\delta = (u_\delta, \theta_\delta)$$

of the energy

$$E = \frac{1}{2} \int \left[|\nabla u|^2 + \theta^2 \right] dx + K(u, \theta) + \int W(u) dx$$

constrained by

$$|C(\mathbf{u})| = \left| \int \theta u \, dx \right| = \sigma_\delta \tag{5.156}$$

where σ_δ is a suitable positive number.

Therefore, for a suitable Lagrange multiplier $\omega_\delta \in \mathbb{R}$, we have that $\mathbf{u}_\delta$ is a critical point of $E_{\omega_\delta}(\mathbf{u}) = E(\mathbf{u}) - \omega_\delta \left(\left| \int \theta u \, dx \right| - \sigma_\delta \right)$.

So, taking the variations of E_{ω_δ} with respect to u and θ, and using (5.122) and (5.123), we get the equations

$$\frac{\partial E_{\omega_\delta}}{\partial u}(\mathbf{u}_\delta) = -\Delta u_\delta + W'(u_\delta) - (q\varphi_\delta - \omega_\delta)\,\theta_\delta = 0 \tag{5.157}$$

$$\frac{\partial E_{\omega_\delta}}{\partial \theta}(\mathbf{u}_\delta) = \theta_\delta - q\varphi_\delta u_\delta + \omega_\delta u_\delta = 0 \tag{5.158}$$

where φ_δ solves the equation $\Delta\varphi_\delta = q u_\delta \theta_\delta$.

If we recall (5.104) and (5.34), we have that

$$\theta_\delta = -\frac{\rho_\delta}{q u_\delta} = \frac{(q\varphi_\delta - \omega_\delta)\, q u_\delta^2}{q u_\delta} = (q\varphi_\delta - \omega_\delta)\, u_\delta$$

and hence replacing θ_δ in (5.157) we get Eq. (5.109):

$$-\Delta u_\delta + W'(u_\delta) - (q\varphi_\delta - \omega_\delta)^2\, u_\delta = 0.$$

Moreover, multiplying (5.158) by u_δ we have

$$u_\delta \theta_\delta + (\omega_\delta - q\varphi_\delta)\, u_\delta^2 = 0. \tag{5.159}$$

By (5.113) and (5.159) we obtain (5.110), namely

$$\Delta\varphi_\delta + (\omega_\delta - q\varphi_\delta)\, q u_\delta^2 = 0.$$

With some abuse of notation we shall continue to denote by $\mathbf{u}_\delta$ also the full set of the canonical variables. It can be directly verified that

$$\mathbf{u}_\delta = (\psi_\delta, \hat{\psi}_\delta, \mathbf{A}_\delta, \hat{\mathbf{A}}_\delta, \varphi_\delta, \hat{\varphi}_\delta) = (u_\delta(x)e^{-i\omega_\delta t},\ D_\varphi(u_\delta(x)e^{-i\omega_\delta t}),\ \mathbf{0}, \mathbf{0}, \varphi_\delta(x), 0)$$

is a solitary wave for the dynamical system described by (5.88). In order to see that it is a soliton, it is sufficient to argue as in the proof of Theorem 33 at page 54.

Finally, in order to prove (a),(b),(c) in Theorem 104 we apply Theorem 34 and Remark 36. So we have to show that (2.25) is satisfied. To this end let

$$\mathbf{u} = (u, \theta) \in H^1\left(\mathbb{R}^3\right) \times L^2\left(\mathbb{R}^3\right)$$

be a solution of $E'(\mathbf{u}) = 0$, then it is easy to see that $\theta = 0$ and that $u \in H^1\left(\mathbb{R}^3\right)$ solves the equation

$$-\Delta u + W'(u) = 0.$$

So, since $W \geq 0$, we have by the Derrick-Pohozaev identity (Theorem 143), that also $u = 0$. We conclude that

$$\mathbf{u} = (u, \theta) = \mathbf{0}.$$

So all the assumptions of the Theorem 34 and Remark 36 are satisfied and also (a), (b) and (c) of Theorem 104 hold.   □

Chapter 6
The Nonlinear Schrödinger-Maxwell Equations

In the preceding chapter we have seen that the use of the covariant derivative provides a very elegant way to combine the action related to the nonlinear Klein-Gordon equation with the action related to the Maxwell equations. It is possible to use this procedure to couple Schrödinger and Maxwell equations. This situation describes the interaction between a charged "matter field" with the electromagnetic field when the relativistic effects are negligible.

6.1 General Feature of the Schrödinger-Maxwell Equations

6.1.1 Construction of the Nonlinear Schrödinger-Maxwell Equations

Consider the nonlinear Schrödinger equation (3.1) and the related Lagrangian

$$\mathcal{L} = \operatorname{Re}\left(i\,\partial_t\psi\overline{\psi}\right) - \frac{1}{2}\left|\nabla\psi\right|^2 - W\left(\psi\right) - V(x)\left|\psi\right|^2.$$

As usual we denote by $\mathbf{E}$, $\mathbf{H}$ the electric and the magnetic field and by φ and $\mathbf{A} = (A_1, A_2, A_3)$ their gauge potentials (see the preceding chapter). As for the nonlinear Klein-Gordon-Maxwell equations, we can couple (3.1) with Maxwell equations by means of the covariant derivatives (5.23). So $\mathcal{L}$ becomes

$$\mathcal{L}_0 = \operatorname{Re}\left(iD_\varphi\psi\overline{\psi}\right) - \frac{1}{2}\left|\mathbf{D_A}\psi\right|^2 - W\left(\psi\right) - V(x)\left|\psi\right|^2$$

© Springer International Publishing Switzerland 2014
V. Benci, D. Fortunato, *Variational Methods in Nonlinear Field Equations*,
Springer Monographs in Mathematics, DOI 10.1007/978-3-319-06914-2_6

where, as in (5.23),

$$\mathbf{D_A}\psi = (D_1\psi, D_2\psi, D_3\psi),$$

$$D_\varphi = \frac{\partial}{\partial t} + iq\varphi, D_j = \frac{\partial}{\partial x^j} - iqA_j$$

denote the covariant derivatives.

Adding to $\mathcal{L}_0$ the Lagrangian relative to the Maxwell equations

$$\mathcal{L}_1 = \frac{1}{2}\left(\left|\frac{\partial\mathbf{A}}{\partial t} + \nabla\varphi\right|^2 - |\nabla\times\mathbf{A}|^2\right)$$

we get the total Lagrangian

$$\mathcal{L}_{tot} = \mathcal{L}_0 + \mathcal{L}_1. \tag{6.1}$$

So the total action is

$$\mathcal{S} = \int \mathcal{L}_{tot}\, dxdt. \tag{6.2}$$

Observe that the Euler-Lagrange equations NSM relative to the action (6.2) are not invariant for the Galileo group and nor for the Poincaré group. In fact these equations are obtained by coupling the Schrödinger equation (which is Galileo invariant if the potential V is constant) with the Maxwell equations (which are Poincarè invariant). Nevertheless the nonlinear Schrödinger-Maxwell equations NSM describe many interesting physical phenomena when the relativistic effects are not relevant (see [129] and its references). Making the variation of $\mathcal{S}$ with respect to ψ, φ and $\mathbf{A}$ we get the following system of equations

$$iD_\varphi\psi = -\frac{1}{2}\mathbf{D_A^2}\psi + \frac{1}{2}W'(\psi) + V(x)\psi \tag{6.3}$$

$$\nabla\cdot\left(\frac{\partial\mathbf{A}}{\partial t} + \nabla\varphi\right) = q\left(\mathrm{Im}\frac{\partial_t\psi}{\psi} + q\varphi\right)|\psi|^2 \tag{6.4}$$

$$\nabla\times(\nabla\times\mathbf{A}) + \frac{\partial}{\partial t}\left(\frac{\partial\mathbf{A}}{\partial t} + \nabla\varphi\right) = q\left(\mathrm{Im}\frac{\nabla\psi}{\psi} - q\mathbf{A}\right)|\psi|^2. \tag{6.5}$$

If we write ψ in polar form

$$\psi(x,t) = u(x,t)\,e^{iS(x,t)}, \quad u \in \mathbb{R}, \quad S \in \mathbb{R}/2\pi\mathbb{Z}$$

the action (6.2) takes the following form

$$S(u, S, \varphi, \mathbf{A}) = - \int \int \left[\frac{1}{2} |\nabla u|^2 + V(x)u^2 + W(u) \right] dxdt +$$

$$- \int \int \left[\left(\frac{\partial S}{\partial t} + q\varphi \right) + \frac{1}{2} |\nabla S - q\mathbf{A}|^2 \right] u^2 dxdt \qquad (6.6)$$

$$+ \frac{1}{2} \int \int \left(\left| \frac{\partial \mathbf{A}}{\partial t} + \nabla \varphi \right|^2 - |\nabla \times \mathbf{A}|^2 \right) dxdt.$$

Making the variations of S with respect $u, S, \varphi, \mathbf{A}$ we get respectively the equations:

$$- \Delta u + 2V(x)u + W'(u) + \left[|\nabla S - q\mathbf{A}|^2 + 2 \left(\frac{\partial S}{\partial t} + q\varphi \right) \right] u = 0 \qquad (6.7)$$

$$\frac{\partial u^2}{\partial t} + \nabla \cdot \left[(\nabla S - q\mathbf{A}) u^2 \right] = 0 \qquad (6.8)$$

$$- \nabla \cdot \left(\frac{\partial \mathbf{A}}{\partial t} + \nabla \varphi \right) = qu^2 \qquad (6.9)$$

$$\nabla \times (\nabla \times \mathbf{A}) + \frac{\partial}{\partial t} \left(\frac{\partial \mathbf{A}}{\partial t} + \nabla \varphi \right) = q \left(\nabla S - q\mathbf{A} \right) u^2. \qquad (6.10)$$

The last two Eqs. (6.9) and (6.10) are the second couple of the Maxwell equations (Gauss and Ampere laws) with respect to a matter distribution whose charge and current density are respectively ρ and $\mathbf{j}$ defined by:

$$\rho = qu^2 \qquad (6.11)$$

$$\mathbf{j} = q \left(\nabla S - q\mathbf{A} \right) u^2. \qquad (6.12)$$

Notice that Eq. (6.8) is a continuity equation. In fact, using (6.11) and (6.12), (6.8) can be written as follows

$$\frac{\partial \rho}{\partial t} + \nabla \cdot \mathbf{j} = 0. \qquad (6.13)$$

Then we deduce the conservation of the electric charge

$$\int \rho dx = q \int u^2 dx \qquad (6.14)$$

and, consequently, of the hylenic charge

$$C = 1/q \int \rho dx = \int u^2 dx.$$

Moreover (6.8) is a consequence of (6.9) and (6.10). In conclusion our system of equations is reduced to (6.7), (6.9), and (6.10).

Observe that in the electrostatic case. i.e. when

$$\frac{\partial u}{\partial t} = 0, \ S = -\omega t, \ \omega \text{ real}, \ \mathbf{A} = 0, \frac{\partial \varphi}{\partial t} = 0$$

the system (6.7), (6.9), (6.10) reduces to two equations

$$- \Delta u + 2V(x)u + W'(u) + 2 (q\varphi - \omega) u = 0 \tag{6.15}$$

$$- \Delta \varphi = qu^2. \tag{6.16}$$

This system, usually called the Schrödinger-Poisson system or Schrödinger-Maxwell system, describes many interesting physical situations (see e.g. [129] and its references). This system of equations has been largely studied by many authors and under various assumptions on W. There is a huge bibliography on this subject and the list of our references is far to be complete. For the existence of solutions we refer to [6,7,20,26,51,55,57,58,64,68,70,95,120,126,128]. However we know only few results (namely [20, 96]) proving the existence of solitons for such equations. For the study of some qualitative properties of the solutions, like the presence of concentration phenomena and the study of semiclassical limits, we refer to [63,66,67,88,125,127].

Finally we observe that, if we consider φ as a scalar field (and not the time-component of a 4-vector) the Schrödinger-Poisson equations are invariant under the Galileo group if V is constant.

6.1.2 The Energy for NSM

In this section we compute the formula of the energy related to the action (6.6).

Theorem 111. *If* $(u, S, \varphi, \mathbf{A})$ *satisfy the Gauss equations (6.9), the energy takes the following form:*

$$E = \int \frac{1}{2} |\nabla u|^2 + V(x)u^2 + W(u) + \frac{1}{2} \int |\nabla S - q\mathbf{A}|^2 u^2 \tag{6.17}$$

$$+ \frac{1}{2} \int \left| \frac{\partial \mathbf{A}}{\partial t} + \nabla \varphi \right|^2 + |\nabla \times \mathbf{A}|^2.$$

Proof. By a direct calculation, we get that the energy related to (6.6) is

$$E = \int \left(\frac{\partial \mathbf{A}}{\partial t} + \nabla \varphi \right) \cdot \frac{\partial \mathbf{A}}{\partial t} + q\varphi u^2 + \frac{1}{2} \left(|\nabla u|^2 + |\nabla S - q\mathbf{A}|^2 u^2 \right) +$$

$$V(x)u^2 + W(u) - \frac{1}{2} \left| \frac{\partial \mathbf{A}}{\partial t} + \nabla \varphi \right|^2 + \frac{1}{2} |\nabla \times \mathbf{A}|^2 . \tag{6.18}$$

By the Gauss equation (6.9), multiplying by φ and integrating, we get

$$\int q\varphi u^2 = \int \nabla \varphi \cdot \left(\frac{\partial \mathbf{A}}{\partial t} + \nabla \varphi \right) . \tag{6.19}$$

The above equality (6.19) easily implies that

$$\int q\varphi u^2 + \left(\frac{\partial \mathbf{A}}{\partial t} + \nabla \varphi \right) \cdot \frac{\partial \mathbf{A}}{\partial t} - \frac{1}{2} \left| \frac{\partial \mathbf{A}}{\partial t} + \nabla \varphi \right|^2 =$$

$$\frac{1}{2} \int \left| \frac{\partial \mathbf{A}}{\partial t} + \nabla \varphi \right|^2 . \tag{6.20}$$

Inserting (6.20) into (6.18) we get the conclusion. □

6.2 NSM as a Dynamical System

6.2.1 The Modified Lagrangian

The Lagrangian in (6.1) has the following form:

$$\mathcal{L} = \mathrm{Re} \left(i D_\varphi \psi \overline{\psi} \right) - \frac{1}{2} |\mathbf{D_A} \psi|^2 - W(\psi) - V(x) |\psi|^2 \tag{6.21}$$

$$+ \frac{1}{2} \left(|\partial_t \mathbf{A} + \nabla \varphi|^2 - |\nabla \times \mathbf{A}|^2 \right) .$$

As in the case of NKGM, $\mathcal{L}$ does not depend on $\partial_t \varphi$ and it is not possible to make the Legendre transformation with respect to $\partial_t \varphi$ and hence to get an Hamiltonian formulation of the dynamics.

In the following, as in Sect. 5.2.1, we will describe NSM as a dynamical system on a suitable manifold. However, unlike the case of NKGM, ψ does not possess a conjugate variable.

We set

$$\mathbf{Q} = (\mathbf{A}, \varphi), \quad \dot{\mathbf{Q}} = (\partial_t \mathbf{A}, \partial_t \varphi)$$

and

$$M_L = \left\{ (\psi, \mathbf{Q}, \dot{\mathbf{Q}}) : \partial_t \varphi + \nabla \cdot \mathbf{A} = 0, \nabla \cdot (\partial_t \mathbf{A} + \nabla \varphi) = q \psi \overline{\psi} \right\}. \tag{6.22}$$

We consider the *modified Lagrangian* defined as follows:

$$\hat{\mathcal{L}} = \text{Re} \left(i D_\varphi \psi \overline{\psi} \right) - \frac{1}{2} |\mathbf{D_A} \psi|^2 - W(\psi) - V(x) |\psi|^2 \tag{6.23}$$

$$+ \frac{1}{2} \left[|\partial_t \mathbf{A}|^2 - |\nabla \mathbf{A}|^2 - (\partial_t \varphi)^2 + |\nabla \varphi|^2 \right].$$

The dynamics induced by $\hat{\mathcal{L}}$ is given by the following equations:

$$i D_\varphi \psi + \frac{1}{2} \mathbf{D_A^2} \psi - \frac{1}{2} W'(\psi) - V(x)\psi = 0 \tag{6.24}$$

$$\Box \mathbf{A} - q \left(\nabla S - q\mathbf{A} \right) \psi \overline{\psi} = 0 \tag{6.25}$$

$$\Box \varphi - q \psi \overline{\psi} = 0. \tag{6.26}$$

Theorem 112. *The set M_L is invariant for the dynamics induced by Eqs. (6.24)–(6.26). Moreover, if the initial data are in M_L and $(\psi, \mathbf{A}, \varphi)$ is a smooth solution of Eqs. (6.24)–(6.26) then it is also a solution of (6.3)–(6.5).*

Proof. Equations (6.12) and (6.11), using the variable ψ, take the form

$$\mathbf{j} = q \left(\nabla S - q\mathbf{A} \right) \psi \overline{\psi}$$

$$\rho = q \psi \overline{\psi}.$$

Let us first show that M_L is invariant for the dynamics induced by (6.24)–(6.26).
   Let

$$(\psi(x, t), \mathbf{A}(x, t), \varphi(x, t))$$

be the solution of (6.24)–(6.26) with initial data

$$(\psi(x, 0), \mathbf{Q}(x, 0), \dot{\mathbf{Q}}(x, 0)) \in M_L.$$

So

$$\partial_t \varphi(x, t)_{t=0} + \nabla \cdot \mathbf{A}(x, 0) = 0, \ \nabla \cdot (\partial_t \mathbf{A}(x, t)_{t=0} + \nabla \varphi(x, 0)) = -\rho_0 \tag{6.27}$$

where

$$\rho_0 = \rho(x, 0) = q \left(\psi \overline{\psi} \right)_{t=0}.$$

We want to show that for all $t \geq 0$

$$(\psi(x,t), \mathbf{Q}(x,t), \dot{\mathbf{Q}}(x,t)) \in M_L$$

namely that for all $t \geq 0$

$$\partial_t \varphi(x,t) + \nabla \cdot \mathbf{A}(x,t) = 0, \tag{6.28}$$

$$\nabla \cdot (\partial_t \mathbf{A}(x,t) + \nabla \varphi(x,t)) = -q \psi \overline{\psi}. \tag{6.29}$$

We set for all x, t

$$F = F(x,t) = \partial_t \varphi(x,t) + \nabla \cdot \mathbf{A}(x,t).$$

Then

$$\begin{aligned}
\Box F &= \partial_t (\Box \varphi) + \nabla \cdot (\Box \mathbf{A}) = \text{(by (6.25), (6.26))} \\
&= \partial_t \rho + \nabla \cdot \mathbf{j} \text{ (by (6.13))} \\
&= 0
\end{aligned}$$

and so

$$\Box F = 0. \tag{6.30}$$

By (6.27)

$$F(x,0) = \partial_t \varphi(x,t)_{t=0} + \nabla \cdot \mathbf{A}(x,0) = 0. \tag{6.31}$$

Moreover

$$\partial_t F = \partial_t^2 \varphi + \nabla \cdot (\partial_t \mathbf{A}) = \text{(by 6.26)} = \nabla \cdot (\nabla \varphi) + \rho + \nabla \cdot (\partial_t \mathbf{A}(x,t)). \tag{6.32}$$

Then

$$\partial_t F(x,0) = \nabla \cdot (\nabla \varphi(x,0) + \partial_t \mathbf{A}(x,0)) + \rho_0 = \text{(by (6.27))} = 0. \tag{6.33}$$

By (6.30), (6.31), and (6.33) we get

$$\Box F = 0$$

$$F = 0 \text{ for } t = 0$$

$$\partial_t F = 0 \text{ for } t = 0.$$

So,

$$F = 0.$$

Then (6.28) is proved. Now in order to prove (6.29), we observe that by (6.28) we get

$$\partial_t F = \partial_t^2 \varphi(x, t) + \nabla \cdot (\partial_t \mathbf{A}(x, t)) = 0 \text{ for all } t.$$

From which, using (6.32), we have

$$\nabla \cdot (\nabla \varphi) + \rho + \nabla \cdot (\partial_t \mathbf{A}(x, t)) = 0 \text{ for all } t.$$

Then (6.29) clearly holds.

In order to prove the second part of the theorem we argue as in the end of Theorem 97 at page 159.   □

6.2.2 The Phase Space of NSM

The Lagrangian (6.23) depends on the variables $(\psi, \mathbf{Q}, \dot{\mathbf{Q}})$ where

$$\mathbf{Q} = (\mathbf{A}, \varphi) \quad \text{and} \quad \dot{\mathbf{Q}} = (\partial_t \mathbf{A}, \partial_t \varphi);$$

and it is possible to define the conjugate variable of $\mathbf{Q}$ via the Legendre transform. Notice that we leave ψ unchanged. We have:

$$\hat{\mathbf{A}} = \frac{\partial \hat{\mathcal{L}}}{\partial (\partial_t \mathbf{A})} = \partial_t \mathbf{A}$$

$$\hat{\varphi} = \frac{\partial \hat{\mathcal{L}}}{\partial (\partial_t \varphi)} = -\partial_t \varphi.$$

We denote by $\mathbf{u}$ the state of our dynamical system described by the canonical variables

$$\mathbf{u} = (\psi, \mathbf{A}, \hat{\mathbf{A}}, \varphi, \hat{\varphi}).$$

The invariant set M_L (see (6.22)), expressed in the canonical variables $(\psi, \mathbf{A}, \hat{\mathbf{A}}, \varphi, \hat{\varphi})$ is given by

$$M_H = \left\{ (\psi, \mathbf{A}, \hat{\mathbf{A}}, \varphi, \hat{\varphi}): -\hat{\varphi} + \nabla \cdot \mathbf{A} = 0, \nabla \cdot \left(\hat{\mathbf{A}} + \nabla \varphi \right) = -q \psi \overline{\psi} \right\}. \quad (6.34)$$

Using canonical variables $(\psi, \mathbf{A}, \hat{\mathbf{A}}, \varphi, \hat{\varphi})$, Eqs. (6.24)–(6.26) become:

$$i\partial_t \psi = -\frac{1}{2}D_\mathbf{A}^2 \psi + \frac{1}{2}W'(\psi) + (V + q\varphi)\psi$$

$$\partial_t \mathbf{A} = \hat{\mathbf{A}}$$

$$\partial_t \hat{\mathbf{A}} = \Delta\mathbf{A} + q\,\mathrm{Re}\left(iD_\mathbf{A}\psi\overline{\psi}\right) \qquad (6.35)$$

$$\partial_t \varphi = -\hat{\varphi}$$

$$\partial_t \hat{\varphi} = -\Delta\varphi - q\psi\overline{\psi}.$$

Remark 113. In the following we shall assume that the Cauchy problem for (6.35) is globally well posed in a suitable function space (namely $\overline{M}_H$ with the distance (6.41), we shall define in page 192). We refer to [86, 113] and [114] for some results in this direction.

Now we want to express the energy (6.17) in the gauge invariant variables (GIV) $u, \rho, \mathbf{j}, \mathbf{E}, \mathbf{H}$ defined by

$$u = |\psi|$$

$$\rho = q\psi\overline{\psi} \qquad (6.36)$$

$$\mathbf{j} = q\,\mathrm{Re}\left(iD_\mathbf{A}\psi\overline{\psi}\right)$$

$$\mathbf{E} = -\left(\nabla\varphi + \hat{\mathbf{A}}\right)$$

$$\mathbf{H} = \nabla \times \mathbf{A}.$$

We will denote by

$$\mathcal{G} : \mathbf{u} = (\psi, \mathbf{A}, \hat{\mathbf{A}}, \varphi, \hat{\varphi}) \mapsto \mathcal{G}(\mathbf{u}) = (u, \rho, \mathbf{j}, \mathbf{E}, \mathbf{H})$$

the transformation (6.36) so that $\mathbf{U} := \mathcal{G}(\mathbf{u})$ represents a state in the gauge invariant variables.

We now set

$$M_\mathcal{G} = \{\mathbf{U} \in C^\infty(\mathbb{R}^3, \mathbb{R}^{11}) \mid \nabla \cdot \mathbf{E} = \rho,\ \nabla \cdot \mathbf{H} = 0,\ E(\mathbf{U}) < +\infty\}$$

where $E(\mathbf{U})$ is the energy as function of $\mathbf{U}$. The expression of the energy is given by the following proposition:

Proposition 114. *If $\mathbf{U} \in M_\mathcal{G}$, then the energy takes the following expression:*

$$E(\mathbf{U}) = \int\left(\frac{1}{2}|\nabla u|^2 + V(x)u^2 + W(u)\right)dx + \frac{1}{2}\int\left(\frac{\mathbf{j}^2}{q^2 u^2} + \mathbf{E}^2 + \mathbf{H}^2\right)dx.$$

Proof. The proof follows from (6.17) and (6.36).   □

As usual, we write W as follows

$$W(s) = E_0 s^2 + N(s), \quad N(s) = o(s^2)$$

and we will assume $E_0 > 0$. Then we have

$$E(\mathbf{U}) = \int \left(\frac{1}{2} |\nabla u|^2 + V(x)u^2 + E_0 u^2 \right) dx$$

$$+ \frac{1}{2} \int \left(\frac{\mathbf{j}^2}{q^2 u^2} + \mathbf{E}^2 + \mathbf{H}^2 \right) dx + \int N(u) dx.$$

The term $\mathbf{j}^2/u^2$ is singular and the energy does not have the form (5.94). In order to avoid this problem, it is convenient to introduce new gauge invariant variables which eliminate this singularity:

$$\Theta = \frac{\mathbf{j}}{qu} = \frac{1}{q} (\nabla S - q\mathbf{A}) u. \tag{6.37}$$

We denote this new change of variables by

$$\mathcal{Q} : (\psi, \mathbf{A}, \hat{\mathbf{A}}, \varphi, \hat{\varphi}) \mapsto (u, \Theta, \mathbf{E}, \mathbf{H})$$

and we set $\mathbf{V} = \mathcal{Q}(\mathbf{u}) = (u, \Theta, \mathbf{E}, \mathbf{H})$.

Using these new variables the energy and the charge take the form:

$$E(\mathbf{V}) = \int \left(\frac{1}{2} |\nabla u|^2 + V(x)u^2 + E_0 u^2 \right) dx$$

$$+ \frac{1}{2} \int \left(\Theta^2 + \mathbf{E}^2 + \mathbf{H}^2 \right) dx + \int N(u) dx. \tag{6.38}$$

$$C(u) = \int u^2 \, dx. \tag{6.39}$$

Thus, we can take a norm having the form (5.95), namely

$$\|\mathbf{V}\| = \left(\int \left[|\nabla u|^2 + 2E_0 u^2 + \Theta^2 + \mathbf{E}^2 + \mathbf{H}^2 \right] dx \right)^{\frac{1}{2}}. \tag{6.40}$$

This norm will induce a distance on the set M_H defined by (6.34):

$$d_{M_H}(\mathbf{u}_1, \mathbf{u}_2) = \|\mathcal{Q}(\mathbf{u}_1) - \mathcal{Q}(\mathbf{u}_2)\|. \tag{6.41}$$

We will denote by $\overline{M}_H$ the completion of M_H with respect to the above distance.

As we pointed out in a preceding remark, we shall assume that the Cauchy problem for Eq. (6.35) is globally well posed in $\overline{M}_H$ with respect to the distance (6.41). So NSM can be considered as a dynamical system (M, γ_t) whose dynamics is given by (6.35).

So the definition of hylomorphic soliton for NSM is the analogous of Definition 103:

Definition 115. A state $\mathbf{u} = (\psi, \mathbf{A}, \hat{\mathbf{A}}, \varphi, \hat{\varphi}) \in \overline{M}_H$ is called hylomorphic soliton with respect to the dynamics given by (6.35) if it satisfies the properties of Definitions 16 and 20 where the linear space X is replaced by $\overline{M}_H$ equipped with the metric (6.41).

6.3 Existence of Charged Solitons for NSM

In this subsection we shall prove the following existence results of hylomorphic solitons for NSM.

Theorem 116. *Let W and V satisfy the assumptions (WB-i)–(WB-iv) and (3.47),(3.48) in Sect. 3.2.2. Then there exists $\bar{q} > 0$ such that for every $q \in [0, \bar{q}]$ the dynamical system described by (6.35) has a family $\mathbf{u}_\delta$ ($\delta \in (0, \delta_\infty)$, $\delta_\infty > 0$) of hylomorphic solitons in the sense of Definition 115. The evolution of $\mathbf{u}_\delta$ is given by*

$$\mathbf{u}_\delta(t) = (\psi_\delta(t), \mathbf{A}_\delta(t), \hat{\mathbf{A}}_\delta(t), \varphi_\delta(t), \hat{\varphi}_\delta(t))$$
$$= (u_\delta(x)e^{-i\delta\omega t}, \mathbf{0}, \mathbf{0}, \varphi_\delta(x), 0).$$

More exactly, $u_\delta(x), \varphi_\delta(x), \omega_\delta$ satisfy Eqs. (6.15), (6.16) and hence, $\mathbf{u}_\delta(t)$ satisfies (6.35). Moreover if $\delta_1 < \delta_2$ we have that

(a) $\Lambda(\mathbf{u}_{\delta_1}) < \Lambda(\mathbf{u}_{\delta_2})$.
(b) $\|u_{\delta_1}\|_{L^2} > \|u_{\delta_2}\|_{L^2}$.

The proof of Theorem 116 is based on the abstract Theorem 34. First of all observe that, since V satisfies (3.48), the energy E is invariant under the raprentation T_z of the group $\mathbb{Z}^3$

$$T_z\mathbf{u}(x) = \mathbf{u}(x + Az), \quad z \in \mathbb{Z}^3$$

where A is as in (3.48)

6.3.1 The Gauge Invariant Variables and the Splitting Property

In the following we assume that W and V satisfy assumptions (WB-i)–(WB-iv) and (V-i), (V-ii) in Sect. 3.2.2.

Consider the set

$$M = \left\{ \mathbf{u} = (u, \Theta, \mathbf{E}, \mathbf{H}) \in H^1\left(\mathbb{R}^3\right) \times L^2\left(\mathbb{R}^3\right)^9 : \nabla \cdot \mathbf{E} = qu^2; \ \nabla \cdot \mathbf{H} = 0 \right\}$$
(6.42)

equipped with the norm (6.40). The energy and the charge are defined by (6.38), (6.39). Unfortunately, as for the NKGM (see the preceding chapter), M is not a linear space and Theorem 34 cannot be applied. However, arguing as for the NKGM, we take

$$\mathbf{H} = 0, \ \mathbf{E} = -\nabla\varphi$$

and hence also $\Theta = 0$. Then the constraint

$$\nabla \cdot \mathbf{E} = qu^2$$

permits to write $\mathbf{E}$ as a function of $u \in H^1\left(\mathbb{R}^3\right)$, namely

$$\mathbf{E} = -\nabla\Phi(u)$$
(6.43)

where $\varphi = \Phi(u)$ solves the equation

$$-\Delta\varphi = qu^2.$$
(6.44)

Clearly we can write also

$$\mathbf{E}(u) = -q\nabla\left(\int \frac{u^2(y)}{|x-y|}\,dy\right).$$

Since we can take $\Theta = \mathbf{H} = \mathbf{0}$ and since we can express the electric field $\mathbf{E}$ as function of u by (6.43), the energy in (6.38) will depend only on the variable u. Then the space of the states X can be identified with $H^1\left(\mathbb{R}^3\right)$.

The space X will be equipped with the norm related to the quadratic part of the energy, namely:

$$\|u\|^2 = \int \left(|\nabla u|^2 + 2E_0 u^2\right)dx$$

where E_0 is defined by (3.43). Then the energy E can be written as follows:

$$E(u) = \frac{1}{2}\|u\|^2 + \int V(x)u^2 dx + \int N(u)dx + K(u)$$
(6.45)

where

$$K(u) = \frac{1}{2} \int \mathbf{E}^2(u) = \frac{1}{2} \int |\nabla \Phi(u)|^2 . \tag{6.46}$$

As in the case of NKGM equation it can be shown that K is differentiable and

$$K'(u) = 2q\Phi(u) = 2q\varphi u. \tag{6.47}$$

We recall (see (6.14)) that the hylenic charge C takes the form

$$C(u) = \int u^2. \tag{6.48}$$

Finally, as usual

$$\Lambda(u) = \frac{E(u)}{C(u)}$$

will denote the hylenic ratio.

In the following, we shall apply the theory of Sect. 2.2.2 to the functions E and C.

Lemma 117. *The energy E given by (6.45) and the hylenic charge C given by (6.48) satisfy the splitting property (EC-2).*

Proof. The hylenic charge C satisfies the splitting property since it is quadratic. In order to prove that the energy

$$E(u) = \frac{1}{2} \|u\|^2 + \int V(x)u^2 + \int N(u) + K(u)$$

satisfies the splitting property it is sufficient to prove that each piece does it. The term $\frac{1}{2}\|u\|^2$ and $\int V(x)u^2$ satisfy the splitting property since they are quadratic; also $\int N(u)dx$ satisfies this property as it has been proved in Lemma 53. It remains to prove that the splitting property is satisfied by the functional

$$K(u) = \frac{1}{2} \int |\nabla \Phi(u)|^2 .$$

So let $\{u_n\} \subset H^1(\mathbb{R}^3)$ be a sequence such that

$$u_n \rightharpoonup u \text{ weakly in } H^1(\mathbb{R}^3)$$

and set

$$\tilde{u}_n = u_n - u.$$

Moreover set $\tilde{\varphi}_n = \Phi(\tilde{u}_n)$, $\varphi_n = \Phi(u_n)$, $\varphi = \Phi(u)$, namely $\tilde{\varphi}_n$, φ_n, φ are the solutions in $\mathcal{D}^{1,2}$ of the following equations:

$$-\Delta\tilde{\varphi}_n = \tilde{u}_n^2$$
$$-\Delta\varphi_n = u_n^2$$
$$-\Delta\varphi = u^2.$$

Notice that here, in order to simplify the notation, we have put $q = 1$.

We need to prove that

$$K(u_n) = K(u) + K(\tilde{u}_n) + o(1).$$

Now let us estimate $K(u_n)$:

$$2K(u_n) = \int |\nabla\Phi(u_n)|^2 = \int |\nabla\varphi_n|^2 = \int -\Delta\varphi_n \cdot \varphi_n = \int u_n^2\varphi_n$$

$$= \int (\tilde{u}_n + u)^2 \varphi_n = \int u^2\varphi_n + 2\int u\tilde{u}_n\varphi_n + \int \tilde{u}_n^2\varphi_n$$

$$= \int u^2\varphi + \int u^2(\varphi_n - \varphi) + 2\int u\tilde{u}_n\varphi_n + \int \tilde{u}_n^2\tilde{\varphi}_n + \int \tilde{u}_n^2(\varphi_n - \tilde{\varphi}_n)$$

$$= 2K(u) + 2K(\tilde{u}_n) + A_n + B_n + C_n$$

where

$$A_n = \int u^2(\varphi_n - \varphi)$$

$$B_n = 2\int u\tilde{u}_n\varphi_n$$

$$C_n = \int \tilde{u}_n^2(\varphi_n - \tilde{\varphi}_n).$$

It remains to show that the last three terms vanish. We estimate each piece separately:

$$A_n = \int u^2(\varphi_n - \varphi) = \int -\Delta\varphi(\varphi_n - \varphi) = -\int \varphi\Delta(\varphi_n - \varphi)$$

$$= \int \varphi(u_n^2 - u^2) = \int \varphi(u_n - u)(u_n + u)$$

$$= \int_{B_R} \varphi(u_n - u)(u_n + u) + \int_{B_R^c} \varphi(u_n - u)(u_n + u).$$

We fix $\varepsilon > 0$. We can choose R so large that $\|\varphi\|_{L^6(B_R^c)} < \varepsilon$, and so

$$\left| \int_{B_R^c} \varphi \, (u_n - u) \, (u_n + u) \right| \le \varepsilon \, \|(u_n - u)\|_{L^2(B_R^c)} \, \|(u_n + u)\|_{L^3(B_R^c)}$$

$$\le c_1 \, \varepsilon \, \|(u_n - u)\|_{H^1(B_R^c)} \, \|(u_n + u)\|_{H^1(B_R^c)} \le c_2 \varepsilon$$

where c_1 and c_2 do not depend on n. Moreover

$$u_n \to u \text{ in } L^3(B_R)$$

and hence

$$\int_{B_R} \varphi \, (u_n - u) \, (u_n + u) = o(1).$$

Then

$$|A_n| \le c_2 \varepsilon + o(1). \tag{6.49}$$

Now let us consider B_n; we have that

$$B_n = 2 \int u \tilde{u}_n \varphi_n = 2 \int_{B_R} u \tilde{u}_n \varphi_n + 2 \int_{B_R^c} u \tilde{u}_n \varphi_n.$$

Observe that u_n^2 is bounded in $L^{\frac{6}{5}}(\mathbb{R}^3)$ which is continuously embedded into $\left(\mathcal{D}^{1,2}\right)'$. Then

$$\varphi_n = (-\Delta)^{-1} u_n^2 \text{ is bounded in } \mathcal{D}^{1,2}.$$

We can choose R so large that $\|u\|_{L^3(B_R^c)} < \varepsilon$, and so

$$\left| \int_{B_R^c} u \tilde{u}_n \varphi_n \right| \le \varepsilon \, \|\tilde{u}_n\|_{L^2(B_R^c)} \, \|\varphi_n\|_{L^6(B_R^c)} \le c_3 \varepsilon$$

where c_3 does not depend on n. Moreover $\tilde{u}_n \to 0$ in $L^2(B_R)$ and so

$$\int_{B_R} u \varphi_n \tilde{u}_n = o(1).$$

Then

$$|B_n| \le c_3 \varepsilon + o(1). \tag{6.50}$$

Let us consider C_n; we have that

$$C_n = \int \tilde{u}_n^2 (\varphi_n - \tilde{\varphi}_n) = \int -\Delta\tilde{\varphi}_n (\varphi_n - \tilde{\varphi}_n) = -\int \tilde{\varphi}_n \Delta (\varphi_n - \tilde{\varphi}_n)$$

$$= \int \tilde{\varphi}_n (u_n^2 - \tilde{u}_n^2) = \int \tilde{\varphi}_n (u_n - \tilde{u}_n)(u_n + \tilde{u}_n) = \int \tilde{\varphi}_n u (u_n + \tilde{u}_n)$$

$$= \int \tilde{\varphi}_n u^2 + \int \tilde{\varphi}_n u(u_n - u) + \int \tilde{\varphi}_n u\tilde{u}_n = \int \tilde{\varphi}_n u^2 + 2\int \tilde{\varphi}_n u\tilde{u}_n$$

$$= \int \tilde{\varphi}_n u^2 + 2\int_{B_R^c} \tilde{\varphi}_n u\tilde{u}_n + 2\int_{B_R} \tilde{\varphi}_n u\tilde{u}_n.$$

Observe that

$$\tilde{\varphi}_n = \Delta^{-1}(\tilde{u}_n^2),$$

where $\tilde{u}_n^2$ is bounded in $L^{\frac{6}{5}}$. Then $\tilde{\varphi}_n$ is bounded in $\mathcal{D}^{1,2}$.

We can choose R so large that $\|u\|_{L^3(B_R^c)} < \varepsilon$, and so

$$\left| 2\int_{B_R^c} \tilde{\varphi}_n u\tilde{u}_n \right| \leq 2\varepsilon \|\tilde{u}_n\|_{L^2(B_R^c)} \|\tilde{\varphi}_n\|_{L^6(B_R^c)}$$

$$\leq 2\varepsilon \|\tilde{u}_n\|_{H^1} \|\tilde{\varphi}_n\|_{\mathcal{D}^{1,2}} \leq c_4\varepsilon$$

where c_4 does not depend on n. Moreover, $u\tilde{u}_n$ converges strongly to 0 in $L^{6/5}(B_R)$ and hence, $\int_{B_R} \tilde{\varphi}_n u\tilde{u}_n = o(1)$. Now we show that

$$\int \tilde{\varphi}_n u^2 = o(1). \tag{6.51}$$

To this end it will be enough to show that

$$\tilde{\varphi}_n \rightharpoonup 0 \text{ in } L^6. \tag{6.52}$$

Up to a subsequence, we have

$$\tilde{u}_n^2 \rightharpoonup \chi \text{ in } L^{\frac{6}{5}}. \tag{6.53}$$

Now $\chi = 0$. In fact, for any ball $B \subset \mathbb{R}^3$, we have

$$\tilde{u}_n \to 0 \text{ in } L^2(B).$$

Then

$$\tilde{u}_n^2 \to 0 \text{ in } L^1(B).$$

So, up to a subsequence,

$$\tilde{u}_n^2 \to 0 \text{ a.e. in } \mathbb{R}^3$$

and, by (6.53), we get

$$\tilde{u}_n^2 \rightharpoonup 0 \text{ in } L^{\frac{6}{5}}.$$

It follows that

$$\tilde{\varphi}_n = \Delta^{-1}(\tilde{u}_n^2) \rightharpoonup 0 \text{ in } L^6.$$

Then (6.51) holds and

$$|C_n| \le c_4 \varepsilon + o(1). \tag{6.54}$$

Finally, by (6.49), (6.50), and (6.54), we have

$$|A_n| + |B_n| + |C_n| \le (c_2 + c_3 + c_4)\varepsilon + o(1)$$

and the conclusion follows by the arbitrariness of ε. $\qquad\qquad\square$

6.3.2 Analysis of the Hylenic Ratio for NSM

For $u \in X$, we set

$$\|u\|_\sharp = \|u\|_{L^t}, \ 2 < t < 6$$

$$\Lambda_0 := \inf\{\liminf \Lambda(u_n) \mid u_n \text{ is a vanishing sequence}\}, \ \Lambda_\sharp = \liminf_{\|u\|_\sharp \to 0} \Lambda(u).$$

Following the same arguments as in proving Lemma 55, it can be shown that

$$(u_n \text{ vanishing sequence}) \Rightarrow (\|u_n\|_\sharp \to 0).$$

Then clearly we have

$$\Lambda_\sharp \le \Lambda_0. \tag{6.55}$$

Moreover, by the same arguments as those used in proving Lemma 56, we get

$$E_0 \leq \Lambda_\sharp. \tag{6.56}$$

Now we are ready to prove the hylomorphy condition, namely the following lemma holds:

Lemma 118. *Let W and V satisfy assumptions (WB-i)–(WB-iv) and (V-i), (V-ii) in Sect. 3.2.2. Then, if q is sufficiently small, the hylomorphy condition (2.16) holds, namely*

$$\inf_{u \in X} \Lambda(u) < \Lambda_0. \tag{6.57}$$

Proof. Clearly, by (6.56) and (6.55), in order to prove (6.57), it will be enough to show that for q sufficiently small we have

$$\inf_{u \in X} \Lambda(u) < E_0. \tag{6.58}$$

So we shall construct $u \in X$ such that $\Lambda(u) < E_0$.

Let $R > 0$ and let u_R be defined as in Lemma 58, namely

$$u_R = \begin{cases} s_0 & if \ |x| < R \\ 0 & if \ |x| > R+1 \\ \frac{|x|}{R}s_0 - (|x| - R)\frac{R+1}{R}s_0 & if \ R < |x| < R+1 \end{cases}.$$

Moreover let $\varphi_R = \Phi(u_R)$ be the solution of the equation

$$-\Delta\varphi_R = qu_R^2. \tag{6.59}$$

By the same arguments used in proving (3.80), we get

$$\Lambda(\mathbf{u}_R) \leq E_0 + V_0 + \frac{N(s_0)}{s_0^2}\left(\frac{R}{R+1}\right)^3 + \frac{c_1}{R} + \frac{\frac{1}{2}\int |\nabla\varphi_R|^2}{\int u_R^2}. \tag{6.60}$$

In order to estimate the term containing φ_R in (6.60), we argue as in the proof of Lemma 110. Observe that u_R^2 has radial symmetry and that the electric field outside any spherically symmetric charge distribution is the same as if all of the charge were concentrated into a point. So $|\nabla\varphi_R(r)|$ corresponds to the strength of an electrostatic field at distance r, created by an electric charge given by

$$C_{el} = \int_{|x| \leq r} qu_R^2 dx = 4\pi \int_0^r qu_R^2 v^2 dv$$

and located at the origin. So we have

$$|\nabla \varphi_R(r)| = \frac{C_{el}}{r^2} \begin{cases} = \frac{4}{3}\pi q s_0^2 r & if \ r < R \\ \leq \frac{4}{3}\pi q s_0^2 \frac{(R+1)^3}{r^2} & if \ r \geq R \end{cases}.$$

Then

$$\int |\nabla \varphi_R|^2 \, dx \leq c_2 q^2 s_0^4 \left(\int_{r<R} r^2 dr + \int_{r>R} \frac{(R+1)^6}{r^2} dr \right)$$

$$\leq c_3 q^2 s_0^4 \left(R^3 + \frac{(R+1)^6}{R} \right) \leq c_4 q^2 s_0^4 R^5.$$

So

$$\frac{\frac{1}{2}\int |\nabla \varphi_R|^2}{\int u_R^2} \leq \frac{c_5 \int |\nabla \varphi_R|^2}{s_0^2 R^3} \leq c_6 q^2 s_0^2 R^2. \tag{6.61}$$

By (6.61) and (6.60), we get

$$\Lambda(\mathbf{u}_R) \leq E_0 + V_0 + \frac{N(s_0)}{s_0^2} \left(\frac{R}{R+1} \right)^3 + \frac{c_1}{R} + c_6 q^2 s_0^2 R^2. \tag{6.62}$$

Since by our assumptions

$$\frac{N(s_0)}{s_0^2} < -V_0$$

for R large we get

$$V_0 + \frac{N(s_0)}{s_0^2} \left(\frac{R}{R+1} \right)^3 + \frac{c_1}{R} < 0. \tag{6.63}$$

So, if q is small enough, by (6.62) and (6.63) we get

$$\Lambda(u_R) < E_0$$

$\square$

Proof of Theorem 116. We shall show that all the assumptions of Theorem 38 are satisfied.

Assumptions (EC-0), (EC-1) , are clearly satisfied. By using the same arguments as in proving Lemma 106, it can be shown that the splitting property (EC-2) holds. Then, since $E(u), C(u)$ satisfy the coercitivity assumption (EC-3*) in the absence of the field (see Lemma 54), we have that $E(u), C(u)$ satisfy (EC-3*) also in the presence of the field.

By Lemma 118 the hylomorphy condition (2.16) holds.

Then let u_δ be a minimizer of the energy

$$E(u) = \int \left[\frac{1}{2} |\nabla u|^2 + V(x)u^2 + W(u) \right] dx + K(u)$$

constrained by

$$C(u) = \int u^2 \, dx = \sigma_\delta. \tag{6.64}$$

Therefore, for a suitable Lagrange multiplier $\omega_\delta \in \mathbb{R}$, we have that u_δ is a critical point of $E_{\omega_\delta}(u) = E(u) - 2\omega_\delta \left(\int u^2 \, dx - \sigma_\delta \right)$.

So, taking the variations of E_ω with respect to u and using (6.47), we get

$$\frac{\partial E_{\omega_\delta}}{\partial u}(u_\delta) = -\Delta u_\delta + 2V(x)u_\delta + W'(u_\delta) + 2q\varphi_\delta u_\delta - 2\omega_\delta u_\delta = 0 \tag{6.65}$$

where $\varphi_\delta = \Phi(u_\delta)$ solves the equation $-\Delta\varphi = qu_\delta^2$.

Denoting with $\mathbf{u}_\delta$ the full set of variables, we get that

$$\mathbf{u}_\delta = (\psi_\delta, \mathbf{A}_\delta, \hat{\mathbf{A}}_\delta, \varphi_\delta, \hat{\varphi}_\delta) = (u_\delta(x)e^{-i\omega_\delta t}, \mathbf{0}, \mathbf{0}, \varphi_\delta(x), 0)$$

is a solitary wave for the dynamical system described by (6.35). In order to see that it is a soliton, it is sufficient to argue as in the proof of Theorem 33 at page 54.

In order to prove (a), (b) as usual, we apply Theorem 34 and Remark 36. The assumption (2.25) is satisfied. In fact

$$C'(u) = 0 \Longrightarrow C'(u) = 0 \Longrightarrow u = 0.$$

$\square$

Chapter 7
The Nonlinear Beam Equation

This chapter is devoted to the nonlinear beam equation. In the last years, this equation has been studied by McKenna, Walter and others as a model for suspended bridges. Among the other things, they discovered by numerical simulations the existence of solitary waves. In this chapter we prove that these solitary waves can be considered hylomorphic solitons.

7.1 General Feature

Let us consider the nonlinear beam equation

$$\frac{\partial^2 u}{\partial t^2} + \frac{\partial^4 u}{\partial x^4} + W'(u) = 0 \tag{7.1}$$

where $u = u(t, x)$ is real function, $t, x \in \mathbb{R}$ and $W \in C^1(\mathbb{R})$. We shall use Theorem 34 to prove that Eq. (7.1) admits hylomorphic soliton solutions provided W satisfies suitable assumptions (see Theorem 119). In particular these assumptions are satisfied by

$$W(s) = \begin{cases} \frac{1}{2}s^2 \ \text{ for } \ s \leq 1 \\ \\ s - \frac{1}{2} \ \text{ for } \ s \geq 1 \end{cases} . \tag{7.2}$$

Equation (7.1) with $W(s)$ as in (7.2) has been proposed as model for a suspension bridge (see [102, 103, 110]). In particular in [111] and [130] the existence of travelling waves has been proved.

Observe that $u(t, x) - 1$ denotes the displacement of the beam from the unloaded state $u(x) \equiv 1$ and the bridge is seen as a vibrating beam supported by cables which

© Springer International Publishing Switzerland 2014
V. Benci, D. Fortunato, *Variational Methods in Nonlinear Field Equations*,
Springer Monographs in Mathematics, DOI 10.1007/978-3-319-06914-2_7

are treated as springs. The force relative to the potential $W(s)$ in (7.2) is given by

$$F(s) = -W'(s) = \begin{cases} -s \text{ for } s \leq 1 \\ -1 \text{ for } s \geq 1 \end{cases},$$

namely, for $s \geq 1$, only the constant gravity force -1 acts; while, for $s \leq 1$, an elastic force (of intensity $1 - s$), due to the suspension cables, must be added to the constant gravity force -1 . We observe that assumptions (W-i)–(W-iii) in Theorem 119 are satisfied also by the potential

$$W(s) = s - 1 + e^{-s} \tag{7.3}$$

which has been considered in [111] and in [130] as an alternative smooth model for a suspension bridge.

7.2   Existence of Solitons

Equation (7.1) has a variational structure, namely it is the Euler-Lagrange equation with respect to the functional

$$S = \frac{1}{2} \int \int \left(u_t^2 - u_{xx}^2 \right) dx \, dt - \int \int W(u) dx \, dt. \tag{7.4}$$

The Lagrangian relative to the action (7.4) is

$$\mathcal{L} = \frac{1}{2} \left(u_t^2 - u_{xx}^2 \right) - W(u). \tag{7.5}$$

This Lagrangian does not depend on t and x. Then, by Noether's Theorem, the energy E and the momentum C defined by

$$E = \int \left(\frac{\partial \mathcal{L}}{\partial u_t} u_t - \mathcal{L} \right) dx = \frac{1}{2} \int \left(u_t^2 + u_{xx}^2 \right) dx + \int W(u) dx$$

$$C = - \int \left(\frac{\partial \mathcal{L}}{\partial u_t} u_x \right) dx = - \int u_t u_x \, dx$$

are constant along the solutions of (7.1).

Equation (7.1), can be rewritten as an Hamiltonian system as follows:

$$\begin{cases} \partial_t u = v \\ \partial_t v = -\partial_x^4 u - W'(u). \end{cases} \tag{7.6}$$

The phase space is given by

$$X = H^2(\mathbb{R}) \times L^2(\mathbb{R})$$

and the generic point in X will be denoted by

$$\mathbf{u} = \begin{bmatrix} u \\ v \end{bmatrix}.$$

Here $H^2(\mathbb{R})$ denotes the usual Sobolev space.

The norm of X is given by

$$\|\mathbf{u}\| = \left(\int \left(v^2 + u_{xx}^2 + u^2 \right) dx \right)^{\frac{1}{2}}. \tag{7.7}$$

The energy and the momentum, as functionals defined on X, take the following form

$$E(\mathbf{u}) = \frac{1}{2} \int \left(v^2 + u_{xx}^2 \right) dx + \int W(u) dx \tag{7.8}$$

$$C(\mathbf{u}) = - \int v u_x \, dx.$$

As usual we set

$$\Lambda = \frac{E}{|C|}.$$

First of all observe that E and C are invariant under translations i. e. under the representation T_z of the group $G = \mathbb{R}$

$$T_z \mathbf{u}(x) = \mathbf{u}(x + z), \quad z \in \mathbb{R}.$$

We will apply the abstract Theorems 33 and 34, where the momentum $C(\mathbf{u})$ plays the role of the hylenic charge.

We make the following assumptions on W:

- (W-i) **(Positivity)** $\exists \eta > 0$ such that $W(s) \geq \eta s^2$ for $|s| \leq 1$ and $W(s) \geq \eta$ for $|s| \geq 1$.
- (W-ii) **(Nondegeneracy at 0)** $W(0) = W'(0) = 0$, $W''(0) = 1$.
- (W-iii) **(Hylomorphy)** $\exists M > 0$, $\exists \alpha \in [0, 2)$, $\forall s \geq 0$,

$$W(s) \leq Ms^\alpha.$$

Here there are some comments on assumptions (W-ii) and (W-iii). (W-ii) The assumption $W''(0) = 1$ can be weakened just assuming the existence of $W''(0)$. In fact, by (W-i) we have $W''(0) > 0$ and we can reduce to the case $W''(0) = 1$, by rescaling space and time. By this assumption we can write

$$W(s) = \frac{1}{2}s^2 + N(s), \quad N(s) = o(s^2). \tag{7.9}$$

(W-iii) This is the crucial assumption which characterizes the potentials which might produce hylomorphic solitons; notice that this assumptions concerns W only for the positive values of s.

We have the following results:

Theorem 119. *Assume that (W-i)–(W-iii) hold, then there exists an open interval* $(0, \delta_\infty)$ *such that, for every* $\delta \in (0, \delta_\infty)$, *there is a hylomorphic soliton* $\mathbf{u}_\delta$ *(Definition 20) for the dynamical system (7.6). Moreover, if we assume also that*

$$W'(s)s \geq 0 \, for \, all \, s \in \mathbb{R}, \tag{7.10}$$

then $\delta_1 < \delta_2$ *implies that*

(a) $E(\mathbf{u}_{\delta_1}) > E(\mathbf{u}_{\delta_2})$.
(b) $\Lambda(\mathbf{u}_{\delta_1}) < \Lambda(\mathbf{u}_{\delta_2})$.
(c) $|C(\mathbf{u}_{\delta_1})| > |C(\mathbf{u}_{\delta 2})|$.

Theorem 120. *Let* $\mathbf{u}_\delta = (u_\delta, v_\delta)$ *be a soliton as in Theorem 119. Then the solution of Eq. (7.1) with initial data* (u_δ, v_δ) *is the travelling wave:*

$$u(t, x) = u_\delta(x - ct)$$

where c is a constant depending on $\mathbf{u}_\delta$. *Moreover* u_δ *solves the following equation*

$$\frac{\partial^4 u_\delta}{\partial x^4} + c^2 \frac{\partial^2 u_\delta}{\partial x^2} + W'(u_\delta) = 0. \tag{7.11}$$

Remark 121. So we get the existence of solutions of (7.1) by a different proof from that in [111] and [130]. We point out that (7.1) could have solutions which are not minimizers. In this case these solutions give rise to solitary waves which are not solitons.

Remark 122. We point out that the first part of Theorem 120 was first proved in [38].

7.2.1   Coercivity and Splitting Property

In order to prove the first part of Theorem 119 we need to show that (W-i)–(W-iii) imply that assumptions (EC-0)–(EC-3) of the abstract Theorem 33 are satisfied. Moreover, in order to prove the second part of Theorem 119, we need to show that, assuming also (7.10), all the assumptions of Theorem 34 are satisfied.

In the next two lemmas we prove the coercitivity of the energy E, namely that E satisfies (EC-3)

Lemma 123. *Let* $M > 0$. *Then there exists a constant* $C > 0$ *such that* $(E(\mathbf{u}) \leq M) \Rightarrow (\|\mathbf{u}\| \leq C)$, *where the norm* $\|\cdot\|$ *is defined in (7.7).*

Proof. Assume that

$$E(\mathbf{u}) = \frac{1}{2} \int \left(v^2 + u_{xx}^2\right) dx + \int W(u) dx \leq M. \tag{7.12}$$

Then, since $W(u) \geq 0$, we have that

$$\int \left(v^2 + u_{xx}^2\right) dx \leq M. \tag{7.13}$$

It remains to prove that also

$$\int u^2 dx \text{ is bounded.} \tag{7.14}$$

We now set

$$\Omega_u^+ = \{x \mid u(x) > 1\}; \ \Omega_u^- = \{x \mid u(x) < -1\}.$$

Since $u \in H^2(\mathbb{R})$, $u(x) \to 0$ as $|x| \to \infty$, then Ω_u^+, Ω_u^- are bounded. By (7.12) and (W-i) we have

$$M \geq \int W(u) dx \geq \int_{\Omega_u^+ \cup \Omega_u^-} W(u) dx \geq \eta \left|\Omega_u^+\right| + \eta \left|\Omega_u^-\right|, \tag{7.15}$$

where $|\Omega|$ denotes the measure of Ω. Now we show that

$$\int_{\Omega_u^+} u^2 dx \text{ is bounded.} \tag{7.16}$$

Set $v = u - 1$, then, since $v = 0$ on $\partial \Omega_u^+$, by the Poincarè inequality, we have

$$\int_{\Omega_u^+} v^2 dx \leq |\Omega_u^+|^2 \int_{\Omega_u^+} v_x^2 dx. \tag{7.17}$$

On the other hand

$$\int_{\Omega_u^+} v_x^2 dx = - \int_{\Omega_u^+} v \, v_{xx} dx \leq \|v\|_{L^2(\Omega_u^+)} \|v_{xx}\|_{L^2(\Omega_u^+)}. \tag{7.18}$$

Then, since $v = u - 1$, by (7.17) and (7.18),

$$\|u - 1\|^2_{L^2(\Omega_u^+)} \leq \left|\Omega_u^+\right|^2 \|u - 1\|_{L^2(\Omega_u^+)} \|u_{xx}\|_{L^2(\Omega_u^+)}$$

which gives

$$\|u\|^2_{L^2(\Omega_u^+)} - 2\left|\Omega_u^+\right|^{\frac{1}{2}} \|u\|_{L^2(\Omega_u^+)} + \left|\Omega_u^+\right| \leq \left|\Omega_u^+\right|^2 \left(\|u\|_{L^2(\Omega_u^+)} + \left|\Omega_u^+\right|^{\frac{1}{2}}\right) \|u_{xx}\|_{L^2(\Omega_u^+)}.$$

$$(7.19)$$

By (7.13) and (7.15) we have

$$\|u_{xx}\|_{L^2(\Omega_u^+)} \leq \sqrt{M}, \quad \left|\Omega_u^+\right| \leq \frac{M}{\eta}.$$ $$(7.20)$$

By (7.19) and (7.20) we get

$$\|u\|^2_{L^2(\Omega_u^+)} - 2\left(\frac{M}{\eta}\right)^{\frac{1}{2}} \|u\|_{L^2(\Omega_u^+)} \leq \left|\Omega_u^+\right|^2 \sqrt{M} \left(\|u\|_{L^2(\Omega_u^+)} + \sqrt{\frac{M}{\eta}}\right)$$

$$\leq \frac{M^{\frac{5}{2}}}{\eta^2} \left(\|u\|_{L^2(\Omega_u^+)} + \sqrt{\frac{M}{\eta}}\right).$$

From which we easily deduce (7.16). Analogously, we get also that

$$\int_{\Omega_u^-} u^2 dx \text{ is bounded.}$$ $$(7.21)$$

By (W-i)

$$M \geq \int W(u)dx = \int_{|u(x)| \leq 1} W(u(x))dx + \int_{\Omega_u^+ \cup \Omega_u^-} W(u(x))dx \geq \eta \int_{|u(x)| \leq 1} u^2 dx.$$

So, by (7.16), (7.21) and the above inequality, there is a constant R such that

$$\int u^2 dx = \int_{|u(x)| \leq 1} u^2 dx + \int_{\Omega_u^+ \cup \Omega_u^-} u^2 dx \leq \frac{M}{\eta} + R.$$

We conclude that $\int u^2 dx$ is bounded.   □

Lemma 124. *Let $\mathbf{u}_n$ be a sequence in X such that*

$$E(\mathbf{u}_n) \to 0.$$ $$(7.22)$$

Then, up to a subsequence, we have $\|\mathbf{u}_n\| \to 0$, where the norm $\|\cdot\|$ is defined in (7.7).

Proof. Let $\mathbf{u}_n = (u_n, v_n)$, $u_n \in H^2(\mathbb{R})$, $v_n \in L^2(\mathbb{R})$, be a sequence such that $E(\mathbf{u}_n) \to 0$. Then clearly $\|v_n\|_{L^2} \to 0$. By Lemma 123, u_n is bounded in $H^2(\mathbb{R})$ and hence, by the Sobolev embedding theorems, u_n is bounded in $L^\infty(\mathbb{R})$, moreover for all n we have $u_n(x) \to 0$ for $|x| \to \infty$.

For each n let τ_n be a maximum point of $|u_n|$ and set

$$u'_n(x) = u_n(\tau_n + x), \quad v'_n(x) = v_n(\tau_n + x),$$

so that

$$|u'_n(0)| = \max |u'_n|. \tag{7.23}$$

Clearly u'_n is bounded in $H^2(\mathbb{R})$, then, up to a subsequence, we get

$$u'_n \rightharpoonup u \text{ weakly in } H^2(\mathbb{R}) \tag{7.24}$$

and consequently

$$\frac{d^2 u'_n}{dx^2} \rightharpoonup \frac{d^2 u}{dx^2} \text{ weakly in } L^2(\mathbb{R}). \tag{7.25}$$

On the other end, since $E(\mathbf{u}_n) \to 0$, we have $\frac{d^2 u_n}{dx^2} \to 0$ in $L^2(\mathbb{R})$. Then also

$$\frac{d^2 u'_n}{dx^2} \to 0 \text{ in } L^2(\mathbb{R}). \tag{7.26}$$

From (7.25) and (7.26) we get

$$\frac{d^2 u}{dx^2} = 0.$$

So $u \in H^2(\mathbb{R})$ is linear and consequently

$$u = 0. \tag{7.27}$$

Now set

$$B_R = \{x \in \mathbb{R} : |x| < R\}, \quad R > 0$$

then, by the compact embedding $H^2(B_R) \subset\subset L^\infty(B_R)$, by (7.24) and (7.27), we get

$$u'_n \to 0 \text{ in } L^\infty(B_R). \tag{7.28}$$

By (7.23) and (7.28) we get

$$\|u'_n\|_{L^\infty(\mathbb{R})} = |u'_n(0)| \to 0.$$

So, if n is sufficiently large, we have $|u'_n(x)| \le 1$ for all x.

Then, setting $\mathbf{u}'_n = (u'_n, v'_n)$, by (W-i), we have, for n large enough, that

$$
\begin{aligned}
E\left(\mathbf{u}'_n\right) &= \int \left(\frac{1}{2}\left(v'^2_n + \left(\partial^2_{xx} u'_n\right)^2\right) + W(u'_n)\right) dx \\
&\ge \int \left(\frac{1}{2}\left(v'^2_n + \left(\partial^2_{xx} u'_n\right)^2\right) + \eta u'^2_n\right) dx \\
&\ge c\left\|\mathbf{u}'_n\right\|^2
\end{aligned}
\tag{7.29}
$$

where c is a positive constant.

Since

$$E\left(\mathbf{u}'_n\right) = E\left(\mathbf{u}_n\right), \; \left\|\mathbf{u}'_n\right\| = \left\|\mathbf{u}_n\right\|,$$

by (7.29) and (7.22) we have

$$\|\mathbf{u}_n\| \to 0.$$

□

Next we prove that E and C satisfy the splitting property, namely that assumption (EC-2) of Theorem 33 is satisfied.

Lemma 125. *Consider any sequence*

$$\mathbf{u}_n = \mathbf{u} + \mathbf{w}_n \in X$$

where $\mathbf{w}_n$ converges weakly to 0. Then

$$E(\mathbf{u}_n) = E(\mathbf{u}) + E(\mathbf{w}_n) + o(1) \tag{7.30}$$

and

$$C(\mathbf{u}_n) = C(\mathbf{u}) + C(\mathbf{w}_n) + o(1). \tag{7.31}$$

Proof. We shall follow the same arguments as those used in the proof of Lemma 53.

First of all we introduce the following notation:

$$K(u) = \int N\left(u\right) dx \text{ and } K_\Omega(u) = \int_\Omega N\left(u\right) dx, \; \Omega \text{ open subset in } \mathbb{R}.$$

Observe that by (7.7)–(7.9) we have for $\mathbf{u} = (u, v) \in X$

$$E(\mathbf{u}) = \frac{1}{2} \|\mathbf{u}\|^2 + \int N(u). \tag{7.32}$$

As usual u, w_n will denote the first components respectively of $\mathbf{u}, \mathbf{w}_n \in H^2(\mathbb{R}) \times L^2(\mathbb{R})$.

We have to show that $\lim\limits_{n \to \infty} |E(\mathbf{u} + \mathbf{w}_n) - E(\mathbf{u}) - E(\mathbf{w}_n)| = 0$. By (7.32) we have that

$$\lim_{n \to \infty} |E(\mathbf{u} + \mathbf{w}_n) - E(\mathbf{u}) - E(\mathbf{w}_n)| \tag{7.33}$$

$$\leq \lim_{n \to \infty} \frac{1}{2} \left| \|\mathbf{u} + \mathbf{w}_n\|^2 - \|\mathbf{u}\|^2 - \|\mathbf{w}_n\|^2 \right|$$

$$+ \lim_{n \to \infty} \left| \int (N(u + w_n) - N(u) - N(w_n)) \, dx \right|.$$

If $(\cdot, \cdot)$ denotes the inner product related to the norm $\|\cdot\|$, we have:

$$\lim_{n \to \infty} \left| \|\mathbf{u} + \mathbf{w}_n\|^2 - \|\mathbf{u}\|^2 - \|\mathbf{w}_n\|^2 \right| = \lim_{n \to \infty} |2(\mathbf{u}, \mathbf{w}_n)| = 0. \tag{7.34}$$

Then by (7.33) and (7.34) we have

$$\lim_{n \to \infty} |E(\mathbf{u} + \mathbf{w}_n) - E(\mathbf{u}) - E(\mathbf{w}_n)| \tag{7.35}$$

$$\leq \lim_{n \to \infty} \left| \int (N(u + w_n) - N(u) - N(w_n)) \, dx \right|. \tag{7.36}$$

Choose $\varepsilon > 0$ and $R = R(\varepsilon) > 0$ such that

$$\left| \int_{B_R^c} N(u) \right| < \varepsilon, \quad \int_{B_R^c} |u|^2 < \varepsilon \tag{7.37}$$

where

$$B_R^c = \mathbb{R} - B_R \text{ and } B_R = \{x \in \mathbb{R} : |x| < R\}.$$

Since $w_n \rightharpoonup 0$ weakly in $H^2(\mathbb{R})$, by usual compactness arguments, we have that

$$K_{B_R}(w_n) \to 0 \text{ and } K_{B_R}(u + w_n) \to K_{B_R}(u). \tag{7.38}$$

Then we have

$$\lim_{n\to\infty} \left| \int [N(u+w_n) - N(u) - N(w_n)] \right|$$

$$= \lim_{n\to\infty} \left| K_{B_R^c}(u+w_n) + K_{B_R}(u+w_n) \right.$$

$$\left. - K_{B_R^c}(u) - K_{B_R}(u) - K_{B_R^c}(w_n) - K_{B_R}(w_n) \right|. \tag{7.39}$$

The above equality with (7.38) and (7.37) gives

$$\lim_{n\to\infty} \left| \int [N(u+w_n) - N(u) - N(w_n)] \right| \tag{7.40}$$

$$= \lim_{n\to\infty} \left| K_{B_R^c}(u+w_n) - K_{B_R^c}(u) - K_{B_R^c}(w_n) \right|$$

$$\leq \lim_{n\to\infty} \left| K_{B_R^c}(u+w_n) - K_{B_R^c}(w_n) \right| + \varepsilon. \tag{7.41}$$

By the intermediate value theorem there are ζ_n in $(0, 1)$ such that

$$\left| K_{B_R^c}(u+w_n) - K_{B_R^c}(w_n) \right| = \left| \int_{B_R^c} N'(\zeta_n u + w_n) \, u dx \right|. \tag{7.42}$$

Since w_n is bounded in $H^2(\mathbb{R})$, $\zeta_n u + w_n$ is bounded in L^∞, so that there exists a positive constant M such that

$$\|\zeta_n u + w_n\|_{L^\infty} \leq M. \tag{7.43}$$

Now by (7.9) there exists a constant $C_M > 0$ depending on M such that

$$\left| N'(s) \right| \leq C_M |s| \quad \text{for } |s| \leq M. \tag{7.44}$$

Then by (7.42)–(7.44) we get

$$\left| K_{B_R^c}(u+w_n) - K_{B_R^c}(w_n) \right| \leq \int_{B_R^c} \left| N'(\zeta_n u + w_n) u \right| dx$$

$$\leq C_M \int_{B_R^c} \left| (\zeta_n u + w_n) u \right| dx$$

$$\leq C_M \left(\|u\|_{L^2(B_R^c)}^2 + \|w_n\|_{L^2(B_R^c)} \|u\|_{L^2(B_R^c)} \right)$$

$$\leq (\text{by } (7.37)) \, C_M (\varepsilon + c_1 \varepsilon) = c_2 \varepsilon$$

where $c_1 = \sup \|w_n\|_{L^2(B_R^c)}$ and $c_2 = C_M (1 + c_1)$.

Then we get

$$\left| K_{B_R^c} (u + w_n) - K_{B_R^c} (w_n) \right| \le c_2 \varepsilon. \tag{7.45}$$

So by (7.41) and (7.45)

$$\lim_{n \to \infty} \left| \int [N (u + w_n) - N (u) - N (w_n)] \, dx \right| \le c_2 \varepsilon + \varepsilon. \tag{7.46}$$

Finally by (7.35) and (7.46) and since ε is arbitrary we get

$$\lim_{n \to \infty} |E (\mathbf{u} + \mathbf{w}_n) - E (\mathbf{u}) - E (\mathbf{w}_n)| = 0$$

and so (7.30) is proved. The proof of (7.31) is immediate. $\qquad \square$

7.2.2 Analysis of the Hylenic Ratio

If $\mathbf{u} \in X$, $\mathbf{u} = (u, \hat{u})$, we set

$$\|\mathbf{u}\|_\sharp = \|u\|_{L^\infty}. \tag{7.47}$$

Lemma 126. *The seminorm* $\|\mathbf{u}\|_\sharp$ *defined in (7.47) satisfies the property (2.19) (see page 37), namely*

$$\{\mathbf{u}_n \text{ is a vanishing sequence}\} \Rightarrow \|\mathbf{u}_n\|_\sharp \to 0.$$

Proof. Let $\mathbf{u}_n \in X$, $\mathbf{u}_n = (u_n, \hat{u}_n)$ be a vanishing sequence (see Definition 28) and, arguing by contradiction, assume that $\|\mathbf{u}_n\|_\sharp$ does not converge to 0. Then, up to a subsequence, we have

$$\|u_n\|_{L^\infty} \ge a > 0.$$

Then for any positive integer n there exists $x_n \in \mathbb{R}$ s. t.

$$|u_n(x_n)| \ge \frac{a}{2}.$$

So, setting $T_{x_n} u_n(x) = u_n(x + x_n)$, we get for any n

$$|T_{x_n} u_n(0)| = |u_n(x_n)| \ge \frac{a}{2} > 0. \tag{7.48}$$

On the other hand, since $\mathbf{u}_n = (u_n, \hat{u}_n)$ is vanishing,

$$T_{x_n} u_n \rightharpoonup 0 \text{ weakly in } H^2(\mathbb{R}).$$

Then, since $H^2(\mathbb{R})$ is compactly embedded into $L^\infty(-1, 1)$, we have

$$T_{x_n} u_n \to 0 \text{ strongly in } L^\infty(-1, 1) \tag{7.49}$$

and clearly (7.49) contradicts (7.48).   □

Now, as usual, we set

$$\Lambda(\mathbf{u}) = \frac{E(\mathbf{u})}{|C(\mathbf{u})|}$$

and

$$\Lambda_* = \inf_{\mathbf{u} \in X} \Lambda(\mathbf{u}) = \inf_{\mathbf{u} \in X} \frac{\frac{1}{2}\|\mathbf{u}\|^2 + \int N(u)dx}{|C(\mathbf{u})|}$$

$$\Lambda_0 := \inf\{\liminf \Lambda(\mathbf{u}_n) \mid \mathbf{u}_n \text{ is a vanishing sequence}\}$$

$$\Lambda_\sharp = \liminf_{\|\mathbf{u}\|_\sharp \to 0} \Lambda(\mathbf{u}) = \lim_{\varepsilon \to 0} \inf\{\Lambda(\varepsilon\mathbf{u}) \mid \|\mathbf{u}\|_\sharp = 1\}.$$

By Lemma 126 and by the definitions of Λ_0 and $\Lambda_\sharp$, we have that

$$\Lambda_0 \geq \Lambda_\sharp. \tag{7.50}$$

So let us evaluate $\Lambda_\sharp$.

Lemma 127. *The following inequality holds:*

$$\Lambda_\sharp \geq 1.$$

Proof. Clearly we have:

$$\Lambda_\sharp = \liminf_{\|\mathbf{u}\|_\sharp \to 0} \Lambda(\mathbf{u}) =$$

$$\lim_{\varepsilon \to 0} \inf\left\{\frac{E(\varepsilon\mathbf{u})}{|C(\varepsilon\mathbf{u})|} \mid \|\mathbf{u}\|_\sharp = 1\right\} =$$

$$\lim_{\varepsilon \to 0} \inf\left\{\frac{\frac{\varepsilon^2}{2}\|\mathbf{u}\|^2 + \int N(\varepsilon u)dx}{|C(\varepsilon\mathbf{u})|} \mid \|u\|_{L^\infty} = 1\right\} \geq$$

$$\lim_{\varepsilon \to 0} \inf \left\{ \frac{\frac{\varepsilon^2}{2}\|\mathbf{u}\|^2 - \int |N(\varepsilon u)|\, dx}{|C(\varepsilon \mathbf{u})|} \ \middle| \ \|u\|_{L^\infty} = 1 \right\} \geq$$

$$\lim_{\varepsilon \to 0} \inf \left\{ \frac{\frac{\varepsilon^2}{2}\|\mathbf{u}\|^2 - \int |N(\varepsilon u)|\, dx}{|C(\varepsilon \mathbf{u})|} \ \middle| \ \|u\|_{L^\infty} = 1 \right\} \geq$$

$$\lim_{\varepsilon \to 0} \inf \left\{ \frac{\frac{\varepsilon^2}{2}\|\mathbf{u}\|^2 - \varepsilon^2 \int g(\varepsilon)\,|u|^2\, dx}{\varepsilon^2 |C(\mathbf{u})|} \ \middle| \ \|u\|_{L^\infty} = 1 \right\}$$

where

$$g(\varepsilon) = \sup \left\{ \frac{|N(\varepsilon u)|}{|\varepsilon u|^2} : \|u\|_{L^\infty} = 1 \right\}.$$

So we get

$$\Lambda_\sharp \geq \lim_{\varepsilon \to 0} \inf \left\{ \frac{\frac{1}{2}\|\mathbf{u}\|^2 - \int g(\varepsilon)\,|u|^2\, dx}{|C(\mathbf{u})|} \ \middle| \ \|u\|_{L^\infty} = 1 \right\}. \tag{7.51a}$$

By (7.9)

$$\frac{N(s)}{s^2} \to 0 \text{ as } s \to 0.$$

Then we have

$$g(\varepsilon) \to 0 \text{ as } \varepsilon \to 0. \tag{7.52}$$

By (7.51a) and (7.52) we have

$$\Lambda_\sharp \geq \lim_{\varepsilon \to 0} \inf \left\{ \frac{\left(\frac{1}{2} - g(\varepsilon)\right)\|\mathbf{u}\|^2}{|C(\mathbf{u})|} \ \middle| \ \|u\|_{L^\infty} = 1 \right\} \geq$$

$$\inf \left\{ \frac{\frac{1}{2}\|\mathbf{u}\|^2}{|C(\mathbf{u})|} : \mathbf{u} \in X \right\}.$$

Then

$$\Lambda_\sharp \geq \inf \left\{ \frac{\frac{1}{2}\|\mathbf{u}\|^2}{|C(\mathbf{u})|} : \mathbf{u} \in X \right\}.$$

So the lemma will be proved if we show that for all $\mathbf{u} \in X$

$$\frac{\frac{1}{2} \|\mathbf{u}\|^2}{|C(\mathbf{u})|} \geq 1. \tag{7.53}$$

To prove (7.53) observe that for $\mathbf{u} = (v, u) \in X$ we have

$$|C(\mathbf{u})| \leq \int |v \partial_x u| \, dx \leq \left(\int v^2 \, dx \right)^{1/2} \cdot \left(\int |\partial_x u|^2 \, dx \right)^{1/2}$$

$$\leq \frac{1}{2} \int v^2 \, dx + \frac{1}{2} \int |\partial_x u|^2 \, dx$$

$$= \frac{1}{2} \int v^2 \, dx - \frac{1}{2} \int u u_{xx} \, dx$$

$$\leq \frac{1}{2} \int v^2 \, dx + \frac{1}{2} \int \frac{1}{2} \left[u^2 + u_{xx}^2 \right] \, dx$$

$$\leq \frac{1}{2} \int \left[v^2 + u_{xx}^2 + u^2 \right] \, dx = \frac{1}{2} \|\mathbf{u}\|^2.$$

$\square$

The next lemma provides a crucial estimate for the existence of solitons:

Lemma 128. *The hylomorphy condition holds, namely*

$$\Lambda_* = \inf_{\mathbf{u} \in X} \Lambda(\mathbf{u}) < \Lambda_0.$$

Proof. By (7.50) and Lemma 127 we get

$$\Lambda_0 \geq 1.$$

So in order to prove the Lemma, it will be enough to prove that

$$\Lambda_* < 1. \tag{7.54}$$

Let $U \in C^2$ be a positive function with compact support such that

$$\frac{\int (U_{xx})^2}{\int (U_x)^2} < \frac{1}{2}. \tag{7.55}$$

Such a function exists; in fact if U_0 is any positive function with compact support, $U(x) = U_0\left(\frac{x}{\lambda}\right)$ satisfies (7.55) for λ sufficiently large. Take

$$\mathbf{u}_R = (u_R, v) = (RU, RU_x).$$

By the definition of X, $\mathbf{u}_R \in X$. Now we can estimate Λ_*:

$$\Lambda_* = \inf_{u \in X} \frac{\frac{1}{2}\|\mathbf{u}\|^2 + \int N(u)dx}{|C(\mathbf{u})|} \le \frac{\frac{1}{2}\|\mathbf{u}_R\|^2 + \int N(u_R)dx}{|C(\mathbf{u}_R)|}$$

$$= \frac{\frac{1}{2}\int\left[(RU_x)^2 + (RU_{xx})^2 + (RU)^2\right]dx + \int N(RU)dx}{\int(RU_x)^2\,dx}$$

$$= \frac{\frac{1}{2}\int\left[(RU_x)^2 + (RU_{xx})^2\right]dx}{\int(RU_x)^2\,dx} + \frac{\int W(RU)dx}{\int(RU_x)^2\,dx}$$

$$= \frac{1}{2} + \frac{1}{2}\frac{\int(U_{xx})^2\,dx}{\int(U_x)^2\,dx} + \frac{\int W(RU)dx}{\int(RU_x)^2\,dx} \quad \text{(by (W-iii))}$$

$$\le \frac{1}{2} + \frac{1}{2}\frac{\int(U_{xx})^2\,dx}{\int(U_x)^2\,dx} + \frac{\int M\,|RU|^\alpha\,dx}{\int(RU_x)^2\,dx} \quad \text{(by (7.55))}$$

$$< \frac{1}{2} + \frac{1}{4} + \frac{M}{R^{2-\alpha}} \cdot \frac{\int|U|^\alpha\,dx}{\int U_x^2\,dx}.$$

Then, for R sufficiently large, we get (7.54) $\qquad\qquad\square$

Proof of Theorem 119. The preceding lemmas show that E and C satisfy (EC-0),...,(EC-3) and the hylomorphy condition (2.16) of Definition 29. So we can apply Theorem 33 in order to get the existence of an open interval $(0, \delta_\infty)$ such that, for every $\delta \in (0, \delta_\infty)$, there is a hylomorphic soliton $\mathbf{u}_\delta$ (Definition 20) for the dynamical system (7.6). Now we assume also that

$$W'(s)s \ge 0 \text{ for all } s \in \mathbb{R}_+. \tag{7.56}$$

Then also assumption (2.25) is satisfied. In fact, let $\mathbf{u} = (u, v) \in X$ such that

$$\|E'(\mathbf{u})\| + \|C'(\mathbf{u})\| = 0.$$

So easy computations show that

$$\frac{\partial^4 u}{\partial x^4} + W'(u) = 0 \text{ and } v = 0. \tag{7.57}$$

From (7.57) and (7.56) we get

$$\mathbf{u} = (u, v) = 0.$$

So assumption (2.25) is satisfied. Then, by Theorem 34 and Remark 36, if $\delta_1 < \delta_2$ we have:

(a) $E(\mathbf{u}_{\delta_1}) > E(\mathbf{u}_{\delta_2})$.
(b) $\Lambda(\mathbf{u}_{\delta_1}) < \Lambda(\mathbf{u}_{\delta_2})$.
(c) $|C(\mathbf{u}_{\delta_1})| > |C(\mathbf{u}_{\delta2})|$.

$\square$

Proof of Theorem 120. Since $\mathbf{u}_\delta = (u_\delta, v_\delta) \in X = H^2(\mathbb{R}) \times L^2(\mathbb{R})$ is a minimizer, we have $J'(\mathbf{u}_\delta) = 0$. Then

$$\frac{E'(\mathbf{u}_\delta)}{C(\mathbf{u}_\delta)} - \frac{E(\mathbf{u}_\delta)}{C(\mathbf{u}_\delta)^2} C'(\mathbf{u}_\delta) + \delta E'(\mathbf{u}_\delta) = 0$$

namely

$$\left(C(\mathbf{u}_\delta) + \delta C(\mathbf{u}_\delta)^2\right) E'(\mathbf{u}_\delta) = E(\mathbf{u}_\delta)C'(\mathbf{u}_\delta).$$

Since we can take $C(\mathbf{u}_\delta) > 0$, then $C(\mathbf{u}_\delta) + \delta C(\mathbf{u}_\delta)^2 > 0$, and hence we can divide both sides by $C(\mathbf{u}_\delta) + \delta C(\mathbf{u}_\delta)^2$ and we get

$$E'(\mathbf{u}_\delta) = c C'(\mathbf{u}_\delta) \tag{7.58}$$

where

$$c = \frac{E(\mathbf{u}_\delta)}{C(\mathbf{u}_\delta) + \delta C(\mathbf{u}_\delta)^2}.$$

If we write (7.58) explicitly, we get for all $\varphi \in H^2(\mathbb{R})$ and all $\psi \in L^2(\mathbb{R})$

$$\int \partial_x^2 u_\delta \partial_x^2 \varphi + W'(u_\delta)\varphi = c \int v_\delta \partial_x \varphi$$

$$\int v_\delta \psi = c \int \psi \partial_x u_\delta$$

namely

$$\partial_x^4 u_\delta + W'(u_\delta) = -c \partial_x v_\delta$$

$$v_\delta = c \partial_x u_\delta$$

and so we get

$$\partial_x^4 u_\delta + c^2 \partial_x^2 u_\delta + W'(u_\delta) = 0.$$

Now we can check directly that

$$u(t, x) = u_\delta(x - ct)$$

solves Eq. (7.1) with initial conditions $(u_\delta(x), -c \partial_x u_\delta(x))$.

$\square$

Chapter 8
Vortices

Let us consider a FT variational dynamical system (see Definition 10). The angular momentum is the constant of the motion due to the invariance of the Lagrangian under space rotations (see Sect. 1.3.2).

A vortex is a finite energy solution with non vanishing angular momentum.

The existence of vortices is an interesting and old issue in many questions of mathematical physics as superconductivity, classical and quantum field theory, string and elementary particle theory (see the pioneering papers [5, 115] and e.g. the more recent ones [94, 141–143, 147] with their references).

From mathematical viewpoint, the existence of vortices for the nonlinear Klein-Gordon equations (NKG), for nonlinear Schrödinger equations (3.1) and for nonlinear Klein-Gordon-Maxwell equations has been studied in some recent papers [11, 12, 16, 18, 19, 32, 33, 35, 45].

In this chapter we shall use Theorem 34 in order to prove the existence of vortices for the nonlinear Schrödinger equation and the nonlinear Klein-Gordon equation.

8.1 Vortices for the Nonlinear Schrödinger Equation

In this section we study the existence of vortices for (3.1).

8.1.1 Statement of the Results

Consider the nonlinear Schrödinger equations (3.1) in three space variables.

As we have seen in Sect. 3.1.1, the energy $E(\psi)$ and the hylenic charge $C(\psi)$ of ψ have respectively the following expressions

© Springer International Publishing Switzerland 2014   219
V. Benci, D. Fortunato, *Variational Methods in Nonlinear Field Equations*,
Springer Monographs in Mathematics, DOI 10.1007/978-3-319-06914-2_8

$$E(\psi) = \int \left[\frac{1}{2} |\nabla \psi|^2 + W(\psi) + V(x) |\psi|^2 \right] dx$$

$$C(\psi) = \int |\psi|^2 \, dx. \tag{8.1}$$

Now assume that the potential V depends only on the third space variable x_3 and that it is periodic

$$V(x_1, x_2, x_3) = V(x_3 + k) \text{ for all } (x_1, x_2, x_3) \in \mathbb{R}^3 \text{ and } k \in \mathbb{Z}. \tag{8.2}$$

Under this assumption the Lagrangian (3.3) is invariant with respect rotations around the x_3 axis. In this case the third component of the angular momentum

$$M_3(\psi) = \text{Re} \int (x_1 \partial_{x_2} \psi - x_2 \partial_{x_1} \psi) \, dx$$

is a constant of the motion. Using the polar form

$$\psi(t, x) = u(t, x) e^{iS(t,x)}, \ u \in \mathbb{R}$$

$M_3(\psi)$ can be written as follows

$$M_3(\psi) = \int (x_1 \partial_{x_2} S - x_2 \partial_{x_1} S) \, u^2 \, dx.$$

A solution of (3.1) is called standing wave (see Sect. 3.2.1) if it has the following form:

$$\psi(t, x) = \psi_0(x) e^{-i\omega t}, \quad \omega \in \mathbb{R}. \tag{8.3}$$

We want to look for vortices of (3.1) which are standing waves.

It is immediate to check that if $\psi_0(x)$ in (8.3) is real valued, the angular momentum $M_3(\psi)$ is trivial. However, if $\psi_0(x)$ is allowed to have complex values, it is possible to have $M_3(\psi) \neq 0$. Thus, we are led to make an ansatz in (3.1) of the following form:

$$\psi(t, x) = u(x) e^{i(\ell\theta(x) - \omega t)}, \quad u(x) \geq 0, \ \omega \in \mathbb{R}, \ \ell \in \mathbb{Z} - \{0\} \tag{8.4}$$

$$\theta(x) = \text{Im} \log(x_1 + i x_2) \in \mathbb{R}/2\pi\mathbb{Z}; \ x = (x_1, x_2, x_3). \tag{8.5}$$

The function θ maps smoothly $\mathbb{R}^3 - \Sigma$ into $\mathbb{R}/2\pi\mathbb{Z}$, where

$$\Sigma = \{(x_1, x_2, x_3) \in \mathbb{R}^3 \mid x_1 = x_2 = 0\}. \tag{8.6}$$

We shall set, with abuse of notation,

$$\nabla\theta = \left(-\frac{x_1}{x_1^2 + x_2^2}, \frac{x_2}{x_1^2 + x_2^2}\right).$$

Moreover, we assume that u has a *cylindrical symmetry*, namely

$$u(x) = u(r, x_3), \text{ where } r = \sqrt{x_1^2 + x_2^2}. \tag{8.7}$$

Inserting ansatz (8.4) in Eq. (3.1), we get the system

$$\begin{cases} -\Delta u + \ell^2 |\nabla\theta|^2 u + W'(u) + 2V(x_3)\psi = 2\omega u \\ u^2 \Delta\theta + 2\nabla u \cdot \nabla\theta = 0. \end{cases}$$

By the definition of θ and (8.7) we have

$$\Delta\theta = 0, \quad \nabla\theta \cdot \nabla u = 0, \quad |\nabla\theta|^2 = \frac{1}{r^2},$$

where the dot $\cdot$ denotes the Euclidean scalar product.

So the above system reduces to find solutions, with symmetry (8.7), of the equation

$$-\Delta u + \frac{\ell^2}{r^2}u + W'(u) + 2V(x)u = 2\omega u. \tag{8.8}$$

Direct computations show that the energy and the third component of the angular momentum of a solution ψ like in (8.4) become

$$E_\ell(\psi) = E_\ell\left(u(x)\, e^{i(\ell\theta(x)-\omega t)}\right)$$

$$= \int_{\mathbb{R}^3}\left[\frac{1}{2}|\nabla u|^2 + \left(\frac{1}{2}\frac{\ell^2}{r^2} + V(x)\right)u^2 + W(u)\right]dx \tag{8.9}$$

$$M_3\left(u(x)\, e^{i(\ell\theta(x)-\omega t)}\right) = \ell\int_{\mathbb{R}^3} u^2 dx. \tag{8.10}$$

We point out that M_3 in (8.10) is nontrivial if and only if both ℓ and u are not zero.

The following theorem holds:

Theorem 129. *Assume that W and V satisfy assumptions (WB-i)–(WB-iv) and (V-i), (V-ii) at page 71. Then for any integer $l \neq 0$ there exists a family ψ_δ ($0 < \delta < \delta_\infty$) of vortices for (3.1)*

$$\psi_\delta = u_\delta\left(x\right)e^{i\left(\ell\theta\left(x\right)-\omega_\delta t\right)},\ u_\delta\left(x\right) \geq 0\ satisfying\ (8.7),\ \omega_\delta \in \mathbb{R}.$$

These vortices have angular momentum

$$M(\psi_\delta) = (0,0,\ell\,\|u_\delta\|_{L^2}^2).$$

Moreover, if $\delta_1 < \delta_2$, u_{δ_1}, u_{δ_2} are distinct, namely we have that

$$\delta_1 < \delta_2 \implies \|u_{\delta_2}\|_{L^2}^2 < \|u_{\delta_1}\|_{L^2}^2.$$

8.1.2   Proof of the Main Result

Let Σ denote the x_3 axis (see (8.6)) and let X be the Hilbert space obtained by the closure of $\mathcal{D}(\mathbb{R}^3 - \Sigma)$, with respect to the norm

$$\|u\|_X^2 = \int_{\mathbb{R}^3}\left[|\nabla u|^2 + \left(\frac{\ell}{r^2}+1\right)u^2\right]dx. \tag{8.11}$$

Comparing (8.8) with (8.9) and (8.1) clearly we have that the solutions of Eq. (8.8) can be obtained as critical points of the functional (8.9) on the manifold

$$\mathfrak{M}_c := \{u \in X \mid C(u) = c\}.$$

Clearly, using this approach, 2ω will be the Lagrange multiplier.

Now consider the action S_θ of the group S^1 on $u(x_1, x_2, x_3) \in X$, defined by

$$S_\theta u = u(R_\theta(x_1,x_2),x_3),\ \ \theta \in \frac{\mathbb{R}}{2\pi\mathbb{Z}}, \tag{8.12}$$

where R_θ denotes the rotation of an angle θ in the plane x_1, x_2. We set

$$X_r = \{u \in X \mid u = u(r,x_3)\},\ \mathfrak{M}_c^r = \mathfrak{M}_c \cap X_r.$$

Observe that V depends only on x_3, then the functional E_ℓ is invariant under the action (8.12). So by the Palais principle of symmetric criticality [118], the critical points of E_ℓ on $\mathfrak{M}_c^r$ are also critical points of E_ℓ on $\mathfrak{M}_c$.

Then we are reduced to find critical points of E_ℓ on $\mathfrak{M}_c^r$. To do this we apply the minimization result of Corollary 35. So we have to prove that E_ℓ and C satisfy assumptions (EC-0)–(EC-2) and (EC-3*), the hylomorphy condition (2.16) and assumption (2.25).

Observe first that, by (8.2), E_ℓ is invariant under the action T_k of the group $G = \mathbb{Z}$ on X_r defined by

$$T_k u(r, x_3) = u(r, x_3 + k), \quad k \in \mathbb{Z}.$$

Then E_ℓ and C satisfy assumptions (EC-0), (EC-1).

Following, with obvious changes, the same arguments used in proving Lemmas 53 and 54, it can be shown that E_ℓ and C satisfy also the splitting property (EC-2) and the coercivity property (EC-3*). Moreover $C'(u) = 0$ means $u = 0$, then also assumption (2.25) is satisfied.

It remains to prove the hylomorphy condition (2.16). To this end, as usual, we shall carry out in the following subsection an analysis of the hylenic ratio.

8.1.3 Analysis of Hylenic Ratio for Vortices

Lemma 130. *If* $2 < t < 6$, *the norm* $\|u\|_{L^t}$ *satisfies the property (2.19), namely*

$$\{u_n \text{ is a vanishing sequence in } X_r\} \Rightarrow \|u_n\|_{L^t} \to 0.$$

Proof. Let u_n be a vanishing sequence in X_r and, arguing by contradiction, assume that $\|u_n\|_{L^t}$ does not converge to 0. Then, up to a subsequence,

$$\|u_n\|_{L^t} \geq a > 0. \tag{8.13}$$

Since u_n is bounded in X_r, we have that for a suitable constant $M > 0$

$$\|u_n\|_{H^1}^2 \leq M. \tag{8.14}$$

Now we set

$$Q_i = \{(x_1, x_2, x_3) : i \leq x_3 < i + 1\}, i \text{ integer.}$$

Clearly

$$\mathbb{R}^3 = \bigcup_{i \in \mathbb{Z}} Q_i.$$

Let C denote the constant for the Sobolev embedding $H^1(Q_i) \subset L^t(Q_i)$, then, by (8.13) and (8.14), we get the following

$$0 < a^t \leq \int |u_n|^t = \sum_i \int_{Q_i} |u_n|^t = \sum_i \|u_n\|_{L^t(Q_i)}^{t-2} \|u_n\|_{L^t(Q_i)}^2$$

$$\leq \left(\sup_i \|u_n\|_{L^t(Q_i)}^{t-2}\right) \cdot \sum_i \|u_n\|_{L^t(Q_i)}^2$$

$$\leq C \left(\sup_i \|u_n\|_{L^t(Q_i)}^{t-2} \right) \cdot \sum_i \|u_n\|_{H^1(Q_i)}^2$$

$$= C \left(\sup_i \|u_n\|_{L^t(Q_i)}^{t-2} \right) \|u_n\|_{H^1}^2 \leq CM \left(\sup_i \|u_n\|_{L^t(Q_i)}^{t-2} \right).$$

Then

$$\left(\sup_i \|u_n\|_{L^t(Q_i)} \right) \geq \left(\frac{a^t}{CM} \right)^{1/(t-2)}.$$

So, for any n, there exists an integer i_n such that

$$\|u_n\|_{L^t(Q_{i_n})} \geq \alpha > 0. \tag{8.15}$$

Then

$$\|T_{i_n} u_n\|_{L^t(Q_0)} = \|u_n\|_{L^t(Q_{i_n})} \geq \alpha > 0.$$

Since u_n is bounded in X_r, $T_{i_n} u_n$ is also bounded. Then, passing eventually to a subsequence, we have that

$$T_{i_n} u_n \rightharpoonup u_0 \text{ weakly in } X_r.$$

Clearly, if we show that $u_0 \neq 0$, we get a contradiction with the assumption that u_n is vanishing.

Now, let $\varphi = \varphi(x_3)$ be a nonnegative, C^∞-function whose value is 1 for $0 < x_3 < 1$ and 0 for $|x_3| > 2$. Then the sequence $\varphi T_{i_n} u_n$ is bounded in $H_0^1(\mathbb{R}^2 \times (-2, 2))$, moreover $\varphi T_{i_n} u_n$ is invariant under the action (8.12). Then, using the compactness result proved in [76], we have

$$\varphi T_{i_n} u_n \to \chi \text{ strongly in } L^t(\mathbb{R}^2 \times (-2, 2)).$$

On the other hand

$$\varphi T_{i_n} u_n \to \varphi u_0 \ a.e. \tag{8.16}$$

Then

$$\varphi T_{i_n} u_n \to \varphi u_0 \text{ strongly in } L^t(\mathbb{R}^2 \times (-2, 2)). \tag{8.17}$$

Moreover

$$\|\varphi T_{i_n} u_n\|_{L^t(\mathbb{R}^2 \times (-2,2))} \geq \|\varphi T_{i_n} u_n\|_{L^t(Q_0)} = \|u_n\|_{L^t(Q_{i_n})} \geq \alpha > 0. \tag{8.18}$$

Then by (8.17) and (8.18)

$$\|\varphi u_0\|_{L^t(\mathbb{R}^2 \times (-2,2))} \geq \alpha > 0.$$

Thus we have that $u_0 \neq 0$. $\qquad\qquad\qquad\qquad\qquad\qquad\qquad\qquad\square$

Lemma 131. *Assume that W and V satisfy assumptions (WB-i)–(WB-iv) and (V-i), (V-ii) at page 71. Then E_l and C, defined in (8.9) and (8.1), satisfy the hylomorphy condition*

$$\inf_{u \in X_r} \frac{E_l(u)}{C(u)} < \Lambda_0.$$

By Lemma 130 and by definition of Λ_0 we have

$$\Lambda_0 \geq \liminf_{u \in X_r, \|u\|_{L^t} \to 0} \Lambda(u). \qquad\qquad (8.19)$$

By the same arguments used in Lemma 56 in Sect. 3.2.4, it can be shown that

$$\liminf_{u \in X_r, \|u\|_{L^t} \to 0} \Lambda(u) \geq E_0. \qquad\qquad (8.20)$$

So, by (8.20) and (8.19), we get

$$\Lambda_0 \geq E_0.$$

Then, in order to prove the hylomorphy condition, it will be enough to construct $u \in X_r$ such that

$$\Lambda(u) < E_0. \qquad\qquad (8.21)$$

The construction of such u needs some work since we require that u belongs to X_r, namely we require that u is invariant under the S^1 action (8.12) and it is 0 near the x_3 axis, so that $\int \frac{u^2}{r^2}$ converges.

For $0 < \mu < \lambda$ we set:

$$T_{\lambda, \mu} = \left\{ (r, x_3) : (r - \lambda)^2 + x_3^2 \leq \mu^2 \right\}$$

and, for $\lambda > 2$, we consider a smooth function u_λ with cylindrical symmetry such that

$$u_\lambda(r, x_3) = \begin{cases} s_0 & \text{if } (r, x_3) \in T_{\lambda, \lambda/2} \\ \\ 0 & \text{if } (r, x_3) \notin T_{\lambda, \lambda/2+1} \end{cases} \qquad\qquad (8.22)$$

where s_0 is such that $\frac{N(s_0)}{s_0^2} < -V_0$ (see (3.44)). So we can take $\eta > 0$ so that

$$\frac{N(s_0)}{s_0^2} < -V_0 - \eta. \tag{8.23}$$

Moreover we may assume that

$$|\nabla u_\lambda (r, x_3)| \le 2 \text{ for } (r, x_3) \in T_{\lambda, \lambda/2+1} \setminus T_{\lambda, \lambda/2}. \tag{8.24}$$

We have

$$\Lambda(u_\lambda) = \frac{\int \left[|\nabla u_\lambda|^2 + \frac{\ell^2 u_\lambda^2}{r^2} + 2V u^2 \right] dx}{2 \int u_\lambda^2} + \frac{\int W(u_\lambda) dx}{\int u_\lambda^2}. \tag{8.25}$$

By (8.24) and (8.22) a direct computation shows that

$$\int |\nabla u_\lambda|^2 \le 4 meas(T_{\lambda, \lambda/2+1} \setminus T_{\lambda, \lambda/2}) \le c_1 \lambda^2 \tag{8.26}$$

$$\int \frac{u_\lambda^2}{r^2} \le \frac{c_2}{\lambda^2} meas(T_{\lambda, \lambda/2+1}) \le c_3 \lambda \tag{8.27}$$

$$\int u_\lambda^2 \ge c_4 meas(T_{\lambda, \lambda/2}) \ge c_5 \lambda^3 \tag{8.28}$$

where $c_1 - c_5$ are positive constants. So that

$$\frac{\int \left[|\nabla u_\lambda|^2 + \frac{\ell^2 u_\lambda^2}{r^2} + 2V u_\lambda^2 \right] dx}{2 \int u_\lambda^2} \le V_0 + O\left(\frac{1}{\lambda}\right). \tag{8.29}$$

By (8.25) and (8.29) we have

$$\Lambda(u_\lambda) \le V_0 + \frac{\int W(u_\lambda) dx}{\int u_\lambda^2} + O\left(\frac{1}{\lambda}\right) = V_0 + E_0 + \frac{\int N(u_\lambda) dx}{\int u_\lambda^2} + O\left(\frac{1}{\lambda}\right). \tag{8.30}$$

Now

$$\frac{\int N(u_\lambda) dx}{\int u_\lambda^2} = \frac{\int_{T_{\lambda, \lambda/2}} N(u_\lambda) dx + \int_{T_{\lambda, \lambda/2+1} \setminus T_{\lambda, \lambda/2}} N(u_\lambda) dx}{\int u_\lambda^2}$$

$$\le \frac{\int_{T_{\lambda, \lambda/2}} N(s_0) dx + \int_{T_{\lambda, \lambda/2+1} \setminus T_{\lambda, \lambda/2}} |N(u_\lambda)| dx}{\int u_\lambda^2}$$

$$\leq \frac{m(T_{\lambda,\lambda/2})N(s_0)}{\int_{T_{\lambda,\lambda/2+1}} s_0^2 dx} + \frac{\int_{T_{\lambda,\lambda/2+1}\backslash T_{\lambda,\lambda/2}} |N(u_\lambda)| dx}{\int u_\lambda^2}$$

$$\leq \frac{N(s_0)}{s_0^2} \left(\frac{\frac{\lambda}{2}}{\frac{\lambda}{2}+1}\right)^2 + \frac{\int_{T_{\lambda,\lambda/2+1}\backslash T_{\lambda,\lambda/2}} |N(u_\lambda)| dx}{\int_{T_{\lambda,\lambda/2}} u_\lambda^2}$$

$$\leq \frac{N(s_0)}{s_0^2} \left(\frac{\frac{\lambda}{2}}{\frac{\lambda}{2}+1}\right)^2 + \frac{c_1 m(T_{\lambda,\lambda/2+1}\backslash T_{\lambda,\lambda/2})}{s_0^2 m(T_{\lambda,\lambda/2})}$$

$$\leq \frac{N(s_0)}{s_0^2} \left(\frac{\frac{\lambda}{2}}{\frac{\lambda}{2}+1}\right)^2 + \frac{c_2}{\lambda}.$$

Then we have

$$\frac{\int N(u_\lambda) dx}{\int u_\lambda^2} \leq \frac{N(s_0)}{s_0^2} \left(\frac{\frac{\lambda}{2}}{\frac{\lambda}{2}+1}\right)^2 + \frac{c_2}{\lambda}. \tag{8.31}$$

So by (8.30) and (8.31) we get

$$\Lambda(u_\lambda) \leq V_0 + E_0 + \frac{N(s_0)}{s_0^2} \left(\frac{\frac{\lambda}{2}}{\frac{\lambda}{2}+1}\right)^2 + O\left(\frac{1}{\lambda}\right).$$

By (8.23)

$$\Lambda(u_\lambda) < V_0 + E_0 + (-V_0 - \eta)\left(\frac{\frac{\lambda}{2}}{\frac{\lambda}{2}+1}\right)^2 + O\left(\frac{1}{\lambda}\right).$$

Then, if λ is large enough,

$$\Lambda(u_\lambda) < E_0.$$

So (8.21) holds.   □

8.1.4 Solutions in the Sense of Distribution

We recall that X is the Hilbert space obtained by the closure of $\mathcal{D}(\mathbb{R}^3 - \Sigma)$ ($\Sigma = \{(x_1, x_2, x_3) \in \mathbb{R}^3 \mid x_1 = x_2 = 0\}$) with respect to the norm (8.11) and that the solutions u we find are critical points of the functional E_l defined by (8.9) on the manifold $\mathfrak{M}_c := \{u \in X \mid C(u) = c\}$. So u solves the equation

$$\int \left(\nabla u \cdot \nabla v + \frac{\ell^2}{r^2} uv + W'(u) v + 2V(x) uv \right) dx = 2\omega \int uv dx \text{ for all } v \in X.$$

$$(8.32)$$

Since $\mathcal{D}(\mathbb{R}^3)$ is not contained in X, a solution of (8.32) need not to be a solution of (8.8) in the sense of distributions in $\mathbb{R}^3$. In fact, since $\frac{\ell^2}{r^2}$ is singular on Σ, it may be that for some test functions $v \in \mathcal{D}(\mathbb{R}^3)$ the integral $\int \frac{\ell^2}{r^2} uv dx$ diverges, unless u is sufficiently small as $x \to \Sigma$. In this subsection we show that this divergence does not occur, namely the singularity is removable in the sense of the following theorem:

Theorem 132. *Let $u_0 \in X$, $u_0 \geq 0$ be a solution of (8.32). Then u_0 is a solution of equation (8.8) in the sense of distribution in $\mathbb{R}^3$, namely*

$$\int \left(\nabla u \cdot \nabla v + \frac{\ell^2}{r^2} uv + W'(u) v + 2V(x) uv \right) dx = 2\omega \int uv dx \quad \forall v \in \mathcal{D}(\mathbb{R}^3).$$

$$(8.33)$$

Let χ_n (n positive integer) be a family of smooth functions depending only on $r = \sqrt{x_1^2 + x_2^2}$ and x_3 and which satisfy the following assumptions:

- $\chi_n(r, x_3) = 1$ for $r \geq \frac{2}{n}$.
- $\chi_n(r, x_3) = 0$ for $r \leq \frac{1}{n}$.
- $|\chi_n(r, x_3)| \leq 1$.
- $|\nabla \chi_n(r, x_3)| \leq 2n$.
- $\chi_{n+1}(r, x_3) \geq \chi_n(r, x_3)$.

Lemma 133. *Let φ be a function in $H^1 \cap L^\infty$ with bounded support and set $\varphi_n = \varphi \cdot \chi_n$. Then, up to a subsequence, we have that*

$$\varphi_n \to \varphi \text{ weakly in } H^1.$$

Proof. Clearly $\varphi_n \to \varphi$ a.e. Then, by standard arguments, the conclusion holds if we show that $\{\varphi_n\}$ is bounded in H^1. Clearly $\{\varphi_n\}$ is bounded in L^2. Let us now prove that

$$\left\{ \int |\nabla \varphi_n|^2 \right\} \text{ is bounded.}$$

We have

$$\int |\nabla \varphi_n|^2 \leq 2 \int |\nabla \varphi \cdot \chi_n|^2 + |\varphi \cdot \nabla \chi_n|^2$$

$$\leq 2\int|\nabla\varphi|^2 + 2\int_{\Gamma_\varepsilon}|\varphi\cdot\nabla\chi_n|^2$$

where

$$\Gamma_\varepsilon = \left\{x\in\mathbb{R}^3 : \varphi\neq 0 \text{ and } |\nabla\chi_n(r,z)|\neq 0\right\}.$$

By our construction, $|\Gamma_\varepsilon|\leq c/n^2$ where c depends only on φ. Thus

$$\int|\nabla\varphi_n|^2 \leq 2\int|\nabla\varphi|^2 + 2\|\varphi\|_{L^\infty}^2\int_{\Gamma_\varepsilon}|\nabla\chi_n|^2$$

$$\leq 2\int|\nabla\varphi|^2 + 2\|\varphi\|_{L^\infty}^2\cdot|\Gamma_\varepsilon|\cdot\|\nabla\chi_n\|_{L^\infty}^2$$

$$\leq 2\int|\nabla\varphi|^2 + 8c\,\|\varphi\|_{L^\infty}^2.$$

Thus φ_n is bounded in H^1 and $\varphi_n\to\varphi$ weakly in H^1. $\qquad\square$

Now we are ready to prove Theorem 132.

Proof of Theorem 132. We take any $v\in\mathcal{D}(\mathbb{R}^3)$ and set $\varphi_n = v^+\chi_n$ where $v^+ = \frac{|v|+v}{2}$. Then, taking φ_n as test function in Eq. (8.32), we have

$$\int\nabla u_0\cdot\nabla\varphi_n + \left(l^2\frac{u_0}{r^2} + W'(u_0) + 2Vu_0 - 2\omega u_0\right)\varphi_n dx = 0. \qquad (8.34)$$

Equation (8.34) can be written as follows

$$A_n + B_n + C_n = 0 \qquad (8.35)$$

where

$$A_n = \int\nabla u_0\cdot\nabla\varphi_n,\quad B_n = \int\left(-2\omega u_0 + W'(u_0) + 2Vu_0\right)\varphi_n dx \qquad (8.36)$$

$$C_n = \int l^2\frac{u_0}{r^2}\varphi_n. \qquad (8.37)$$

By Lemma 133

$$\varphi_n\to v^+ \text{ weakly in } H^1. \qquad (8.38)$$

Then we have

$$A_n\to\int\nabla u_0\cdot\nabla v^+. \qquad (8.39)$$

Now

$$\left(-2\omega u_0 + W'(u_0) + 2V u_0\right) \in L^{6/5} = \left(L^6\right)'.$$

Then, using again (8.38) and by the embedding $H^1 \subset L^6$, we have

$$B_n \to \int \left(-2\omega u_0 + W'(u_0) + 2V u_0\right) v^+ < \infty. \tag{8.40}$$

Finally we prove that

$$C_n \to \int l^2 \frac{u_0}{r^2} v^+ < \infty. \tag{8.41}$$

By (8.35), (8.40), and (8.39) we have that

$$C_n = \int l^2 \frac{u_0}{r^2} \varphi_n \text{ is bounded.} \tag{8.42}$$

The sequence $l^2 \frac{u_0}{r^2} \varphi_n$ is monotone and it converges a.e. to $l^2 \frac{u_0}{r^2} v^+$. Then, by the monotone convergence theorem, we get

$$\int l^2 \frac{u_0}{r^2} \varphi_n dx \to \int l^2 \frac{u_0}{r^2} v^+ dx. \tag{8.43}$$

By (8.42) and (8.43) we get (8.41).

Taking the limit in (8.35) and by using (8.39)–(8.41) we have

$$\int \nabla u_0 \cdot \nabla v^+ + \left(l^2 \frac{u_0}{r^2} + W'(u_0) + 2V u_0 - 2\omega u_0\right) v^+ = 0. \tag{8.44}$$

Now, taking $\varphi_n = v^- \chi_n$, $v^- = \frac{|v|-v}{2}$, and arguing in the same way as before, we get

$$\int \nabla u_0 \cdot \nabla v^- + \left(l^2 \frac{u_0}{r^2} + W'(u_0) + 2V u_0 - 2\omega u_0\right) v^- = 0. \tag{8.45}$$

Then by (8.44) and (8.45) we get

$$\int \nabla u_0 \cdot \nabla v + \left(l^2 \frac{u_0}{r^2} + W'(u_0) + 2V u_0 - 2\omega u_0\right) v = 0.$$

Since $v \in \mathcal{D}(\mathbb{R}^3)$ is arbitrary, we get that Eq. (8.33) is solved in the sense of distribution in $\mathbb{R}^3$. $\square$

8.2   Vortices for the Nonlinear Klein-Gordon Equation

Consider the nonlinear Klein-Gordon equation

$$\Box\psi + W'(\psi) = 0. \tag{8.46}$$

We look for vortices, i.e. finite energy solutions of (8.46) having non trivial angular momentum.

As in the case of the Schrödinger equation, we look for solutions of the type

$$\psi(t,x) = u(x) e^{i(\ell\theta(x)-\omega t)}, \quad \omega \in \mathbb{R}, \ \ell \in \mathbb{Z} - \{0\} \tag{8.47}$$

where

$$\theta(x) = \mathrm{Im}\log(x_1 + ix_2) \in \mathbb{R}/2\pi\mathbb{Z}; \quad x = (x_1, x_2, x_3).$$

and $u(x) \geq 0$ has cylindrical symmetry, namely it satisfies (8.7). Easy calculations show that the angular momentum of such $\psi(t,x)$ is given by

$$M(\psi) = \ell\omega(0,0,\|u\|_{L^2}^2).$$

We point out that $M(\psi)$ is nontrivial if and only if ℓ, ω and u are not zero. Inserting (8.47) in (8.46) we get

$$\begin{cases} -\Delta u + \ell^2 |\nabla\theta|^2 u + W'(u) = \omega^2 u \\ \nabla\cdot(u^2\nabla\theta) = 0. \end{cases}$$

By the definition of θ and (8.7) we have

$$\Delta\theta = 0, \quad \nabla\theta\cdot\nabla u = 0, \quad |\nabla\theta|^2 = \frac{1}{r^2}$$

where the dot $\cdot$ denotes the Euclidean scalar product.

So the above system reduces to find non trivial solutions u, with the symmetry property (8.7), of the equation

$$-\Delta u + \frac{\ell^2}{r^2}u + W'(u) = \omega^2 u \quad . \tag{8.48}$$

Observe that, if in (8.8) we take $V = 0$ and replace 2ω with ω^2, we get Eq. (8.48). So we can use exactly the same arguments as those used in proving Theorem 129. In this case the lagrangian multiplier ω^2 is positive, then, by (8.48), we need to assume also that

$$W'(s)s \geq 0 \text{ for } s \geq 0. \tag{8.49}$$

The following theorem holds:

Theorem 134. *Assume that W satisfies assumptions (WC-0)–(WC-ii) (page 120) and (WC-iii') (page 122). Moreover assume that (8.49) holds. Then for any integer $l \neq 0$ there exists a family ψ_δ $(0 < \delta < \delta_\infty)$ of vortices for (8.46)*

$$\psi_\delta = u_\delta(x)\, e^{i(\ell\theta(x) - \omega_\delta t)}, \ u_\delta(x) \geq 0 \text{ satisfying (8.7), } \omega_\delta \in \mathbb{R}.$$

These vortices have angular momentum

$$M(\psi_\delta) = \omega_\delta l(0, 0, \|u_\delta\|_{L^2}^2).$$

Moreover, if $\delta_1 < \delta_2$, $u_{\delta_1}, u_{\delta_2}$ are distinct, namely we have that

$$\delta_1 < \delta_2 \implies \|u_{\delta_2}\|_{L^2}^2 < \|u_{\delta_1}\|_{L^2}^2.$$

Finally observe that, as for (8.8), the functions $u_\delta(x)$ in Theorem 134 satisfy (8.48) in the sense of distributions in all $\mathbb{R}^3$(see Sect. 8.1.4).

8.3   Vortices for Nonlinear Klein-Gordon-Maxwell Equations

We look for stationary solutions of the system (5.28)–(5.31), namely solutions of the form

$$\psi(t, x) = u(x)\, e^{iS(x,t)}, \ u \in \mathbb{R}^+, \ \omega \in \mathbb{R}, \ S = S_0(x) - \omega t \in \frac{\mathbb{R}}{2\pi\mathbb{Z}} \quad (8.50)$$

$$\partial_t \mathbf{A} = 0, \ \partial_t \varphi = 0. \quad (8.51)$$

Substituting (8.50) and (8.51) in (5.28)–(5.31), we get the following equations:

$$-\Delta u + \left[|\nabla S_0 - q\mathbf{A}|^2 - (\omega - q\varphi)^2 \right] u + W'(u) = 0 \quad (8.52)$$

$$-\nabla \cdot \left[(\nabla S_0 - q\mathbf{A})\, u^2 \right] = 0 \quad (8.53)$$

$$-\Delta\varphi = q\, (\omega - q\varphi)\, u^2 \quad (8.54)$$

$$\nabla \times (\nabla \times \mathbf{A}) = q\, (\nabla S_0 - q\mathbf{A})\, u^2. \quad (8.55)$$

Observe that Eq. (8.53) easily follows from Eq. (8.55). Then we are reduced to study the system (8.52), (8.54), and (8.55). The energy of a solution of Eqs. (8.52), (8.54), and (8.55) has the following expression

$$
\mathcal{E} = \frac{1}{2} \int \left(|\nabla u|^2 + |\nabla \varphi|^2 + |\nabla \times \mathbf{A}|^2 + (|\nabla S_0 - q\mathbf{A}|^2 + (\omega - q\varphi)^2) u^2 \right)
$$
$$
+ \int W(u). \tag{8.56}
$$

Moreover the (electric) charge is given by

$$
Q = q\sigma \tag{8.57}
$$

where

$$
\sigma = \int (\omega - q\varphi) u^2 dx. \tag{8.58}
$$

Clearly, when $u = 0$, the only finite energy gauge potentials which solve (8.54), (8.55) are the trivial ones.

It is possible to have three types of finite energy stationary non trivial solutions:

- Electrostatic solutions: $\mathbf{A} = 0$, $\varphi \neq 0$.
- Magnetostatic solutions: $\mathbf{A} \neq 0$, $\varphi = 0$.
- Electro-magneto-static solutions: $\mathbf{A} \neq 0$, $\varphi \neq 0$.

Under suitable assumptions, all these types of solutions exist. The existence and stability of electrostatic solutions for Eqs. (8.52)–(8.54) have been proved in Sect. 5.3. Now, we are interested in electro-magneto-static solutions. It will be seen that these solutions are vortices.

We use the notation (8.5) and we take $S_0(x) = \ell\theta(x)$, ℓ integer.

In this case, Eqs. (8.52), (8.54), and (8.55) become

$$
-\Delta u + \left[|\ell\nabla\theta - q\mathbf{A}|^2 - (\omega - q\varphi)^2 \right] u + W'(u) = 0 \tag{8.59}
$$

$$
-\Delta\varphi = q (\omega - q\varphi) u^2 \tag{8.60}
$$

$$
\nabla \times (\nabla \times \mathbf{A}) = q (\ell\nabla\theta - q\mathbf{A}) u^2 \tag{8.61}
$$

and the function ψ can be written in the following form

$$
\psi(t, x) = u(x) e^{i(\ell\theta(x) - \omega t)}; \ \ell \in \mathbb{Z} - \{0\}. \tag{8.62}
$$

If $\ell \neq 0$, a finite energy solution $(u, \omega, \varphi, \mathbf{A})$ of Eqs. (8.52), (8.54), and (8.55) is a vortex. In fact, we shall see (Proposition 139) that the angular momentum $\mathbf{M}_m$ of the matter field does not vanish.

8.3.1   The Main Existence Result

By rescaling time and space we can assume without loss of generality

$$m^2 = 1.$$

In the following we denote by e_1, e_2, e_3 the standard basis for the Euclidean space.
We can state the main existence result.

Theorem 135. *Assume that W satisfies assumptions (WC-0)–(WC-ii) (page 120)
and (WC-iii') (page 122). Then for all $\ell \in \mathbb{Z}$ there exists $\bar{q} > 0$ such that for
every $0 \le q \le \bar{q}$ the Eqs. (8.59)–(8.61) admit a finite energy solution in the sense
of distributions (u, ω, φ, A), $u \ne 0$, $\omega > 0$. The maps u, φ depend only on the
variables $r = \sqrt{x_1^2 + x_2^2}$ and x_3*

$$u = u(r, x_3), \quad \varphi = \varphi(r, x_3)$$

and the magnetic potential A has the following form

$$\mathbf{A} = a(r, x_3)\nabla\theta = a(r, x_3)\left(\frac{x_2}{r^2}\mathbf{e}_1 - \frac{x_1}{r^2}\mathbf{e}_2\right). \tag{8.63}$$

*If $q = 0$, then $\varphi = 0, \mathbf{A} = 0$. If $q > 0$ then $\varphi \ne 0$. Moreover $\mathbf{A} \ne 0$ if and only
if $\ell \ne 0$.*

Proof. The proof of this theorem is rather technical and we refer to the paper [35].
□

Remark 136. When there is no coupling with the electromagnetic field, i.e. $q = 0$,
Eqs. (8.59)–(8.61) reduce to find vortices to the nonlinear Klein-Gordon equation
and we get the same result of Sect. 8.2.

Remark 137. When $\ell = 0$ and $q > 0$ the last part of Theorem 135 states the
existence of electrostatic solutions, namely finite energy solutions with $u \ne 0$, $\varphi
\ne 0$ and $\mathbf{A} = 0$. This result is contained in Theorem 104.

Remark 138. By the presence of the term $\nabla\theta$ Eqs. (8.59) and (8.61) are not
invariant under the $O(3)$ group action as it happens for the Eqs. (5.28)–(5.30) we
started from. Indeed there is a breaking of radial symmetry and the solutions u, φ,
A in Theorem 135 have only an S^1 symmetry.

Proposition 139. *Let (u, ω, φ, A) be a non trivial, finite energy solution of
Eqs. (8.59)–(8.61) as in Theorem 135. Then the angular momentum $\mathbf{M}_m$ has the
following expression*

$$\mathbf{M}_m = -\left[\int (\ell - qa)(\omega - q\varphi) u^2 dx\right]\mathbf{e}_3 \tag{8.64}$$

and, if $\ell \ne 0$, it does not vanish.

Proof. By (5.53), (8.62), (8.58), (5.35) and (8.63), we have that

$$\mathbf{M}_m = \int \mathbf{x} \times \nabla \theta \, (\ell - qa) \, (\omega - q\varphi) \, u^2 dx.$$

Let us compute

$$\mathbf{x} \times \nabla \theta = (x_1 \mathbf{e}_1 + x_2 \mathbf{e}_2 + x_3 \mathbf{e}_3) \times \left(\frac{x_2}{r^2} \mathbf{e}_1 - \frac{x_1}{r^2} \mathbf{e}_2 \right)$$

$$= -\frac{x_1^2}{r^2} \mathbf{e}_3 - \frac{x_2^2}{r^2} \mathbf{e}_3 + \frac{x_2 x_3}{r^2} \mathbf{e}_2 + \frac{x_1 x_3}{r^2} \mathbf{e}_1$$

$$= \frac{x_1 x_3}{r^2} \mathbf{e}_1 + \frac{x_2 x_3}{r^2} \mathbf{e}_2 - \mathbf{e}_3.$$

Then

$$\mathbf{M}_m (\psi) = \int \left(\frac{x_1 x_3}{r^2} \mathbf{e}_1 + \frac{x_2 x_3}{r^2} \mathbf{e}_2 - \mathbf{e}_3 \right) (\ell - qa) \, (\omega - q\varphi) \, u^2 dx. \qquad (8.65)$$

On the other hand, since the functions $x_1 x_3 \frac{(\ell-qa)(\omega-q\varphi)u^2}{r^2}$ and $x_2 x_3 \frac{(\ell-qa)(\omega-q\varphi)u^2}{r^2}$ are odd in x_1 and x_2 respectively, we have

$$\int x_1 x_3 \frac{(\ell - qa) \, (\omega - q\varphi) \, u^2}{r^2} = \int x_2 x_3 \frac{(\ell - qa) \, (\omega - q\varphi) \, u^2}{r^2} = 0. \qquad (8.66)$$

Then (8.64) follows from (8.65) and (8.66). Now let $\ell \neq 0$. In order to see that $\mathbf{M}_m \neq 0$, it is sufficient to prove that

$$(\ell - qa) \, (\omega - q\varphi) > 0 \qquad (8.67)$$

or that

$$(\ell - qa) \, (\omega - q\varphi) < 0. \qquad (8.68)$$

Clearly, since $\ell, \omega \neq 0$ (8.67) or (8.68) are satisfied when $q = 0$. Now let $q > 0$. Assume that $\ell > 0$ and we show that (8.67) is verified. The case $\ell < 0$ can be treated analogously.

By (8.54) we have that

$$-\Delta \varphi + q^2 u^2 \varphi = q\omega u^2.$$

Since ω/q is a supersolution, by the maximum principle, $\varphi < \omega/q$ and hence $\omega - q\varphi > 0$. So, in order to prove (8.67), it remains to show that

$$\ell - qa > 0. \qquad (8.69)$$

By (8.55) we have that

$$\nabla \times (\nabla \times \mathbf{A}) = q \, (\ell \nabla \theta - q \mathbf{A}) \, u^2. \tag{8.70}$$

Now a straight computation shows that,

$$\nabla \times (\nabla \times a \nabla \theta) = b \, \nabla \theta \tag{8.71}$$

where

$$b = -\frac{\partial^2 a}{\partial r^2} + \frac{1}{2} \frac{\partial a}{\partial r} - \frac{\partial^2 a}{\partial x_3^2}.$$

Then, setting $\mathbf{A} = a \nabla \theta$ in (8.70) and using (8.71), we have

$$-\frac{\partial^2 a}{\partial r^2} + \frac{1}{2} \frac{\partial a}{\partial r} - \frac{\partial^2 a}{\partial x_3^2} = q \, (\ell - qa) \, u^2.$$

Since ℓ/q is a supersolution, by the maximum principle, $a < \ell/q$ and hence (8.69) is proved. $\square$

Finally let us observe that under general assumptions on W, magnetostatic solutions (i.e. with $\omega = \varphi = 0$) do not exist. In fact the following proposition holds:

Proposition 140. *Assume that W satisfies the assumptions $W(0) = 0$ and $W'(s)s \geq 0$. Then (8.59)–(8.61) has no solutions with $\omega = \varphi = 0$.*

Proof. Set $\omega = 0$, $\varphi = 0$ in (8.59) and we get

$$-\Delta u + |\ell \nabla \theta - q \mathbf{A}|^2 \, u + W'(u) = 0.$$

Then, multiplying by u and integrating, we get

$$\int |\nabla u|^2 + |\ell \nabla \theta - q \mathbf{A}|^2 \, u^2 + W'(u)u = 0.$$

So, since $W'(s)s \geq 0$, we get $u = 0$. $\square$

Appendix

A.1 Some Inequalities

Lemma 141 (Strauss inequality [139]). *There is a constant $C > 0$ such that for all $u \in \mathcal{D}\left(\mathbb{R}^N\right)$, radially symmetric, we have*

$$|u(x)| \leq C \frac{\|u\|_{H^1}}{|x|^{\frac{N-1}{2}}}.$$

Proof. We have that

$$\frac{d}{dr}\left(r^{N-1}u^2\right) = 2r^{N-1}u\frac{du}{dr} + (N-1)\,r^{N-2}u^2 \geq 2r^{N-1}u\frac{du}{dr}.$$

Then, integrating over $(R, +\infty)$ we get

$$-R^{N-1}u\,(R)^2 \geq 2\int_R^{+\infty} r^{N-1}u\frac{du}{dr}dr$$

and so

$$R^{N-1}u\,(R)^2 \leq 2\int_R^{+\infty}\left|u\frac{du}{dr}\right|r^{N-1}dr \leq \int_R^{+\infty}\left(\left|\frac{du}{dr}\right|^2 + u^2\right)r^{N-1}dr \leq C\,\|u\|_{H^1}^2\,.$$

$\square$

Theorem 142 (Strauss embedding theorem [139]). *Let $H_r^1 = \{u \in H^1(\mathbb{R}^N),$ u radially symmetric$\}$ and $N \geq 2$. Then for $2 < p < 2^*$ $(2^* = +\infty$ if $N = 2)$ the embedding*

© Springer International Publishing Switzerland 2014
V. Benci, D. Fortunato, *Variational Methods in Nonlinear Field Equations*,
Springer Monographs in Mathematics, DOI 10.1007/978-3-319-06914-2

$$H_r^1 \to L^p$$

is compact.

Proof. Let $u_n \rightharpoonup 0$ weakly in H_r^1; we need to prove that $u_n \longrightarrow 0$ strongly in L^p. Since $u_n \rightharpoonup 0$, then there is a constant M such that $\|u_n\|_{H^1} \leq M$; so by Lemma 141, we have that

$$\int_{\mathbb{R}^N - B_R} |u_n|^p \, dx \leq \|u_n\|_{L^\infty(\mathbb{R}^N - B_R)}^{p-2} \int_{\mathbb{R}^N - B_R} |u_n|^2 \, dx$$

$$\leq \left(C \frac{\|u_n\|_{H^1}}{R^{\frac{N-1}{2}}} \right)^{p-2} \|u_n\|_{L^2}^2 \leq \frac{C^{p-2} M^p}{R^\alpha} \tag{A.1}$$

where

$$B_R = \{x \in \mathbb{R}^N : |x| < R\} \text{ and } \alpha = \frac{(N-1)(p-2)}{2} > 0.$$

Since $u_n \longrightarrow 0$ strongly in $L^p(B_R)$, by (A.1), we have that

$$\limsup_{n \to \infty} \int |u_n|^p \, dx = \lim_{n \to \infty} \|u_n\|_{L^p(B_R)}^p + \limsup_{n \to \infty} \|u_n\|_{L^p(\mathbb{R}^N - B_R)}^p \leq \frac{C^{p-2} M^p}{R^\alpha}.$$

By the arbitrariness of R, it follows that $u_n \longrightarrow 0$ strongly in L^p. $\square$

A.2   Pohozhev-Derrick Theorem

Theorem 143 (Pohozaev-Derrik theorem [72, 121]). *Let $u \in H^1(\mathbb{R}^N)$ be a regular solution of*

$$-\Delta u + h(u) = 0 \tag{A.2}$$

where h is a continuous, real function s.t. $h(0) = 0$. Assume that

$$H(u) \in L^1(\mathbb{R}^N)$$

where

$$H(t) = \int_0^t h(s) \, ds.$$

Then the following equalities hold

$$\int H(u)dx = \left(\frac{1}{N} - \frac{1}{2}\right)\int |\nabla u|^2 dx; \tag{A.3}$$

$$\int |\partial_i u|^2 dx = \frac{1}{N}\int |\nabla u|^2 dx, \ i = 1, \dots, N. \tag{A.4}$$

Proof. Here we prove (A.3) and (A.4) under the additional assumption

$$x_i \partial_i u \in H^1(\mathbb{R}^N). \tag{A.5}$$

Observe that this assumption could be avoided by following a different proof. Here we prefer to use a rescaling argument which requires (A.5). On the other hand this argument is quite natural in this context.

Let $u \in H^1(\mathbb{R}^N)$ be a solution of (A.2). Then it is a critical point of the functional

$$J(u) = \frac{1}{2}\int |\nabla u|^2 dx + \int H(u)dx.$$

Rescale the x_1 variable and set for $\lambda \neq 0$

$$u_\lambda(x_1, \dots, x_N) = u\left(\frac{x_1}{\lambda}, x_2, \dots, x_N\right).$$

Then

$$\frac{d}{d\lambda}J(u_\lambda) = dJ(u_\lambda)\left[\frac{d}{d\lambda}u_\lambda\right] = \frac{1}{\lambda}dJ(u_\lambda)\left[x_1\frac{\partial u_\lambda}{\partial x_1}\right]. \tag{A.6}$$

If $\lambda = 1$ we have $u_\lambda = u$. Then, since $dJ(u) = 0$ and $x_1\frac{\partial u}{\partial x_1}$ belongs to $H^1(\mathbb{R}^N)$, we have from (A.6)

$$\frac{d}{d\lambda}J(u_\lambda) = 0 \text{ for } \lambda = 1. \tag{A.7}$$

Now

$$J(u_\lambda) = \frac{1}{2}\int \left(\frac{1}{\lambda}|\partial_1 u|^2 + \sum_{i \neq 1}\lambda |\partial_i u|^2\right)dx + \lambda \int H(u)dx.$$

Then from (A.7) we get

$$\frac{d}{d\lambda}J(u_\lambda)\bigg|_{\lambda=1} = \frac{1}{2}\int \left(-|\partial_1 u|^2 + \sum_{i \neq 1}|\partial_i u|^2\right)dx + \int H(u)dx = 0,$$

then

$$\int \left(-|\partial_1 u|^2 + \frac{1}{2}|\nabla u|^2\right) dx + \int H(u)dx = 0. \qquad (A.8)$$

Analogous equalities can be obtained for $i = 2,\ldots,N$. So, by adding on $i = 1,\ldots,N$, we obtain

$$\int \left(-|\nabla u|^2 + \frac{N}{2}|\nabla u|^2\right) dx + N \int H(u)dx = 0,$$

from which (A.3). □

Now write (A.8) for $x_j,\ j \neq 1$

$$\int (-|\partial_j u|^2 + \frac{1}{2}|\nabla u|^2)dx + \int H(u)dx = 0, \qquad (A.9)$$

then subtract (A.8) from (A.9). So we get

$$\int |\partial_1 u|^2 \, dx = \int |\partial_j u|^2 \, dx, \ j \neq 1$$

from which, adding for $j = 1,\ldots,n$, we get

$$N \int |\partial_1 u|^2 \, dx = \int |\nabla u|^2 \, dx,$$

from which (A.4). □

A.3   An Existence Result for an Elliptic Equation

Many existence theorems of solitary waves reduce to the following elliptic equation in $\mathbb{R}^N$

$$\begin{cases} -\Delta u + G'(u) = 0 \\ u > 0 \end{cases} \qquad (A.10)$$

where G is a C^1-function with $G(0) = 0$.

This equation has been studied by many authors (see e.g. [46, 59, 139] and their bibliography). In particular, in [46], there are sufficient and "almost necessary" conditions for the existence of "finite energy" solutions, i.e. solutions $u \in H^1(\mathbb{R}^N)$.

Since this equation is the basic equation for the existence of solitary waves, we give an existence proof. This proof is a variant of the proof in [46]. It is simpler but it uses slightly more restrictive assumptions.

Equation (A.10) is the Euler-Lagrange equation relative to the functional

$$J(u) = \frac{1}{2} \int |\nabla u|^2 \, dx + \int G(u) dx.$$

Let $N \geq 3$ and $G : \mathbb{R}^+ \to \mathbb{R}$ be a C^1 function satisfying the following assumptions:

(G-i) $G(0) = G'(0) = 0, \lim\sup\limits_{s \to 0^+} \frac{G'(s)}{s} < \infty.$

(G-ii) $G'(s) \geq c_1 s - c_2 s^{p-1}, 2 < p < 2^* = \frac{2N}{N-2}, s > 0, c_1, c_2 > 0.$

(G-iii) $\exists s_0 \in \mathbb{R}^+ : G(s_0) < 0.$

Theorem 144 (Beresticki-Lions theorem). *Assume $N \geq 3$ and that G satisfies (G-i)–(G-iii). Then Eq. (A.10) has a nontrivial finite energy solution.*

In order to prove the above theorem, first of all we define an auxiliary function $\bar{G}$ distinguishing two cases (a) and (b) as follows:

Case (a) If there exist $c_3, c_4 > 0$ such that

$$|G'(s)| \leq c_3 s + c_4 s^{p-1} \text{ for all } s > 0 \tag{A.11}$$

(where p is defined by (G-ii)), then we set

$$\bar{G}(s) = \begin{cases} G(s) \text{ for } s \geq 0 \\ 0 \quad\quad \text{for } s \leq 0 \end{cases}. \tag{A.12}$$

In order to define $\bar{G}$ in the other case (b), we show first that, if G does not satisfy (A.11), then there exists s_1 such that

$$s_1 > s_0 \text{ and } G'(s_1) > 0. \tag{A.13}$$

In fact, if G does not satisfy (A.11), then for all $c_3, c_4 > 0$ there exists $s_1 > 0$ such that

$$|G'(s_1)| > c_3 s_1 + c_4 s_1^{p-1}. \tag{A.14}$$

Then, taking in (A.14) $c_4 = c_2$ and $c_3 = M = \sup \left\{ \frac{|G'(s)|}{s} : s \in (0, s_0] \right\}$ (M is finite by (G-i) and since G is C^1), we have

$$|G'(s_1)| > M s_1 + c_2 s_1^{p-1}. \tag{A.15}$$

Then $\frac{|G'(s_1)|}{s_1} > M$. So, by definition of M, we have $s_1 > s_0$. Now we prove that $G'(s_1) > 0$. Arguing by contradiction, assume that $G'(s_1) \leq 0$, then by (A.15)

$$G'(s_1) < -M s_1 - c_2 s_1^{p-1}. \tag{A.16}$$

By (G-ii) and (A.16) we get

$$-Ms_1 - c_2 s_1^{p-1} < c_1 s_1 - c_2 s_1^{p-1} \le G'(s_1) < -Ms_1 - c_2 s_1^{p-1}.$$

So we get a contradiction. Then $G'(s_1) > 0$ and we conclude that (A.13) holds.

Case (b) In the case in which G does not satisfy (A.11), we set

$$\bar{G}(s) = \begin{cases} G(s) & \text{for } 0 \le s \le s_1; \\ G(s_1) + G'(s_1)(s - s_1) & \text{for } s \ge s_1; \\ 0 & \text{for } s \le 0. \end{cases} \tag{A.17}$$

So we conclude that in any case, $\bar{G}$ satisfies the assumptions (G-i), (G-ii) (G-iii) and the following ones:

(G-iv) $|\bar{G}'(s)| \le c_3 s + c_4 s^{p-1}$, $s \ge 0$, $2 \le p < 2^*$.
(G-v) $\forall s < 0$, $\bar{G}(s) = 0$.

Lemma 145. *Let $u \in H^1(\mathbb{R}^N)$ be a solution of the following equation:*

$$-\Delta u + \bar{G}'(u) = 0 \tag{A.18}$$

where $\bar{G}$ is defined by (A.17). Then, u is positive and it is a solution of (A.10).

Proof. The fact that u is positive is a straightforward consequence of the maximum principle and the fact that $\bar{G}'(u) = 0$ for $u \le 0$. Then, if (A.11) holds, $\bar{G}$ is defined by (A.12) and $\bar{G}'(u) = G'(u)$ and hence u is a solution of (A.10). If (A.11) does not hold, $\bar{G}$ is defined by (A.17) and, by the maximum principle, it follows that $u(x) \le s_1$; then, also in this case $\bar{G}'(u) = G'(u)$ and u is a solution of (A.10).   □

Thus, by the above lemma it is not restrictive to assume that G satisfies (G-i) and (G-iii)–(G-v), since otherwise we can work with $\bar{G}$.

We set

$$H_r^1 = \left\{ u \in H^1(\mathbb{R}^N) : u = u(|x|) \right\}.$$

Lemma 146. *There exists $\breve{u} \in H_r^1$ such that*

$$\int G(\breve{u})dx < 0.$$

Proof. We set

$$u_R(x) = \begin{cases} s_0 & \text{for } |x| < R; \\ s_0 - s_0(|x| - R) & \text{for } R \le |x| \le R + 1 \\ 0 & \text{for } |x| > R + 1. \end{cases} \tag{A.19}$$

Thus we have

$$\int_{\mathbb{R}^N} G(u_R)dx$$

$$\leq C_1 \left(\left[\max_{|s| \in [0,s_0]} G(s) \right] \int_R^{R+1} r^{N-1}dr + \int_0^R G(s_0)r^{N-1}dr \right)$$

$$\leq C_2 \left(\left[(R+1)^N - R^N \right] + \frac{1}{N}G(s_0)R^N \right)$$

$$\leq C_3 \left(R^{N-1} + \frac{1}{N}G(s_0)R^N \right),$$

where C_1, C_2, C_3 are positive constants. By (G-iii), $G(s_0) < 0$; hence, for R sufficiently large, $\int G(u_R)dx < 0$. $\square$

Proof of Theorem 144. Take a function $\beta \in C^1(\mathbb{R})$ such that

- (β-i) $\forall s \in \mathbb{R}$, $0 \leq \beta(s) \leq 1$.
- (β-ii) $\forall s < 0$, $\beta(s) > 0$.
- (β-iii) $\forall s \geq 0$, $\beta(s) = 0$.
- (β-iv) $\forall s \in \mathbb{R}$, $s < 0 :$ $\beta'(s) < 0$.

Now we define a C^1-functional on H_r^1 as follows:

$$F(u) = \frac{1}{2} \int |\nabla u|^2 dx - b\beta \left(\int G(u)dx \right);$$

here b is a positive constant defined by

$$b = \frac{\frac{1}{2}\int |\nabla \breve{u}|^2 dx + 1}{\beta \left(\int G(\breve{u})dx \right)}$$

and $\breve{u}$ is defined by Lemma 146. This choice of b, implies that

$$F(\breve{u}) = -1.$$

Clearly F is a C^1 functional on H_r^1.

Since β is bounded, then $F(u)$ is bounded below and its infimum is a number less or equal to -1. Let u_n be a minimizing sequence for F. Also, β bounded implies $\frac{1}{2}\int |\nabla u_n|^2 dx$ bounded and hence, up to a subsequence, $u_n \rightharpoonup w$ weakly in $\mathcal{D}_r^{1,2}$.[1] Since the infimum is negative we have that $\beta \left(\int G(u_n)dx \right) > 0$, and hence, by ($\beta$-ii),

---

[1] $\mathcal{D}_r^{1,2}$ denotes the closure of the set of radially symmetric C^∞-functions with compact support with respect to the norm $\sqrt{\int |\nabla u|^2 dx}$.

$$\int G(u_n)dx < 0. \tag{A.20}$$

We show that $\|u_n\|_{H^1}$ is bounded. Since $N \geq 3$, by (G-ii), we have that there are constants $c_5, c_6 > 0$ such that, for $s \geq 0$

$$G(s) = \int_0^s G'(t)dt \geq \int_0^s \left(c_1 t - c_2 t^{p-1}\right) dt$$

$$\geq \frac{1}{2}c_1 s^2 - \frac{c_2}{p} s^p \geq c_5 s^2 - c_6 s^{2^*}.$$

By this inequality, by (A.20) and by the Sobolev inequality, we get

$$0 > \int G(u_n)dx \geq c_5 \int u_n^2 dx - c_6 \int |u_n|^{2^*} dx$$

$$\geq c_5 \|u_n\|_{L^2}^2 - c_7 \|\nabla u_n\|_{L^2}^{2^*}, \ c_7 > 0.$$

Thus, since $\|\nabla u_n\|_{L^2}$ is bounded, by the above inequality, also $\|u_n\|_{L^2}$ is bounded and hence also $\|u_n\|_{H^1}$ is bounded.

So, by Theorem 142, up to a subsequence, $u_n \to w$ in L^p strongly, where p is defined by (G-ii). By our assumptions, the functional

$$u \to \int G(u)dx$$

is continuous in L^p. Then we have that $\int G(u_n)dx \to \int G(w)dx$ and hence $\beta \left(\int G(u_n)dx \right) \to \beta \left(\int G(w)dx \right)$. Since $u \to \int |\nabla u|^2 dx$ is weakly l.s.c., it follows that w is a minimizer of F. Thus w is different from 0 and it satisfies the equation $F'(w) = 0$, i.e.

$$-\Delta w + \lambda G'(w) = 0$$

where

$$\lambda = -b\beta' \left(\int G(w)dx \right).$$

By (β-iv), we have that $\lambda \geq 0$, and since $w \neq 0$, we have that $\lambda > 0$. Now set

$$u(x) = w \left(\frac{x}{\sqrt{\lambda}} \right).$$

Clearly u satisfies Eq. (A.10). $\square$

References

1. A. Abbondandolo, V. Benci, Solitary waves and Bohmian mechanics. Proc. Natl. Acad. Sci. **99**(24), 15257–15261 (2002)
2. W.K. Abou Salem, Solitary wave dynamics in time-dependent potentials. J. Math. Phys. **49**(3), 032101, 29pp (2008)
3. W.K. Abou Salem, Effective dynamics of solitons in the presence of rough nonlinear perturbations. Nonlinearity **22**(4), 747–763 (2009)
4. W.K. Abou Salem, J. Fröhlich, I.M. Sigal, Colliding solitons for the nonlinear Schrödinger equation. Commun. Math. Phys. **291**(1), 151–176 (2009)
5. A.A. Abrikosov, On the magnetic properties of superconductors of the second group. Sov. Phys. JETP **5**, 1174–1182 (1957)
6. A. Ambrosetti, The Schrödinger-Poisson systems. Milan J. Math. **76**(1), 257–274 (2008). doi:10.1007/s00032-008-0094-z
7. A. Ambrosetti, D. Ruiz, Multiple bound states for the Schrödinger-Poisson problem. Commun. Contemp. Math. **10**, 391–404 (2008)
8. A. Azzollini, A. Pomponio, Ground state solutions for the nonlinear Klein-Gordon-Maxwell equations. Topol. Methods Nonlinear Anal. **35**, 33–42 (2010)
9. A. Azzollini, L. Pisani, A. Pomponio, Improved estimates and a limit case for the electrostatic Klein-Gordon-Maxwell system. Proc. R. Soc. Edinb. Sect. A **141**, 449–463 (2011)
10. M. Badiale, V. Benci, S. Rolando, Solitary waves: physical aspects and Mathematical results. Rend. Sem. Mat. Univ. Pol. Torino **62**(2), 107–154 (2004)
11. M.Badiale, V.Benci, S.Rolando, A nonlinear elliptic equation with singular potential and applications to nonlinear field equations. J. Eur. Math. Soc. **9**, 355–381 (2007)
12. M. Badiale, V. Benci S. Rolando, Three dimensional vortices in the nonlinear wave equation. Boll. U.M.I., Ser. 9 **2**, 105–134 (2009)
13. D. Bambusi, T. Penati, Continuous approximation of breathers in one- and two-dimensional DNLS lattices. Nonlinearity **23**(1), 143–157 (2010)
14. J. Bellazzini, V. Benci, C. Bonanno, A.M. Micheletti, Solitons for the nonlinear Klein-Gordon-equation. Adv. Nonlinear Stud. **10**, 481–500 (2010)
15. J. Bellazzini, V. Benci, C. Bonanno, E. Sinibaldi, Hylomorphic solitons in the nonlinear Klein-Gordon-equation. Dyn. Part. Differ. Equ. **6**, 311–333 (2009)
16. J. Bellazzini, V. Benci, C. Bonanno, E. Sinibaldi, On the existence of hylomorphic vortices in the nonlinear Klein-Gordon equation. Dyn. Part. Differ. Equ. **10**, 1–24 (2013). arXiv:1211.5553
17. J. Bellazzini, V. Benci, M. Ghimenti, A.M. Micheletti, On the existence of the fundamental eigenvalue of an elliptic problem in $\mathbb{R}^N$. Adv. Nonlinear Stud. **7**, 439–458 (2007)

18. J. Bellazzini, C. Bonanno, Nonlinear Schrödinger equations with strongly singular potentials. Proc. R. Soc. Edinb. Sect. A Mathematics **140**, 707–721 (2010)
19. J. Bellazzini, C. Bonanno, G. Siciliano, Magnetostatic vortices in two dimensional Abelian gauge theory. Mediterr. J. Math. **6**, 347–366 (2009)
20. J. Bellazzini, G. Siciliano, Stable standing waves for a class of nonlinear Schrödinger-Poisson equations. Z. Angew. Math. Phys. **62**(2), 267–280 (2011). doi:10.1007/s00033-010-0092-1
21. V. Benci, Hylomorphic solitons. Milan J. Math. **77**, 271–332 (2009)
22. V. Benci, G. Cerami, Positive solutions of some nonlinear elliptic problems in exterior domains. Arch. Rational Mech. Anal. **99**(4), 283–300 (1987)
23. V. Benci, D. Fortunato, L.Pisani, Remarks on topological solitons. Topol. Methods Nonlinear Anal. **7**, 349–367 (1996)
24. V. Benci, D. Fortunato, A new variational principle for the fundamental equations of classical physics. Found. Phys. **28**, 333–352 (1998)
25. V. Benci. D. Fortunato, L. Pisani, Soliton like solutions of a Lorentz invariant equation in dimension 3. Rev. Math. Phys. **3**, 315–344 (1998)
26. V. Benci, D. Fortunato, An eigenvalue problem for the Schrödinger-Maxwell equations. Topol. Methods Nonlinear Anal. **11**, 283–293 (1998)
27. V. Benci, D. Fortunato, A. Masiello, L. Pisani, Solitons and electromagnetic field. Math. Zeit. **232**, 73–102 (1999)
28. V. Benci, D. Fortunato, Solitary waves of the nonlinear Klein-Gordon field equation coupled with the Maxwell equations. Rev. Math. Phys. **14**, 409–420 (2002)
29. V. Benci, D. Fortunato, Solitary waves in classical field theory, in *Nonlinear Analysis and Applications to Physical Sciences*, ed. by V. Benci, A. Masiello (Springer, Milano, 2004), pp. 1–50
30. V. Benci, D. Fortunato, Solitary waves in the nonlinear wave equation and in gauge theories. J. Fixed Point Theory Appl. **1**, 61–86 (2007)
31. V. Benci, D. Fortunato, Solitary waves in Abelian gauge theories. Adv. Nonlinear Stud. **3**, 327–352 (2008)
32. V. Benci, D. Fortunato, Three dimensional vortices in Abelian Gauge Theories. Nonlinear Anal. T.M.A. **70**, 4402–4421 (2009)
33. V. Benci, D. Fortunato, Existence of hylomorphic solitary waves in Klein-Gordon and in Klein-Gordon-Maxwell equations. Rend. Lincei Mat. Appl. **20**, 243–279 (2009). arXiv:0903.3508
34. V. Benci, D. Fortunato, Hylomorphic solitons on lattices. Discret. Contin. Dyn. Syst. **28**, 875–879 (2010)
35. V. Benci, D. Fortunato, Spinning Q-balls for the Klein-Gordon-Maxwell equations. Commun. Math. Phys. **295**, 639–668 (2010). doi:10.1007/s00220-010-0985-z
36. V. Benci, D. Fortunato, On the existence of stable charged Q-balls. J. Math. Phys. **52**, (2011). doi:10.1063/1.3629848
37. V. Benci, D. Fortunato, Hamiltonian formulation of the Klein-Gordon-Maxwell equations. Rend. Lincei Mat. Appl. **22**, 1–22 (2011)
38. V. Benci, D. Fortunato, Existence of Solitons in the nonlinear beam equation. J. Fixed Point Theory Appl. **11**, 261–278 (2012). arXiv:1102.5315v1(math.AP)
39. V. Benci, D. Fortunato, A minimization method and applications to the study of solitons. Nonlinear Anal. **75**, 4398–4421 (2012)
40. V. Benci, D. Fortunato, Hylomorphic solitons and charged Q-balls: existence and stability. Chaos Solitons Fractals **58**, 1–15 (2014). arXiv:1212.3236
41. V. Benci, D. Fortunato, Solitons in Schrödinger-Maxwell equations. J. Fixed Point Theory Appl. **15**, (2014). arXiv:1303.1415
42. V. Benci, M. Ghimenti, A.M. Micheletti, The nonlinear Schrödinger equation: solitons dynamics. J. Diff. Equ. **249**, 3312–3340 (2010)
43. V. Benci, M. Ghimenti, A.M. Micheletti, On the dynamics of solitons in the nonlinear Schrödinger equation. Arch. Ration. Mech. Anal. **205**, 467–492 (2012). doi:10.1007/s00205-012-0510-y

44. V. Benci, P. Freguglia, *Modelli e realtà* (Bollati Boringhieri, Torino, 2011)
45. V. Benci, N. Visciglia, Solitary waves with non vanishing angular momentum. Adv. Nonlinear Stud. **3**, 151–160 (2003)
46. H. Berestycki, P.L. Lions, Nonlinear scalar field equations, I – existence of a ground state. Arch. Ration. Mech. Anal. **82**(4), 313–345 (1983)
47. C. Bonanno, Existence and multiplicity of stable bound states for the nonlinear Klein-Gordon equation. Nonlinear Anal. Theory Methods Appl. **72**, 2031–2046 (2010)
48. H. Brezis, E.H. Lieb, Minimum action solutions of some vector field equations. Commun. Math. Phys. **96**, 97–113 (1984)
49. J.C. Bronski, R.L. Jerrard, Soliton dynamics in a potential. Math. Res. Lett. **7**(2–3), 329–342 (2000)
50. V.S. Buslaev, C. Sulem, On asymptotic stability of solitary waves for nonlinear Schrödinger equations. Annales de l'institut Henri Poincaré (C) Analyse non linéaire **20**, 419–475 (2003)
51. A.M. Candela, A. Salvatore, Multiple solitary waves for non-homogeneous Schrödinger-Maxwell equations. Mediterr. J. Math. **3**, 483–493 (2006)
52. D. Cassani, Existence and non-existence of solitary waves for the critical Klein-Gordon equation coupled with Maxwell's equations. Nonlinear Anal. **58**, 733–747 (2004)
53. T. Cazenave, *Semilinear Schrödinger Equations*. Courant Lecture Notes in Mathematics, vol. 10 (New York University Courant Institute of Mathematical Sciences, New York, 2003)
54. T. Cazenave, P.L. Lions, Orbital stability of standing waves for some nonlinear Schrödinger equations. Commun. Math. Phys. **85**, 549–561 (1982)
55. G. Cerami, G. Vaira, Positive solutions for some non-autonomous Schrödinger–Poisson systems. J. Differ. Equ. **248**, 521–543 (2010). doi:10.1016/j.jde.2009.06.017
56. J. Chabrowski, *Weak Convergence Methods for Semilinear Elliptic Equations* (World Scientific, Singapore, 1999)
57. G.M. Coclite, A multiplicity result for the nonlinear Schrödinger-Maxwell equations. Commun. Appl. Anal. **7**, 417–42 (2003)
58. G.M. Coclite, V. Georgiev, Solitary waves for Maxwell-Schrödinger equations. Electron. J. Differ. Equ. **94**, 1–31 (2004)
59. S. Coleman, V. Glaser, A. Martin, Action minima among solutions to a class of euclidean scalar field equation. Commun. Math. Phys. **58**, 211–221 (1978)
60. S. Coleman, Q-Balls Nucl. Phys. **B262**, 263–283 (1985) Erratum **B269**, 744–745 (1986)
61. S. Cuccagna, On asymptotic stability in 3D of kinks for the φ^4 model. Trans. Am. Math. Soc. **360**(5), 2581–2614 (1986)
62. S. Cuccagna, T. Mizumachi, On asymptotically stability in energy space of ground states for nonlinear Schrödinger equations. Commun. Math. Phys. **284**(1), 51–77 (2008)
63. T. D'Aprile, Semiclassical states for the nonlinear Schrödinger equations with the electromagnetic field. Nonlinear Differ. Equ. Appl. **13**, 526–549 (2006)
64. T. D'Aprile, D. Mugnai,Solitary waves for nonlinear Klein-Gordon-Maxwell and Schrödinger -Maxwell equations. Proc. R. Soc. Edinb. Sect A Math **134**, 893–906 (2004)
65. T. D'Aprile, D. Mugnai, Non-existence results for the coupled Klein-Gordon- Maxwell equations. Adv. Nonlinear Stud. **4**, 307–322 (2004)
66. T. D'Aprile, J. Wei, On bound states concentrating on spheres for the Maxwell-Schrödinger equation. SIAM J. Math. Anal. **37**, 321–342 (2005)
67. T. D'Aprile, J. Wei, Standing waves in the Maxwell-Schrödinger equation and an optimal configuration problem. Calc. Var. Part. Differ. Equ. **25**(1), 105–137 (2006)
68. P. D'Avenia, A. Pomponio, G. Vaira, Infinitely many positive solutions for a Schrödinger-Poisson system. Nonlinear Anal. Theory Methods Appl. **74**, 5705–5721 (2011)
69. P. D'Avenia, L. Pisani, Nonlinear Klein-Gordon equations coupled with Born-Infeld Equations. Electron. J. Differ. Equ. **26**, 1–13 (2002)
70. P. D'Avenia, Non-radially symmetric solutions of nonlinear Schrödinger equation coupled with Maxwell equations. Adv. Nonlinear Stud. **2** (2), 177–192 (2002)

71. L. De Broglie, *Un tentative d'interprétation causale et non linéaire de la Mécanique ondulatoire: la théorie de la double solution* (Gauthier-Villars, Paris, 1958) English traslation: *Non-Linear Wave Mechanics, A Causal Interpretation* (Elsevier, Amsterdam, 1960)

72. C.H. Derrick, Comments on nonlinear wave equations as model for elementary particles. J. Math. Phys. **5**, 1252–1254 (1964)

73. R.K. Dodd, J.C. Eilebeck, J.D. Gibbon, H.C. Morris, *Solitons and Nonlinear Wave Equations* (Academic, London/New York, 1982)

74. S. Dodelson, L. Widrow, Baryon symmetric baryogenesis. Phys. Rev. Lett. **64**, 340–343 (1990)

75. D. Eardley, V. Moncrief, The global existence of Yang-Mills-Higgs fields in $\mathbf{R}^{3+1}$. Commun. Math. Phys. **83**, 171–212 (1982)

76. M. Esteban, P.L. Lions, A compactness lemma. Nonlinear Anal. **7**, 381–385 (1983)

77. J. Fröhlich, S. Gustafson, B.L.G. Jonsson, I.M. Sigal, Solitary wave dynamics in an external potential. Commun. Math. Phys. **250**(3), 613–642 (2004)

78. J. Fröhlich, S. Gustafson, B.L.G. Jonsson, I.M. Sigal, Long time motion of NLS solitary waves in a confining potential. Ann. Henri Poincaré **7**(4), 621–660 (2006)

79. Z. Gang, M.I. Weinstein, Dynamics of nonlinear Schrödinger/Gross-Pitaevskii equations: mass transfer in systems with solitons and degenerate neutral modes. Anal. PDE **1**(3), 267–322 (2008)

80. I.M. Gelfand, S.V. Fomin, *Calculus of Variations* (Prentice-Hall, Englewood Cliffs, 1963)

81. B.Gidas, W.M.Ni, L.Nirenberg, Symmetry and related properties via the maximum principle. Commun. Math. Phys. **68**, 209–243 (1979)

82. J. Ginibre, G.Velo, On a class of nonlinear Schrödinger equations. II. Scattering theory, general case. J. Funct. Anal. **32**(1), 33–71 (1979)

83. J. Ginibre, G. Velo, On a class of nonlinear Schrödinger equations. III. Special theories in dimensions 1, 2 and 3. Ann. I.H.P. **28**(3), 287–316 (1978)

84. M. Grillakis, J. Shatah, W. Strauss, Stability theory of solitary waves in the presence of symmetry, I. J. Funct. Anal. **74**, 160–197 (1987)

85. M. Grillakis, J. Shatah, W. Strauss, Stability theory of solitary waves in the presence of symmetry. II. J. Funct. Anal. **94**(2), 308–348 (1990)

86. Y. Guo, K. Nakamitsu, W. Strauss, Global finite energy solutions of the Maxwell-Schrödinger system. Commun. Math. Phys. **170**, 181–196 (1995)

87. J. Holmer, M. Zworski, Soliton interaction with slowly varying potentials. Int. Math. Res. Not. IMRN 2008, no. 10, Art. ID rnn026, 36pp

88. I. Ianni, G. Vaira, concentration of positive bound states for the Schrödinger-Poisson problem with potential. Adv. Nonlinear Stud. **8**, 573–595 (2008)

89. J.D. Jackson, *Classical Electrodynamics* (Wiley, New York, 1962)

90. T. Kato, *Nonlinear Schrödinger Equations, Schrödinger Operators (Sønderborg, 1988)*. Lecture Notes in Physics, vol. 345 (Springer, Berlin, 1989), pp. 218–263

91. S. Keraani, Semiclassical limit of a class of Schrödinger equations with potential. Commun. Partial Diff. Eq. **27**, 693–704 (2002)

92. S. Keraani, Semiclassical limit of a class of Schrödinger equations with potential II. Asymptot. Anal. **47**, 171–186 (2006)

93. S. Kichenassamy, *Nonlinear wave equations* (Marcel Dekker, New York, 1996)

94. C. Kim, S. Kim, Y. Kim, Global nontopological vortices. Phys. Rev. D **47**, 5434–5443 (1985)

95. H. Kikuchi, On the existence of solution for elliptic system related to the Maxwell-Schrödinger equations. Nonlinear Anal. **67**, 1445–1456 (2007)

96. H. Kikuki, Existence and stability of standing waves for Schrödinger-Poisson-Slater equation. Adv. Nonlinear Stud. **7**, 403–437 (2007)

97. S. Klainerman, M. Machedon, On the Maxwell-Klein-Gordon equation with finite energy. Duke Math. J. **74**, 19–44 (1994)

98. A. Komech, B. Vainberg, On asymptotic stability of stationary solutions to nonlinear wave and Klein-Gordon equations. Arch. Ration. Mech. Anal. **134**(3), 227–248 (1996)

99. A. Kusenko, M. Shaposhnikov, Supersymmetric Q balls as dark matter. Phys. Lett. B **418**, 46–54 (1998). arXiv:hep-ph/9709492

100. L. Landau, E. Lifchitz, *Mécanique* (Editions Mir, Moscow, 1966)

101. L.Landau, E.Lifchitz, *Théorie du Champ* (Editions Mir, Moscow, 1966)

102. A.C. Lazer, P.J. McKenna Large scale oscillating behaviour in loaded asymmetric systems. Ann. Inst. H. Poincaré, Anal. Nonlineaire, **4**, 244–274 (1987)

103. A.C. Lazer, P.J. McKenna, Large amplitude periodic oscillations in suspension bridge: some new connection with nonlinear analysis. SIAM Rev. **32**, 537–578 (1990)

104. T.D. Lee, Y. Pang, Nontopological solitons. Phys. Rep. **221**, 251–350 (1992)

105. E.H. Lieb, On the lowest eigenvalue of the Laplacian for the intersection of two domains. Invent. Math. **74**, 441–448 (1983)

106. P.-L. Lions, The concentration-compactness principle in the calculus of variations. The locally compact case. I. Ann. Inst. H. Poincaré Anal. Non Linéaire **1**(2), 109–145 (1984)

107. P.L. Lions, The concentration-compactness principle in the calculus of variations. The locally compact case. II. Ann. Inst. H. Poincaré Anal. Non Linéaire **1**(4), 223–283 (1984)

108. E. Long, Existence and stability of solitary waves in non-linear Klein-Gordon-Maxwell equations. Rev. Math. Phys. **18**, 747–779 (2006)

109. N. Manton, P. Sutcliffe, *Topological solitons* (Cambridge Univerity Press, Cambridge, 2004)

110. P.J. McKenna, W. Walter, Nonlinear oscillations in a suspension bridge. Arch. Ration. Mech. Anal. **98**, 167–177 (1987)

111. P.J. McKenna, W. Walter, Travelling waves in a suspension bridge. SIAM J. Appl. Math. **3**(50), 703–715 (1990)

112. D. Mugnai, Solitary waves in Abelian gauge theories with strongly nonlinear potentials. Ann. Inst. H. Poincaré Anal. Non Linéaire **27**, 1055–1071 (2010)

113. M. Nakamitsu, M. Tsutsumi, The cauchy problem for the coupled Maxwell-Schrödinger equations. J. Math. Phys. **27**, 211–216 (1986)

114. M. Nakamura, T. Wada, Global existence and uniqueness of solutions to the Maxwell-Schrödinger equations. Commun. Math. Phys. **276**, 315–339 (2007)

115. H. Nielsen, P. Olesen, Vortex-line models for dual strings. Nucl. Phys. B **61**, 45–61 (1973)

116. L. Nirenberg, On elliptic partial differential equations. Ann. Sc. Norm. Sup. Pisa **13**, 115–162 (1959)

117. E.Noether, Invariante Variationsprobleme. Nachr.kgl. ges. Wiss. Göttingen math.phys. Kl. S. 253–257 (1918)

118. R.S. Palais, The principle of symmetric criticality. Commun. Math. Phys. **69**, 19–30 (1979)

119. D.M. Petrescu, Time decay of solutions of coupled Maxwell-Klein-Gordon equations. Commun. Math. Phys. **179**, 11–24 (1996)

120. L. Pisani, G. Siciliano, Newmann conditions in the Schrödinger-Maxwell system. Topol. Methods Nonlinear Anal. **27**, 251–264 (2007)

121. S.I. Pohozaev Eigenfunctions of the equation $\Delta u + \lambda f(u) = 0$. Sov. Math. Dokl. **165**, 1408–1412 (1965)

122. R. Rajaraman, *Solitons and Instantons* (North Holland, Amsterdam/Oxford/New York/Tokio, 1988)

123. G. Rosen Particle-like solutions to nonlinear complex scalar field theories with positive-definite energy densities. J. Math. Phys. **9**, 996–998 (1968)

124. V. Rubakov, *Classical Theory of Gauge Fields* (Princeton University press, Princeton, 2002)

125. D. Ruiz, Semiclassical states for coupled Schrödinger-Maxwell equations: concentration around a sphere. Math. Models Methods Appl. Sci. **15**, 141–164 (2005)

126. D. Ruiz, The Schrödinger–Poisson equation under the effect of a nonlinear local term. J. Funct. Anal. **237**, 655–674 (2006)

127. D. Ruiz, G. Vaira, Cluster solutions for the Schrödinger-Poisson-Slater problem around a local minimum of the potential. Rev. Mat. Iberoam. **27**, 253–271 (2011)

128. A. Salvatore, Multiple solitary for a non-homogeneous Schrödinger -Maxwell system in $\mathbb{R}^3$. Adv. Nonlinear Stud. **6**, 157–169 (2006)

129. O. Sanchez, J. Soler, Long time dynamics of Schrödinger- Poisson-Slater systems. J. Stat. Phys. **114**, 179–204 (2004)
130. S. Santra, J. Wei, Homoclinic solutions for fourth order traveling wave equations. SIAM J. Math. Anal. **41**(5), 2038–2056 (2009)
131. W. Schlag, Stable manifolds for an orbitally unstable nonlinear Schrödinger equation. Ann. Math. (2) **169** (1), 139–227 (2009)
132. J.S. Russell, *Report of the Fourteenth Meeting of the British Association for the Advancement of Science*, York, Sept 1844, London, 1845, pp. 311–390
133. A. Selvitella, Semiclassical evolution of two rotating solitons for the nonlinear Schrödinger equation with electric potential. Adv. Differ. Equ. **15**(3–4), 315–348 (2010)
134. R. Servadei, M. Squassina, Soliton dynamics for a general class of Schrödinger equations. J. Math. Anal. Appl. **365**, 776–796 (2010)
135. J. Shatah, Stable Standing waves of nonlinear Klein-Gordon equations. Commun. Math. Phys. **91**, 313–327 (1983)
136. J. Shatah, W. Strauss, Instability of nonlinear bound states. Commun. Math. Phys. **100**, 173–190 (1985)
137. M. Squassina, Soliton dynamics for nonlinear Schrödinger equation with magnetic field. Manuscr. Math. **130**, 461–494 (2009). arXiv:0811.2584
138. C. Sulem, P.L. Sulem *The Nonlinear Schrödinger Equation* (Springer, New York, 1999)
139. W.A. Strauss, Existence of solitary waves in higher dimensions. Commun. Math. Phys. **55**, 149–162 (1977)
140. K. Tintarev, K.H. Fieseler, *Concentration Compactness* (Imperial College Press, London, 2007)
141. A. Vilenkinnon, E.P.S. Shellard, *Cosmic Strings and Other Topological Defects* (Cambrige University press, Cambridge, 1994)
142. M.S. Volkov, E. Wöhnert, Spinning Q-balls. Phys. Rev. D **66**, 085003 (2002)
143. M.S. Volkov, Existence of spinning solitons in field theory (2004). arXiv:hep-th/0401030
144. M.I. Weinstein, Modulational stability of ground states of nonlinear Schrödinger equations. SIAM J. Math. Anal. **16**(3), 472–491 (1985)
145. M.I. Weinstein, Lyapunov stability of ground states of nonlinear dispersive evolution equations. Commun. Pure Appl. Math. **39**(1), 51–67 (1986)
146. G.B. Witham, *Linear and Nonlinear Waves* (Wiley, New York, 1974)
147. Y. Yang, *Solitons in Field Theory and Nonlinear Analysis* (Springer, New York/Berlin, 2000)

Index

© Springer International Publishing Switzerland 2014

251

V. Benci, D. Fortunato, *Variational Methods in Nonlinear Field Equations*,
Springer Monographs in Mathematics, DOI 10.1007/978-3-319-06914-2